METHODOLOGICAL APPROACHES FOR SLEEP AND VIGILANCE RESEARCH

METHODOLOGICAL APPROACHES FOR SLEEP AND VIGILANCE RESEARCH

Edited by

ERIC MURILLO-RODRIGUEZ

Universidad Anáhuac Mayab. Mérida, Yucatán. México

Academic Press is an imprint of Elsevier
125 London Wall, London EC2Y 5AS, United Kingdom
525 B Street, Suite 1650, San Diego, CA 92101, United States
50 Hampshire Street, 5th Floor, Cambridge, MA 02139, United States
The Boulevard, Langford Lane, Kidlington, Oxford OX5 1GB, United Kingdom

Library of Congress Cataloging-in-Publication Data
A catalog record for this book is available from the Library of Congress

British Library Cataloguing-in-Publication Data
A catalogue record for this book is available from the British Library

ISBN: 978-0-323-85235-7

For information on all Academic Press publications visit our website at https://www.elsevier.com/books-and-journals

Publisher: Nikki Levy
Acquisitions Editor: Natalie Farra
Editorial Project Manager: Timothy Bennett
Production Project Manager: Maria Bernard
Cover Designer: Victoria Pearson

Typeset by TNQ Technologies

This volume is dedicated to my family.

Eric Murillo-Rodríguez

Contents

Contributors xi
Foreword xv
Acknowledgments xix

1. Definitions and measurements of the states of vigilance **1**
Alejandra Mondino, Pablo Torterolo *and* Giancarlo Vanini

The sleep-wake cycle 1
General anesthesia 5
States of hypometabolism 9
Conclusion 10
Acknowledgments 10
References 10

2. Polysomnography in humans and animal models: basic procedures and analysis **17**
Pablo Torterolo, Joaquín Gonzalez, Santiago Castro-Zaballa, Matías Cavelli, Alejandra Mondino, Claudia Pascovich, Nicolás Rubido, Eric Murillo-Rodríguez *and* Giancarlo Vanini

Introduction 17
Polysomnography 17
Polysomnography in humans 18
Polysomnography in animal models 21
Quantitative electroencephalogram analysis 24
Conclusion 30
Acknowledgments 30
References 30

3. Electrophysiological studies and sleep-wake cycle **33**
Md Aftab Alam, Andrey Kostin *and* Md Noor Alam

Introduction 33
Electrophysiology and sleep-wake regulatory systems 34
Extracellular recording: basic procedure 36
Juxtacellular recording-labeling 48
Basic procedure 49
Juxtacellular recording-labeling and sleep-wake studies 52

Conclusions 53
References 54

4. Physiologic systems dynamics, coupling and network interactions across the sleep-wake cycle **59**

Plamen Ch. Ivanov *and* Ronny P. Bartsch

Physiologic dynamics across sleep stages and circadian phases 59
Physiological systems interactions during wake and sleep 80
Acknowledgments 93
References 93

5. Deep brain stimulation for understanding the sleep-wake phenomena **101**

Francisco J. Urbano *and* Edgar Garcia-Rill

Deep brain stimulation for the treatment of sleep disorder in several thalamocortical dysrhythmia pathologies 101
Neuronal mechanisms underlying deep bran stimulation 103
Pharmacological intervention of histone deacetylase enzymes for the treatment of sleep disorders during deep brain stimulation 105
Acknowledgments 106
References 107

6. Electroencephalography power spectra and electroencephalography functional connectivity in sleep **111**

Chiara Massullo, Giuseppe A. Carbone, Eric Murillo-Rodríguez, Sérgio Machado, Henning Budde, Tetsuya Yamamoto *and* Claudio Imperatori

Quantitative electroencephalography 111
Quantitative electroencephalography in normal sleep 117
Quantitative electroencephalography in abnormal sleep 119
References 128

7. Optogenetics in sleep and integrative systems research **135**

Brook L.W. Sweeten, Laurie L. Wellman *and* Larry D. Sanford

Background 135
Basics of optogenetics 136
Optogenetics versus chemical or electrical stimulation techniques 138
Potential disadvantages of optogenetic 138
Applications of optogenetics in sleep research 139
Surgical procedures 142

Sample protocol for optogenetics in sleep research 146
Experimental variants 149
Conclusions 151
Acknowledgment 151
References 151

8. Immunohistochemical analysis and sleep studies: some recommendations to improve analysis data 155

Fabio García-García, Luis Beltrán Parrazal *and* Armando Jesús Martínez

Introduction 155
Cell markers and sleep 156
Immunohistochemistry and sleep studies 157
Statistical considerations when analyzing immunoreactive cells 161
Pseudoreplication 163
Conclusions 165
Acknowledgments 165
References 165

9. Wireless vigilance state monitoring 171

Paul-Antoine Libourel

Measuring states of vigilance 171
Wireless monitoring of vigilance states 174
New opportunities offered by the wireless monitoring of the vigilance states 183
References 184

10. Wearable and nonwearable sleep-tracking devices 191

Laronda Hollimon, Ellita T. Williams, Iredia M. Olaye, Jesse Moore, Daniel Volshteyn, Debbie P. Chung, Janna Garcia Torres, Girardin Jean-Louis *and* Azizi A. Seixas

Introduction 191
Features 193
Reliability, validity, and accuracy 202
Utility and applications of wearables and nonwearables 208
Assessing sleep health parameters 208
Diagnosis of diseases and health conditions 209
Using wearables to deliver interventions 210
The use of wearables and nonwearables across the lifespan 210
Conclusion 211

Acknowledgment 211
References 212

11. Clinical trial: imaging techniques in sleep studies **215**
Luigi Ferini-Strambi, Andrea Galbiati *and* **Maria Salsone**

Imaging in normal human sleep 215
Imaging in nonrapid eye movement sleep 217
Imaging in rapid eye movement sleep 225
Conclusions and future directions 230
Conflict of interests 230
References 231

12. Objective questionnaires for assessment sleep quality **235**
David L. Streiner

Introduction 235
Steps in scale construction 236
A soupçon of test theory 242
Reliability 242
Validity 248
Summary 251
References 252

13. Clinical psychoinformatics: a novel approach to behavioral states and mental health care driven by machine learning **255**
Tetsuya Yamamoto, Junichiro Yoshimoto, Jocelyne Alcaraz-Silva, Eric Murillo-Rodríguez, Claudio Imperatori, Sérgio Machado *and* **Henning Budde**

Introduction 255
Overview of machine learning 256
Clinical applications of machine learning approaches in the mental health field 265
Limitations of machine learning approaches 269
Future perspectives on the use of machine learning approaches 270
Conclusions 273
References 274

Index *281*

Contributors

Md Aftab Alam
Research Service (151A3), Veterans Affairs Greater Los Angeles Healthcare System, Sepulveda, CA, United States; Department of Psychiatry, University of California, Los Angeles, CA, United States

Md Noor Alam
Research Service (151A3), Veterans Affairs Greater Los Angeles Healthcare System, Sepulveda, CA, United States; Department of Medicine, David Geffen School of Medicine, University of California, Los Angeles, CA, United States

Jocelyne Alcaraz-Silva
Laboratorio de Neurociencias Moleculares e Integrativas, Escuela de Medicina, División Ciencias de la Salud, Universidad Anáhuac Mayab, Mérida, Yucatán, México

Ronny P. Bartsch
Department of Physics, Bar-Ilan University, Ramat Gan, Israel

Henning Budde
Intercontinental Neuroscience Research Group, Rome, Italy; Faculty of Human Sciences, Medical School Hamburg, Hamburg, Germany; Intercontinental Neuroscience Research Group, Tokushima, Tokushima, Japan

Giuseppe A. Carbone
Cognitive and Clinical Psychology Laboratory, Department of Human Science, European University of Rome, Rome, Italy

Santiago Castro-Zaballa
Laboratorio de Neurobiología del Sueño, Departamento de Fisiología, Facultad de Medicina, Universidad de la República, Montevideo, Uruguay

Matías Cavelli
Laboratorio de Neurobiología del Sueño, Departamento de Fisiología, Facultad de Medicina, Universidad de la República, Montevideo, Uruguay; Department of Psychiatry, University of Wisconsin, Madison, WI, United States

Debbie P. Chung
NYU Grossman School of Medicine, New York, NY, United States

Luigi Ferini-Strambi
Vita-Salute San Raffaele University, Milan, Italy; IRCCS San Raffaele Scientific Institute, Department of Clinical Neurosciences, Neurology-Sleep Disorder Center, Milan, Italy

Andrea Galbiati
Vita-Salute San Raffaele University, Milan, Italy; IRCCS San Raffaele Scientific Institute, Department of Clinical Neurosciences, Neurology-Sleep Disorder Center, Milan, Italy

Edgar Garcia-Rill
Emeritus, Center for Translational Neuroscience, Department of Neurobiology and Developmental Sciences, College of Medicine, University of Arkansas for Medical Sciences. Little Rock, AR, United States

Fabio García-García
Biomedicine Department, Health Science Institute, Veracruzana University, Xalapa, Veracruz, Mexico

Joaquín Gonzalez
Laboratorio de Neurobiología del Sueño, Departamento de Fisiología, Facultad de Medicina, Universidad de la República, Montevideo, Uruguay

Laronda Hollimon
NYU Grossman School of Medicine, New York, NY, United States

Claudio Imperatori
Intercontinental Neuroscience Research Group, Tokushima, Tokushima, Japan; Cognitive and Clinical Psychology Laboratory, Department of Human Science, European University of Rome, Rome, Italy; Faculty of Human Sciences, Medical School Hamburg, Hamburg, Germany

Plamen Ch. Ivanov
Keck Laboratory for Network Physiology, Department of Physics, Boston University, Boston, MA, United States; Division of Sleep Medicine, Brigham and Women's Hospital, Harvard Medical School, Boston, MA, United States; Institute of Solid State Physics, Bulgarian Academy of Sciences, Sofia, Bulgaria

Girardin Jean-Louis
NYU Grossman School of Medicine, New York, NY, United States

Andrey Kostin
Research Service (151A3), Veterans Affairs Greater Los Angeles Healthcare System, Sepulveda, CA, United States

Paul-Antoine Libourel
Neurosciences Research Center of Lyon, Inserm U1028 – CNRS UMR5292 – UCBL, Bron, France

Sérgio Machado
Intercontinental Neuroscience Research Group, Rome, Italy; Department of Sports Methods and Techniques, Federal University of Santa Maria, Santa Maria, Brazil; Laboratory of Physical Activity Neuroscience, Neurodiversity Institute, Queimados-RJ, Brazil; Intercontinental Neuroscience Research Group, Tokushima, Tokushima, Japan; Laboratory of Physical Activity Neuroscience, Neurodiveristy Institute, Queimados, Rio de Janeiro, Brazil

Armando Jesús Martínez
Neuroethology Institute, Veracruzana University, Xalapa, Veracruz, Mexico

Chiara Massullo
Cognitive and Clinical Psychology Laboratory, Department of Human Science, European University of Rome, Rome, Italy

Alejandra Mondino
Department of Anesthesiology, University of Michigan, Ann Arbor, MI, United States; Laboratorio de Neurobiología del Sueño, Departamento de Fisiología, Facultad de Medicina, Universidad de la República, Montevideo, Uruguay

Jesse Moore
NYU Grossman School of Medicine, New York, NY, United States

Eric Murillo-Rodríguez
Intercontinental Neuroscience Research Group, Rome, Italy; Laboratorio de Neurociencias Moleculares e Integrativas Escuela de Medicina, División Ciencias de la Salud, Universidad Anáhuac Mayab Mérida, Mérida, Yucatán, México; Intercontinental Neuroscience Research Group, Tokushima, Tokushima, Japan; Intercontinental Neuroscience Research Group, Mérida, Yucatán, México

Iredia M. Olaye
Weill Cornell Medicine of Cornell University, New York, NY, United States

Luis Beltrán Parrazal
Brain Research Center, Veracruzana University, Xalapa, Veracruz, Mexico

Claudia Pascovich
Laboratorio de Neurobiología del Sueño, Departamento de Fisiología, Facultad de Medicina, Universidad de la República, Montevideo, Uruguay; Consciousness and Cognition Laboratory, Department of Psychology, University of Cambridge, Cambridge, United Kingdom

Nicolás Rubido
Aberdeen Biomedical Imaging Centre, University of Aberdeen, Aberdeen, United Kingdom; Instituto de Física, Facultad de Ciencias, Universidad de la República, Montevideo, Uruguay

Maria Salsone
IRCCS San Raffaele Scientific Institute, Department of Clinical Neurosciences, Neurology-Sleep Disorder Center, Milan, Italy; Institute of Molecular Bioimaging and Physiology, National Research Council, Segrate, Italy

Larry D. Sanford
Sleep Research Laboratory, Center for Integrative Neuroscience and Inflammatory Diseases, Department of Pathology and Anatomy, Eastern Virginia Medical School, Norfolk, VA, United States

Azizi A. Seixas
NYU Grossman School of Medicine, New York, NY, United States

David L. Streiner
Department of Psychiatry and Behavioural Neurosciences, McMaster University, Hamilton, ON, Canada

Brook L.W. Sweeten
Sleep Research Laboratory, Center for Integrative Neuroscience and Inflammatory Diseases, Department of Pathology and Anatomy, Eastern Virginia Medical School, Norfolk, VA, United States

Janna Garcia Torres
NYU Grossman School of Medicine, New York, NY, United States

Pablo Torterolo
Laboratorio de Neurobiología del Sueño, Departamento de Fisiología, Facultad de Medicina, Universidad de la República, Montevideo, Uruguay; Intercontinental Neuroscience Research Group, Mérida, Yucatán, México

Francisco J. Urbano
Universidad de Buenos Aires, Facultad de Ciencias Exactas y Naturales, Departamento de Fisiología, Biología Molecular y Celular "Dr. Héctor Maldonado", Ciudad de Buenos Aires, Argentina; CONICET- Instituto de Fisiología, Biología Molecular y Neurociencias (IFIBYNE), Ciudad Universitaria, Ciudad Autónoma de Buenos Aires, Argentina

Giancarlo Vanini
Department of Anesthesiology, University of Michigan, Ann Arbor, MI, United States

Daniel Volshteyn
Cornell University, New York, NY, United States

Laurie L. Wellman
Sleep Research Laboratory, Center for Integrative Neuroscience and Inflammatory Diseases, Department of Pathology and Anatomy, Eastern Virginia Medical School, Norfolk, VA, United States

Ellita T. Williams
NYU Grossman School of Medicine, New York, NY, United States

Tetsuya Yamamoto
Graduate School of Technology, Industrial and Social Sciences Tokushima University, Tokushima, Tokushima, Japan; Intercontinental Neuroscience Research Group, Tokushima, Tokushima, Japan; Intercontinental Neuroscience Research Group, Rome, Italy

Junichiro Yoshimoto
Graduate School of Science and Technology, NAIST, Ikoma, Nara, Japan

Foreword

The study of sleep and the methods for doing so are a rapidly moving and evolving field. As the chapters in this book will attest, investigators have used an enormous array of methods to try to understand the phenomenology, mechanisms, and purpose of sleep. These range from very invasive methods, such as optogenetics, which are currently employed only in cutting-edge basic science laboratories in experimental animals, to questionnaires used as instruments to quantify human sleep experience.

The Sleep Research Society publishes a manual called the *Basics of Sleep Guide*, which is meant as an introductory text for basic laboratory and human studies investigators to learn about the processes of sleep and their mechanisms. While this text goes a long way toward educating new students about sleep, it does not go into the details of methodology. A book that covers this territory succinctly but with sufficient detail to get a new investigator going has long been lacking, and the current volume fills that niche.

The current volume can be roughly divided into three sections. The first section, which encompasses Chapters 1–6 (and to a lesser extent Chapters 9, 10), provides a guide to the electrophysiological measurement of sleep. Since the 1950s, when electroencephalography (EEG) was first applied to the study of sleep, the use of the polysomnogram (EEG, electromyogram, and often respiration or eye movements) has been used as the fundamental method for measuring sleep, dividing it into various physiological stages and quantifying them. Because the EEG as typically used essentially reflects local field potentials at the cortical surface, this has been complemented by more invasive measurements of local field potentials or single neuron recordings deep in the brain. At the same time, there have been major advances in analyzing the electrophysiological data, including power spectrum, coupling dynamics, network interactions, functional connectivity, and in the near future, artificial intelligence. With the miniaturization of electronics, these methods of measurement have reached the point of wireless application to animals and wearable technology for humans, areas that are covered in this volume.

The second section, including Chapters 7–10, deals with cutting-edge methods that are mainly of interest in animal studies of sleep. This is important because the use of experimental animals allows us to dissect the mechanisms that control wake-sleep down to neural circuits and molecular events and to interrogate the brain about how these circuits function, and how their function can be altered. The use of optogenetic (and their alter ego, chemogenetic) methods has matured over the last decade, and these are now standard laboratory techniques. Not only can we activate or inhibit neurons but we can also do so with remarkable anatomical and biochemical specificity. The methods for establishing the different chemical phenotypes of neurons in these brain circuits, such as immunohistochemistry, in situ hybridization, and use of transgenic Cre-expresser mice, also is allowing us to establish neural connectivity, and suggesting ways to gain genetic access to different circuits. This work is moving the field forward at a remarkable rate.

The third section, Chapters 11–13 on human studies, brings the rigorous study of sleep to human sleep and sleep disorders. Measuring human sleep has always raised the problem that the more invasive the measurement, the less likely it is to reflect more natural sleep in the home environment. The early sleep studies in humans were done in a sleep laboratory, an environment that is not conducive to normal sleep behavior as expressed in the home environment. To overcome this obstacle as well as the cost of doing in-laboratory sleep studies, one approach has been to miniaturize the recording devices and make them wearable, so that the method itself will have the least intrusion on the sleep process and can be taken into the home environment. Another approach has been to use questionnaires to provide subjective but rigorously quantifiable measures of sleep quality, daytime sleepiness, and other sleep experiences. The application of artificial intelligence (machine learning) to human sleep measurements has provided an exciting glimpse of how the data that are acquired from human studies can be used to identify patterns that are beyond the ability of humans to discern.

Sleep is by nature a multidisciplinary field. Many people, myself included, originally trained in a different field, but entered into sleep research because it is so fascinating and intersects with so many other biological processes. There is a constant need for the education of these interdisciplinary investigators who enter the world of sleep research. This volume will provide a convenient method of entrée to the methodology used in modern sleep research, and so catalyze the entry of new colleagues who bring with them advanced methods from other fields. For that reason, we look forward to the armamentarium of sleep research methods continuing to expand. In that context, this volume will provide an important entry point for those coming into the field.

Clifford B. Saper, MD, PhD
Beth Israel Deaconess Medical Center, Harvard Medical School
Boston, MA, United States
March 1, 2021

Acknowledgments

I would like to express my gratitude to my family, with special love to Omar, my sister Linda, and my lovely niece Shauly.

Thanks to my students, colleagues, and friends for allowing me to share with them my happiness when starting this project.

An extra thanks to the collaborators for trusting in the project, as well as for your magnificent contributions to this book. Thanks to Natalie Farra and Tim Bennett as well as Elsevier staff for believing in this project.

I would like to offer special thanks to the many people who provided support and assistance in editing, proofreading, and designing the book.

Last but not the least, I beg forgiveness of all those who have been with me over the course of the years and whose names I have failed to mention.

Eric Murillo-Rodríguez

CHAPTER 1

Definitions and measurements of the states of vigilance

Alejandra Mondino[1,2], Pablo Torterolo[2], Giancarlo Vanini[1]
[1]Department of Anesthesiology, University of Michigan, Ann Arbor, MI, United States; [2]Laboratorio de Neurobiología del Sueño, Departamento de Fisiología, Facultad de Medicina, Universidad de la República, Montevideo, Uruguay

The sleep-wake cycle

The sleep-wake cycle is a physiological phenomenon that occurs in virtually all animal species, and is the most evident circadian rhythm in mammals and birds (Ray and Reddy, 2016). The alternation between states of sleep and wakefulness (and alertness levels) is precisely regulated in relation to the light-dark cycle by interacting homeostatic and circadian processes (Borbely et al., 2016). Wakefulness is a state characterized by increased alertness and goal-oriented motor behaviors that are either internally generated or induced in response to sensory inputs from the environment. On the other hand, sleep is a behavioral state that is actively generated by the brain and is characterized by a reduced interaction with the environment, reduced muscular tone and body movement, closed eyes, a typical posture adopted to conserve body temperature, and an increase in the arousal threshold in response to external stimuli. It is important to note that ample evidence from several independent groups shows that invertebrates have a sleep-like state that satisfies all the key criteria used to define vertebrate sleep. In invertebrates, sleep is mainly defined by a reversible period of inactivity, a specific body posture, elevated arousal thresholds to stimuli applied during sleep, and a sleep homeostatic response after sleep deprivation (Shaw et al., 2000; Raizen et al., 2008; Vorster et al., 2014; Iglesias et al., 2019). In contrast with anesthesia or coma, sleep can be easily reversed by sensory stimuli and without any pharmacological manipulation (Schwartz and Klerman, 2019). The mammalian sleep-wake cycle is organized in three different states, wakefulness, nonrapid eye movement sleep (NREM; also called slow wave sleep), and rapid eye movement (REM) sleep (McCarley, 2007; Saper et al., 2010). Each of these states is characterized by distinctive behavioral, autonomic, and electroencephalographic signatures. In addition, cognitive function significantly varies in a state-specific manner across the

Methodological Approaches for Sleep and Vigilance Research
ISBN 978-0-323-85235-7
https://doi.org/10.1016/B978-0-323-85235-7.00002-8

sleep-wake cycle. For example, consciousness is lost during deep NREM sleep, and re-emerges in a distorted manner during REM sleep, when most oneiric activity takes place (Tononi and Laureys, 2009; Torterolo et al., 2019).

Electroencephalographic signatures of sleep and wakefulness

In clinical practice and laboratory research, polysomnography is the gold standard for the objective identification and study of arousal states of sleep and wakefulness (Marino et al., 2013). A basic polysomnography consists in the simultaneous recording of the electroencephalogram (EEG), electrooculogram, and electromyogram (EMG). In clinical settings, the use of polysomnography for diagnostic purposes typically includes the recording of the electrocardiogram, pulse oximetry, and airflow and respiratory effort (Haba-Rubio and Krieger, 2012). Using the most basic array that combines EEG and EMG signals, wakefulness is identified by low-voltage (in the range of μV), high-frequency oscillations, associated with high muscle tone and movements that are evidenced by varying degrees of EMG activity (Vanderwolf, 1969; Winson, 1974; Achermann, 2009; Torterolo et al., 2019). In humans, alpha rhythms are present during quite wakefulness with eyes closed (Gupta et al., 2018). Both theta and alpha rhythms are prominent in the occipital cortex and are typically observed in the "raw", unprocessed EEG recordings (Gupta et al., 2018; Torterolo et al., 2019). During the transition to NREM sleep, EEG frequencies become slower, and the amplitude of the oscillations gradually increases indicating higher synchronization of local neural activity.

NREM sleep is defined by high-voltage and low frequencies, with prominent delta frequencies (0.5—4.0 Hz) and frequent sleep spindles (a burst of 11—15 Hz oscillations with a duration of at least 0.4 s) (Achermann, 2009; Sullivan et al., 2014; Gupta et al., 2018). Relative to wakefulness, NREM sleep is characterized by a marked reduction in muscle tone (i.e., lower EMG amplitude). While in laboratory animals, NREM sleep can be sub-divided into light and deep sleep, in relation to the amount of EEG slow wave (delta) activity in each epoch, in humans, NREM is classified into three sub-stages. N1 is a transitional state from wakefulness, N2 is characterized by k-complexes typically followed by sleep spindles, and N3 is defined by delta waveforms (Carskadon and Dement, 2001). Human sleep alternates between NREM and REM sleep for about 90 min, and each cycle repeats four to five times during the night (Carskadon and Dement, 2001). Laboratory animals are typically nocturnal (rats and mice are the most commonly used species in sleep research) and have multiple brief sleep cycles that occur predominantly during the light (daytime) period. In all cases, REM sleep is always preceded by a NREM sleep bout.

During REM sleep, EEG activity is characterized by low-voltage and fast frequency oscillations, similar to that one during wakefulness. In addition to the theta "saw-tooth" waveforms, REMs and a sustained muscle atonia are key hallmark traits of REM sleep (Carskadon and Dement, 2001). In rodents, EEG theta activity (4.0–9.0 Hz) appears during the transition between NREM and REM sleep, and is maintained throughout the entire REM sleep bout. Because of the similarity between the desynchronized EEG during wakefulness and REM sleep, the latter is also known as "paradoxical sleep".

Ample evidence demonstrates that gamma oscillations (35–100 Hz) and high-frequency oscillations (HFOs, up to 200 Hz) play a role in cognition. During arousal states where cognitive processing occurs (i.e., wakefulness and REM sleep), the power of these frequencies is significantly higher than during NREM sleep, a state where oneiric activity occurs but is less frequent and qualitatively different than in REM sleep (Maloney et al., 1997; Uhlhaas et al., 2011; Mondino et al., 2020). However, HFOs during wakefulness and REM sleep are quantitatively and qualitatively different. During REM sleep, there is a prominent peak in EEG power between 100 and 200 Hz, which it is not detectable during wakefulness (Cavelli et al., 2018; Mondino et al., 2020). In addition, the power of EEG frequencies between 200 and 300 Hz is greater during wakefulness than during REM sleep (Silva-Perez et al., 2020). Based on these differences between REM sleep and wakefulness, and the lower power of these oscillations during NREM sleep, Silva-Pérez et al. (2020) developed a high frequency index "HiFi" calculated as the ratio between the amplitude of the EEG signal between 110–200 Hz and 200–300 Hz. Lower HiFi values were found during wakefulness (because of the high amplitude of frequencies between 200 and 300 Hz), higher during REM sleep (because of the amplitude peak between 110 and 200 Hz), and intermediate values during NREM sleep. The synchronization of neuronal activity within gamma and HFO frequencies is also different between wakefulness and REM sleep. By means of coherence and symbolic transfer entropy analysis (measures of undirected and directed cortical connectivity, respectively), is has been shown that high synchronization within gamma and HFO bands characterizes wakefulness, while it is significantly reduced during both NREM and REM sleep (Castro et al., 2013; Pal et al., 2016; Cavelli et al., 2018; Mondino et al., 2020). Interestingly, interhemispheric gamma coherence is lower during REM sleep compared to NREM sleep (Mondino et al., 2020). It has been proposed that synchronization of neuronal activity in the gamma range is necessary for the integration of fragmentary neural events within the brain to have integrated perceptual experiences (Torterolo et al., 2019). Therefore, the lack of this synchronization during REM sleep may explain the bizarre content of consciousness during dreams (Castro et al., 2013; Rosen, 2018). In addition, the complexity of the signal of the EEG can be correlated with behavioral states. In this regard, by means of different complexity measures such as Lempel Ziv Complexity and Permutation Entropy, it has been shown that EEG signals during wakefulness have higher levels of complexity than during NREM sleep and REM sleep (Schartner et al., 2017; Gonzalez et al., 2019).

Autonomic function during sleep-wake states

Changes in the autonomic nervous system function have been described during the sleep-wake cycle. During sleep, there is a reduction in hearth rate (HR) and blood pressure (BP). The lowest values of HR and BP are seen during NREM sleep, while during REM sleep, HR and BP levels increase up to the level of wakefulness (Schechtman et al., 1985; Rowe et al., 1999; Trinder et al., 2001). The heart rate variability (HRV) has been used to determine the influence of the sympathetic and parasympathetic tone in autonomic changes that occur during sleep. The HRV analysis is a noninvasive procedure based on the electrocardiogram that evaluates the balance between the sympathetic and parasympathetic tone (Deutschman et al., 1994; Sztajzel, 2004). It measures the variation between R-R intervals by quantifying its low frequency (LF) and HFOs. The LF (0.045−0.15 Hz) is modulated by the parasympathetic and sympathetic nervous system, with higher LF indicating greater sympathetic activity. The HF (0.15−0.4 Hz) is, on the other hand, mediated mainly by the parasympathetic system (Pichon et al., 2006; Mazzeo et al., 2011; Ernst, 2017). Conventionally, this analysis has been performed by means of a Fast Fourier Transformation to obtain the power spectral density of each frequency band (Sztajzel, 2004; Pichon et al., 2006). However, this approach has been challenged because it does not allow the investigator to determine the temporal localization of instantaneous changes in the R-R intervals.

Therefore, the wavelet transform analysis has been proposed as a more precise analysis to assess the autonomic tone with time—frequency localization (Pichot et al., 1999; Lotric et al., 2000). By means of this analysis, it has been shown that during NREM sleep, there is an increase in the HF and a decrease in the LF component of the HRV, indicating an increase in parasympathetic tone during this sleep stage (Vaughn et al., 1995). Conversely, REM sleep is characterized by an augmented sympathetic tone, revealed by a higher LF component and a higher LF/HF ratio (Méndez et al., 2006; Cabiddu et al., 2012).

Identification and quantification of sleep-wake states in invertebrate species

As described above, sleep is present in virtually all animals. Importantly, recent research has shown that even animals with a simple nervous system that lacks brain cephalization such as cnidarias have a sleep-like state. In jellyfish, there is a quiescent state during the night revealed by the reduction in its bell pulsation, a behavior used to generate currents of fluid for feeding and expulsion of byproducts (Nath et al., 2017; Jha and Jha, 2020). Similarly, prolonged periods of behavioral quiescence have also been shown in nematodes (Iwanir et al., 2013) and annelids (Morrison, 2013). In animal species that lack a thalamocortical system do not have any of the EEG features that define sleep in birds or mammals. However, in some of them, changes in the neuronal activity have been demonstrated between wakefulness and sleep-like states. In *Drosophila* (the fruit fly), studies using recordings of local field potentials (LFPs) from the medial part of the brain have shown that, relative to wakefulness, there is a significant reduction in spike-like potentials during the quiescent state (Nitz et al., 2002; Cirelli and Bushey, 2008). Moreover, electrophysiological recordings from the protocerebrum of crayfish revealed high-amplitude and slow-frequency waves (8 Hz) during a sleep-like state (Ramón et al., 2004). Collectively, ample evidence demonstrates that sleep is universally present in animal organisms and has a common behavioral signature (quiescence) and species-specific electrographic characteristics.

General anesthesia

The American Society of Anesthesiologists defines general anesthesia as a drug-induced loss of consciousness during which patients are not arousable, even by painful stimulation (American-Society-of-Anesthesiologists, 2018). In addition, hypnosis (i.e., the loss of consciousness), immobility, amnesia, and a partial or complete loss of protective reflexes are key traits of the anesthetized state.

Behavioral assessment of anesthetic-induced loss of consciousness in laboratory animals

Studies using rodents (rats and mice) to investigate mechanisms of general anesthetics use the time to loss and resumption of righting response as validated surrogate measures for the time required to loss and regain consciousness, respectively (Tung et al., 2002; Alkire et al., 2007; Kelz et al., 2008; Vanini et al., 2008, 2014, 2020; Pal et al., 2015b, 2018; Taylor et al., 2016; Wasilczuk et al., 2020). For quantification of the time to the loss of consciousness (i.e., induction time), the animal receives an intraperitoneal or intravenous injection (propofol, ketamine, etomidate) or is exposed to inhalational anesthetic vapors within a transparent induction chamber. In most protocols, anesthetic-induced loss of consciousness is defined as the time when the animal remains in dorsal recumbency for at least 30 or 60 s. The recovery from general anesthesia (i.e., recovery of consciousness) is defined as the time needed to a complete resumption of the righting response, and the animal adopts a normal posture after all four paws are placed on the floor. In the case of intravenous anesthetics, the recovery time is quantified as the time between drug injection ($t = 0$) to the resumption of righting, whereas for inhalational anesthetics corresponds to the time at which the delivery of the anesthetic vapor is discontinued until the resumption of consciousness.

Electroencephalographic signatures of general anesthesia

The analysis of the electroencephalographic features is a feasible and practical approach to objectively monitor the depth of general anesthesia; some EEG features are drug-specific and others are independent of the anesthetic agent (Zoughi et al., 2012). In practice, anesthesiologists rarely interpret raw EEG traces during surgery and rely on processed signals calculated by and displayed on EEG monitors. In general, the EEG frequency decreases, and the amplitude increases as a function of anesthetic dose and depth of anesthesia. From anesthetic induction to surgical anesthesia, most anesthetic drugs show a consistent progression pattern starting with dominant alpha activity. This alpha activity is then followed by continued delta waveforms that, at a deeper anesthetic plane, shifts into a burst-suppression pattern, or a flat-line EEG when brain activity is completely suppressed.

Specific patterns during the induction phase include biphasic changes, characterized by an initial increase, followed by a decrease in alpha and beta activity, and an increase in delta. In the case of propofol, etomidate and thiopental, there is a subsequent reduction in delta activity (Kuizenga et al., 2001; Koskinen et al., 2005; Kortelainen et al., 2007). During surgical anesthesia, there is an increase in the power of delta oscillations and the shift in alpha rhythms from occipital to frontal regions is established (Purdon et al., 2013; Hagihira, 2015). This shift in the alpha rhythm is known as "alpha anteriorization," and it refers to an anesthesia-induced reduction in alpha power (8–13 Hz) in the occipital cortex, with a simultaneous increase in the frontal cortex (Purdon et al., 2013; Vijayan et al., 2013). Moreover, the coherence of slow oscillations (<1.5 Hz) in frontal and occipital regions, as well as between both regions, decreases with propofol anesthesia (Wang et al., 2014). As mentioned above, a burst suppression EEG pattern is produced by a high concentration of general anesthetics (Akrawi et al., 1996; Lukatch et al., 2005). This pattern is characterized by "bursts" of high-voltage activity (150–350 μV amplitude) alternating with periods of depressed (less than 25 μV amplitude) or absent electrographic activity (Japaridze et al., 2015). Although this pattern is seen with most of the anesthetics (or their combination), the duration and amplitude of the bursts vary with different drugs (Steriade et al., 1994; Akrawi et al., 1996).

General anesthetics effectively block the synchronization of activity between distant brain regions. Specifically, by means of coherence and transfer entropy analysis, it has been demonstrated that levels of synchronization and directional connectivity between frontal and parietal cortex (especially in the high-gamma frequency band) correlate with the loss of consciousness produced by general anesthetics (Boly et al., 2012; Pal et al., 2016). Interestingly, anesthetics preferentially inhibit the frontoparietal feedback connectivity while preserving feedforward connectivity (Lee et al., 2009; Ku et al., 2011). Frontoparietal feedback refers to the flow of information from the frontal cortex to posterior areas of the brain, which correlates with consciousness, while feedforward connectivity refers to the information that flows in the opposite direction, and it has been associated with sensory processing (Lamme et al., 1998).

Remarkably, during wakefulness, feedback is higher than feedforward connectivity, and this feedback dominance is significantly reduced during anesthesia (Lee et al., 2009). The complexity of the signal of the EEG can also be useful to predict the depth of anesthesia (Zhang et al., 2001). Studies using Lempel Ziv Complexity analysis (and other algorithms to study brain entropy) revealed that general anesthesia decreases EEG signal complexity (Liang et al., 2015; Hudetz et al., 2016).

The EEG signatures described above apply to most anesthetics. However, some anesthetics such as ketamine do not share these characteristics. Ketamine is a dissociative anesthetic (Domino et al., 1965) that decreases the activity of thalamocortical structures while activating the limbic system and hippocampus (Sinner and Graf, 2008). After induction with ketamine, the EEG is characterized by a "gamma burst" activity, i.e., gamma activity that alternates with slow oscillations. This phase is followed by a stable phase during which the EEG signal is comprised of high gamma and theta, and low-beta activity (Akeju et al., 2016). In addition, ketamine produces a global reduction in coherence (i.e., a measure of undirected cortical connectivity) in the 0.5 to 250 Hz frequency range (Pal et al., 2015a). Although ketamine has distinctive EEG traits, it also reduces feedback connectivity, which is a common correlate of anesthetic-induced unconsciousness across different anesthetics (Lee et al., 2013).

The knowledge about EEG signatures of general anesthesia provided the neuroscientific foundation to develop methods to monitor the depth of the anesthetized state during surgery. The bispectral index (BIS; Aspect Medical Systems, Newton, MA) was approved by the FDA for clinical use in 1996 and is one of the most popular clinical monitors (Gan et al., 1997; O'Connor et al., 2001; Gu et al., 2019). BIS uses a Fourier transformation and bispectral analysis to compute a number (the BIS) between 0 and 100 that represents different depths of anesthesia (80—100: wakefulness; 60—80: light anesthesia; 40—60: general anesthesia; less than 40: deep anesthesia) (Denman et al., 2000; Gu et al., 2019). BIS has been shown to be useful for assessment of anesthetic depth in both adults and children (Denman et al., 2000). It has been hypothesized that the use of BIS can help reduce the risk of awareness during anesthesia, however, clinical studies did not detect differences in the incidence of awareness based on BIS monitoring in comparison with routine care (Avidan et al., 2011; Bottros et al., 2011; Mashour et al., 2012).

Autonomic function is impaired during general anesthesia

Anesthetic drugs blunt autonomic nervous system responses by, directly or indirectly, acting on specific brain regions associated with parasympathetic and sympathetic control such as the dorsal motor nucleus of the vagus nerve and nucleus of the tractus solitarius, respectively (Polk et al., 2019). Thermoregulation is also inhibited by general anesthetics in a dose-dependent manner, increasing the threshold for warm responses such as vasoconstriction and shivering (Xiong et al., 1996; Lenhardt et al., 2009; Alfonsi et al., 2014; Lenhardt, 2018). Therefore, anesthesia rapidly causes hypothermia if the patient is not warmed adequately (Murphy and Murnane, 2019). In addition, general anesthetics decrease heart rate and arterial BP. In fact, it has been shown that the evaluation of the autonomic nervous system by means of the HRV and mean arterial pressure is useful to determine the depth of anesthesia and discriminate between general anesthesia and wake states (Mahfouf et al., 2003; Neukirchen and Kienbaum, 2008; Shalbaf et al., 2015; Polk et al., 2019). Both HF and LF components of the HRV are altered by general anesthesia (Kato et al., 1992; Jeanne et al., 2009), but the degree of changes in these parameters depends on the drug administered. For example, Kanaya and colleagues (Kanaya et al., 2003) showed that propofol decreases HF with almost no effects on LF, while sevoflurane significantly decreases LF.

States of hypometabolism

Several species of mammals and birds lower their metabolic rate and body temperature as an energy saving adaptation to extreme environmental temperatures and reduced food availability (Sonntag and Arendt, 2019). In some species, this process is called daily torpor, which is encompassed within the circadian cycle and last less than 24 h. In contrast, some animals maintain the hypometabolism for several weeks, which is known as hibernation (Geiser, 1994; Royo et al., 2019). The behavior during these hypometabolic states resembles sleep, i.e., animals adopt a sleep-like posture, remain inactive and show substantially higher arousal thresholds (Heller and Ruby, 2004). However, there are several differences between states of torpor and hibernation. The brain activity during torpor and hibernation is dramatically reduced, with isoelectric EEG patterns except for intermittent bursts of spindles (Walker et al., 1977; Royo et al., 2019). All through these states sleep is virtually absent. However, it has been found in hibernating primates that during high-ambient temperatures, periods of REM sleep-like activity occurs (Krystal et al., 2013), and in ground squirrels, there are periods of NREM sleep with temperatures higher than 10°C (Walker et al., 1977). Interestingly, torpor and hibernation are usually followed by sleep, and such sleep is characterized by an increase in slow oscillations amplitude and incidence (Deboer and Tobler, 1994; Cerri et al., 2013; Vyazovskiy et al., 2017). Similar changes in the EEG are seen after sleep deprivation, and therefore, these changes in EEG and the homeostatic sleep need after hypometabolic states support the idea that torpor or hibernation do not have the restorative function of sleep.

Conclusion

In the present chapter, we reviewed the main characteristics of vigilance states that are actively generated by the brain (sleep-wake, torpor, and hibernation) or pharmacologically induced (general anesthesia) in clinical and research settings. In addition, we introduced some of the most used analytical tools in the study of these vigilance states, with an emphasis on EEG analysis. These methods will be described in more detail in the next chapter.

Acknowledgments

This work was supported by the Department of Anesthesiology, University of Michigan, USA.

References

Achermann, P., 2009. EEG analysis applied to sleep. Epileptologie 26, 28–33.

Akeju, O., Song, A.H., Hamilos, A.E., Pavone, K.J., Flores, F.J., Brown, E.N., Purdon, P.L., 2016. Electroencephalogram signatures of ketamine anesthesia-induced unconsciousness. Clin. Neurophysiol. 127, 2414–2422.

Akrawi, W., Drummond, J., Kalkman, C.J., Patel, P., 1996. A comparison of electrophysiologicl characteristics of EEG burst-suppresion as produced by isoflurane, thiopental, etomidate, and propofol. J. Neurosurg. Anesthesiol. 8, 40–46.

Alfonsi, P., Passard, A., Guignard, B., Chauvin, M., Sessler, D.I., 2014. Nefopam and meperidine are infra-additive on the shivering threshold in humans. Anesth. Analg. 119, 58–63.

Alkire, M.T., McReynolds, J.R., Hahn, E.L., Trivedi, A.N., 2007. Thalamic microinjection of nicotine reverses sevoflurane-induced loss of righting reflex in the rat. Anesthesiology 107, 264–272.

American-Society-of-Anesthesiologists, 2018. Continuum of depth of sedation: definition of general anesthesia and levels of sedation/analgesia. In: Continuum of Depth of Sedation: Definition of General Anesthesia and Levels of Sedation/Analgesia. Committee on Quality Management and Departmental Administration. https://www.asahq.org/standards-and-guidelines/continuum-of-depth-of-sedation-definition-of-general-anesthesia-and-levels-of-sedationanalgesia.

Avidan, M.S., Jacobsohn, E., Glick, D., Burnside, B.A., Zhang, L., Villafranca, A., Karl, L., Kamal, S., Torres, B., O'Connor, M., Evers, A.S., Gradwohl, S., Lin, N., Palanca, B.J., Mashour, G.A., 2011. Prevention of intraoperative awareness in a high-risk surgical population. N. Engl. J. Med. 365, 591–600.

Boly, M., Moran, R., Murphy, M., Boveroux, P., Bruno, M.A., Noirhomme, Q., Ledoux, D., Bonhomme, V., Brichant, J.F., Tononi, G., Laureys, S., Friston, K., 2012. Connectivity changes underlying spectral EEG changes during propofol-induced loss of consciousness. J. Neurosci. 32, 7082–7090.

Borbely, A.A., Daan, S., Wirz-Justice, A., Deboer, T., 2016. The two-process model of sleep regulation: a reappraisal. J. Sleep Res. 25, 131–143.

Bottros, M.M., Palanca, B.J., Mashour, G.A., Patel, A., Butler, C., Taylor, A., Lin, N., Avidan, M.S., 2011. Estimation of the bispectral index by anesthesiologists: an inverse turing test. Anesthesiology 114, 1093–1101.

Cabiddu, R., Cerutti, S., Viardot, G., Werner, S., Bianchi, A.M., 2012. Modulation of the sympatho-vagal balance during sleep: frequency domain study of heart rate variability and respiration. Front. Physiol. 3, 45.

Carskadon, M.A., Dement, W., 2001. Normal Human Sleep: An Overview. Elsevier-Saunders, Philadelphia.

Castro, S., Falconi, A., Chase, M.H., Torterolo, P., 2013. Coherent neocortical 40-Hz oscillations are not present during REM sleep. Eur. J. Neurosci. 37, 1330–1339.

Cavelli, M., Rojas-Libano, D., Schwarzkopf, N., Castro-Zaballa, S., Gonzalez, J., Mondino, A., Santana, N., Benedetto, L., Falconi, A., Torterolo, P., 2018. Power and coherence of cortical high-frequency oscillations during wakefulness and sleep. Eur. J. Neurosci. 48, 2728–2737.

Cerri, M., Mastrotto, M., Tupone, D., Martelli, D., Luppi, M., Perez, E., Zamboni, G., Amici, R., 2013. The inhibition of neurons in the central nervous pathways for thermoregulatory cold defense induces a suspended animation state in the rat. J. Neurosci. 33, 2984–2993.

Cirelli, C., Bushey, D., 2008. Sleep and wakefulness in *Drosophila melanogaster*. Ann. N Y Acad. Sci. 1129, 323–329.

Deboer, T., Tobler, I., 1994. Sleep EEG after daily torpor in the Djungarian hamster: similarity to the effects of sleep deprivation. Neurosci. Lett. 166, 35–38.

Denman, W., Swanson, E., Rosow, D., Ezbicki, K., Connors, P., Rosow, C., 2000. Pediatric evaluation of the bispectral index (BIS) monitor and correlation of BIS with end-tidal sevoflurane concentration in infants and children. Anesth. Analg. 90, 872–877.

Deutschman, C., Harris, A.P., Fleisher, L.A., 1994. Changes in heart rate variability under propofol anesthesia: a possible explanation for propofol-induced bradycardia. Anesth. Analg. 79.

Domino, E.F., Chodoff, P., Corssen, G., 1965. Pharmacologic effects of CI-581, a new dissociative anesthetic, in man. Clin. Pharmacol. Ther. 6, 279–291.

Ernst, G., 2017. Heart-rate variability-more than heart beats? Front. Public Health 5, 240–240.

Gan, T., Glass, P.S., Windsor, A., Payne, F., Rosow, C., Sebel, P., Manberg, P., group, B.U.S., 1997. Bispectral index monitoring allows faster emergence and improved recovery from propofol, alfentanil, and nitrous oxide anesthesia. Anesthesiology 87, 808–815.

Geiser, F., 1994. Hibernation and daily torpor in marsupials: a review. Aust. J. Zool. 42, 1–16.

Gonzalez, J., Cavelli, M., Mondino, A., Pascovich, C., Castro-Zaballa, S., Torterolo, P., Rubido, N., 2019. Decreased electrocortical temporal complexity distinguishes sleep from wakefulness. Sci. Rep. 9, 18457.

Gu, Y., Liang, Z., Hagihira, S., 2019. Use of multiple EEG features and artificial neural network to monitor the depth of anesthesia. Sensors (Basel) 19.

Gupta, R., Pandi-Perumal, S.R., BaHa, A.S., 2018. Clinical Atlas of Polysomnography. Apple Academic Press, Oakville, ON, Canada.

Haba-Rubio, J., Krieger, J., 2012. Evaluation instruments for sleep disorders: a brief history of polysomnography and sleep medicine. In: Chiang, R.P., Kang, S. (Eds.), Introduction to Modern Sleep Technology. Springer.

Hagihira, S., 2015. Changes in the electroencephalogram during anaesthesia and their physiological basis. Br. J. Anaesth. 115 (Suppl. 1), i27–i31.

Heller, H.C., Ruby, N.F., 2004. Sleep and circadian rhythms in mammalian torpor. Annu. Rev. Physiol. 66, 275–289.

Hudetz, A.G., Liu, X., Pillay, S., Boly, M., Tononi, G., 2016. Propofol anesthesia reduces Lempel-Ziv complexity of spontaneous brain activity in rats. Neurosci. Lett. 628, 132–135.

Iglesias, T.L., Boal, J.G., Frank, M.G., Zeil, J., Hanlon, R.T., 2019. Cyclic nature of the REM sleep-like state in the cuttlefish Sepia officinalis. J. Exp. Biol. 222.

Iwanir, S., Tramm, N., Nagy, S., Wright, C., Ish, D., Biron, D., 2013. The microarchitecture of *C. elegans* behavior during lethargus: homeostatic bout dynamics, a typical body posture, and regulation by a central neuron. Sleep 36, 385–395.

Japaridze, N., Muthuraman, M., Reinicke, C., Moeller, F., Anwar, A.R., Mideksa, K.G., Pressler, R., Deuschl, G., Stephani, U., Siniatchkin, M., 2015. Neuronal networks during burst suppression as revealed by source analysis. PLoS One 10, e0123807.

Jeanne, M., Logier, R., De Jonckheere, J., Tavernier, B., 2009. Validation of a graphic measurement of heart rate variability to assess analgesia/nociception balance during general anesthesia. In: 31st Annual International Conference of the IEEE EMBS. Minneapolis, Minnesota: IEEE.

Jha, V.M., Jha, S.K., 2020. Sleep: findings in invertebrates and lower vertebrates. In: Jha, V.M., Jha, S.K. (Eds.), Sleep: Evolution and Functions. Springer Nature Singapore Pte Ltd, Singapore.

Kanaya, N., Hirata, N., Kurosawa, S., Nakayama, M., Namiki, A., 2003. Differential effects of propofol and sevoflurane on heart rate variability. Anesthesiology 98, 34–40.

Kato, M., Komatsu, T., Kimura, T., Sugiyama, F., Nakashima, K., Shimada, Y., 1992. Spectral analysis of heart rate variability during isoflurane anesthesia. Anesthesiology 77, 669–674.

Kelz, M.B., Sun, Y., Chen, J., Cheng Meng, Q., Moore, J.T., Veasey, S.C., Dixon, S., Thornton, M., Funato, H., Yanagisawa, M., 2008. An essential role for orexins in emergence from general anesthesia. Proc. Natl. Acad. Sci. U S A 105, 1309–1314.

Kortelainen, J., Koskinen, M., Mustola, S., Seppanen, T., 2007. EEG frequency progression during induction of anesthesia: from start of infusion to onset of burst suppression pattern. In: Proceedings of the 29th Annual International Conference of the IEEE EMBS. Cité Internationale, Lyon, France: IEEE.

Koskinen, M., Mustola, S., Seppanen, T., 2005. Relation of EEG spectrum progression to loss of responsiveness during induction of anesthesia with propofol. Clin. Neurophysiol. 116, 2069–2076.
Krystal, A.D., Schopler, B., Kobbe, S., Williams, C., Rakatondrainibe, H., Yoder, A.D., Klopfer, P., 2013. The relationship of sleep with temperature and metabolic rate in a hibernating primate. PLoS One 8, e69914.
Ku, S.W., Lee, U., Noh, G.J., Jun, I.G., Mashour, G.A., 2011. Preferential inhibition of frontal-to-parietal feedback connectivity is a neurophysiologic correlate of general anesthesia in surgical patients. PLoS One 6, e25155.
Kuizenga, K., Wierda, J.M., Kalkman, C.J., 2001. Biphasic EEG changes in relation to loss of consciousness during induction with thiopental, propofol, etomidate, midazolam or sevoflurane. Br. J. Anesth. 86, 354–360.
Lamme, V., Super, H., Spekreijse, H., 1998. Feedforward, horizontal, and feedback processing in the visual cortex. Curr. Opin. Neurobiol. 8, 529–535.
Lee, U., Kim, S., Noh, G.J., Choi, B.M., Hwang, E., Mashour, G.A., 2009. The directionality and functional organization of frontoparietal connectivity during consciousness and anesthesia in humans. Conscious. Cognit. 18, 1069–1078.
Lee, U., Ku, S.W., Noh, G.J., Baek, S., Choi, B.M., Mashour, G.A., 2013. Disruption of frontal–parietal communication by ketamine, propofol, and sevoflurane. Anesthesiology 118, 1264–1275.
Lenhardt, R., 2018. Body temperature regulation and anesthesia. In: Romanovsky, A.A. (Ed.), Handbook of Clinical Neurology. Elseiver, pp. 635–644.
Lenhardt, R., Orhan-Sungur, M., Komatsu, R., Govinda, R., Kasuya, Y., Sessler, D.I., Wadwha, A., 2009. Suppression of shivering during hypothermia using a novel drug combination in healthy volunteers. Anesthesiology 111, 110–115.
Liang, Z., Wang, Y., Sun, X., Li, D., Voss, L.J., Sleigh, J.W., Hagihira, S., Li, X., 2015. EEG entropy measures in anesthesia. Front. Comput. Neurosci. 9, 16.
Lotric, M., Stefanovska, A., Stajer, D., Urbancic-Rovan, V., 2000. Spectral components of heart rate variability determined by wavelet analysis. Physiol. Meas. 21, 441–457.
Lukatch, H.S., Kiddoo, C.E., Maciver, M.B., 2005. Anesthetic-induced burst suppression EEG activity requires glutamate-mediated excitatory synaptic transmission. Cerebr. Cortex 15, 1322–1331.
Mahfouf, M., Asbury, A.J., Linkens, D.A., 2003. Unconstrained and constrained generalised predictive control of depth of anaesthesia during surgery. Contr. Eng. Pract. 11, 1501–1515.
Maloney, K.J., Cape, E.G., Jones, B.E., 1997. High-frequency electroencephalogram activity in association with sleep-wake states and spontaneous behaviors in the rat. Neuroscience 76, 541–555.
Marino, M., Li, Y., Rueschman, M.N., Winkelman, J.W., Ellenbogen, J.M., Solet, J.M., Dulin, H., Berkman, L.F., Buxton, O.M., 2013. Measuring sleep: accuracy, sensitivity, and specificity of wrist actigraphy compared to polysomnography. Sleep 36, 1747–1755.
Mashour, G.A., Shanks, A., Tremper, K.K., Kheterpal, S., Turner, C.R., Ramachandran, S.K., Picton, P., Schueller, C., Morris, M., Vandervest, J.C., Lin, N., Avidan, M.S., 2012. Prevention of intraoperative awareness with explicit recall in an unselected surgical population: a randomized comparative effectiveness trial. Anesthesiology 117, 717–725.
Mazzeo, A.T., La Monaca, E., Di Leo, R., Vita, G., Santamaria, L.B., 2011. Heart rate variability: a diagnostic and prognostic tool in anesthesia and intensive care. Acta Anaesthesiol. Scand. 55, 797–811.
McCarley, R.W., 2007. Neurobiology of REM and NREM sleep. Sleep Med. 8, 302–330.

Méndez, M., Bianchi, A.M., Villantieri, O., Cerutti, S., 2006. Time-varying analysis of the heart rate variability during REM and non REM sleep stages. In: EMBS Annual International Conference. IEEE, New York City, USA, pp. 3576–3579.
Mondino, A., Cavelli, M., Gonzalez, J., Osorio, L., Castro-Zaballa, S., Costa, A., Vanini, G., Torterolo, P., 2020. Power and coherence in the EEG of the rat: impact of behavioral states, cortical area, lateralization and light/dark phases. Clocks Sleep 2, 536–556.
Morrison, K., 2013. Differential circadian behaviors in aquatic annelids. In: Biology. The State University of New Jersey.
Murphy, T.J., Murnane, K.S., 2019. The serotonin 2C receptor agonist WAY-163909 attenuates ketamine-induced hypothermia in mice. Eur. J. Pharmacol. 842, 255–261.
Nath, R.D., Bedbrook, C.N., Abrams, M.J., Basinger, T., Bois, J.S., Prober, D.A., Sternberg, P.W., Gradinaru, V., Goentoro, L., 2017. The jellyfish Cassiopea exhibits a sleep-like state. Curr. Biol. 27, 2984–2990 e2983.
Neukirchen, M., Kienbaum, P., 2008. Sympathetic nervous system evaluation and importance for clinical general anesthesia. Anesthesiology 109, 1113–1131.
Nitz, D.A., van Swinderen, B., Tononi, G., Greenspan, R., 2002. Electrophysiological correlates of rest and activity in *Drosophila melanogaster*. Curr. Biol. 12, 1934–1940.
O'Connor, M.F., Daves, S.N., Tung, A., Cook, R.I., Thisted, R., Apfelbaum, J., 2001. BIS monitoring to prevent awareness during general anesthesia. Anesthesiology 94, 520–522.
Pal, D., Hambrecht-Wiedbusch, V.S., Silverstein, B.H., Mashour, G.A., 2015a. Electroencephalographic coherence and cortical acetylcholine during ketamine-induced unconsciousness. Br. J. Anaesth. 114, 979–989.
Pal, D., Silverstein, B.H., Lee, H., Mashour, G.A., 2016. Neural correlates of wakefulness, sleep, and general anesthesia: an experimental study in rat. Anesthesiology 125, 929–942.
Pal, D., Jones, J.M., Wisidagamage, S., Meisler, M.H., Mashour, G.A., 2015b. Reduced Nav1.6 sodium channel activity in mice increases in vivo sensitivity to volatile anesthetics. PLoS One 10, e0134960.
Pal, D., Dean, J.G., Liu, T., Li, D., Watson, C.J., Hudetz, A.G., Mashour, G.A., 2018. Differential role of prefrontal and parietal cortices in controlling level of consciousness. Curr. Biol. 28, 2145–2152.e2145.
Pichon, A., Roulaud, M., Antoine-Jonville, S., de Bisschop, C., Denjean, A., 2006. Spectral analysis of heart rate variability: interchangeability between autoregressive analysis and fast Fourier transform. J. Electrocardiol. 39, 31–37.
Pichot, V., Gaspoz, J.M., Molliex, S., Antoniadis, A., Busso, T., Roche, F., Costes, F., Quintin, L., Lacour, J., Barthélémy, J., 1999. Wavelet transform to quantify heart rate variability and to assess its instantaneous changes. J. Appl. Physiol. 1081–1091.
Polk, S., Kashkooli, K., Nagaraj, S., Chamadia, S., Murphy, J.M., Sun, H., Westover, M.B., Barbieri, R., Akeju, O., 2019. Automatic detection of general anesthetic-states using ECGDerived autonomic nervous system features. In: 41st Annual International Conference of the IEEE Engineering in Medicine and Biology Society (EMBC). IEEE, Berlin, Germany, pp. 2019–2022.
Purdon, P.L., Pierce, E.T., Mukamel, E.A., Prerau, M.J., Walsh, J.L., Wong, K.F., Salazar-Gomez, A.F., Harrell, P.G., Sampson, A.L., Cimenser, A., Ching, S., Kopell, N.J., Tavares-Stoeckel, C., Habeeb, K., Merhar, R., Brown, E.N., 2013. Electroencephalogram signatures of loss and recovery of consciousness from propofol. Proc. Natl. Acad. Sci. U S A 110, E1142–E1151.
Raizen, D.M., Zimmerman, J.E., Maycock, M.H., Ta, U.D., You, Y-j, Sundaram, M.V., Pack, A.I., 2008. Lethargus is a *Caenorhabditis elegans* sleep-like state. Nature 451, 569–572.

Ramón, F., Hernández-Falcón, J., Nguyen, B., Bullock, T., 2004. Slow wave sleep in crayfish. Proc. Natl. Acad. Sci. U S A 101, 11857–11861.

Ray, S., Reddy, A.B., 2016. Cross-talk between circadian clocks, sleep-wake cycles, and metabolic networks: dispelling the darkness. Bioessays 38, 394–405.

Rosen, M.G., 2018. How bizarre? A pluralist approach to dream content. Conscious. Cognit. 62, 148–162.

Rowe, K., Moreno, R., Lau, T.R., Wallooppillai, U., Nearing, B., Kocsis, B., Quattrochi, J., Hobson, J.A., Verrier, R.L., 1999. Heart rate surges during REM sleep are associated with theta rhythm and PGO activity in cats. Am. J. Physiol. 277, 843–849.

Royo, J., Aujard, F., Pifferi, F., 2019. Daily torpor and sleep in a non-human primate, the gray mouse lemur (*Microcebus murinus*). Front. Neuroanat. 13, 87.

Saper, C.B., Fuller, P.M., Pedersen, N.P., Lu, J., Scammell, T.E., 2010. Sleep state switching. Neuron 68, 1023–1042.

Schartner, M.M., Pigorini, A., Gibbs, S.A., Arnulfo, G., Sarasso, S., Barnett, L., Nobili, L., Massimini, M., Seth, A.K., Barrett, A.B., 2017. Global and local complexity of intracranial EEG decreases during NREM sleep. Neurosci. Conscious. 2017, niw022.

Schechtman, V.L., Harper, R.M., Lucas, E., Taube, D.M., Chase, M.H., 1985. Sleep-waking modulation of respiratory sinus arrhythmia in aged and young adult cats. Exp. Neurol. 88, 234–239.

Schwartz, W.J., Klerman, E.B., 2019. Circadian neurobiology and the physiologic regulation of sleep and wakefulness. Neurol. Clin. 37, 475–486.

Shalbaf, R., Behnam, H., Jelveh Moghadam, H., 2015. Monitoring depth of anesthesia using combination of EEG measure and hemodynamic variables. Cogn. Neurodyn. 9, 41–51.

Shaw, P.J., Cirelli, C., Greenspan, J., Tononi, G., 2000. Correlates of sleep and waking in *Drosophila melanogaster*. Science 287, 1834–1837.

Silva-Perez, M., Sanchez-Lopez, A., Pompa-Del-Toro, N., Escudero, M., 2020. Identification of the sleep-wake states in rats using the high-frequency activity of the electroencephalogram. J. Sleep Res. e13233.

Sinner, B., Graf, B.M., 2008. Ketamine. In: Schüttler, J., Schwilden, H. (Eds.), Modern Anesthetics. Springer Berlin Heidelberg, Berlin, Heidelberg, pp. 313–333.

Sonntag, M., Arendt, T., 2019. Neuronal activity in the hibernating brain. Front. Neuroanat. 13, 71.

Steriade, M., Amzica, F., Contreras, D., 1994. Cortical and thalamic cellular correlates of electroencephalographic burst-suppression. Electroencephalogr. Clin. Neurophysiol. 90, 1–16.

Sullivan, D., Mizuseki, K., Sorgi, A., Buzsaki, G., 2014. Comparison of sleep spindles and theta oscillations in the hippocampus. J. Neurosci. 34, 662–674.

Sztajzel, J., 2004. Heart rate variability: a noninvasive electrocardiographic method to measure the autonomic nervous system. Swiss Med. Wkly. 134, 514–522.

Taylor, N.E., Van Dort, C.J., Kenny, J.D., Pei, J., Guidera, J.A., Vlasov, K.Y., Lee, J.T., Boyden, E.S., Brown, E.N., Solt, K., 2016. Optogenetic activation of dopamine neurons in the ventral tegmental area induces reanimation from general anesthesia. Proc. Natl. Acad. Sci. U S A 113, 12826–12831.

Tononi, G., Laureys, S., 2009. The neurology of consciousness: an overview. In: Laureys, S., Tononi, G. (Eds.), The Neurology of Consciousness: Cognitive Neuroscience and Neuropathology. Elsevier, San Diego, pp. 375–412.

Torterolo, P., Castro-Zaballa, S., Cavelli, M., Gonzalez, J., 2019. Arousal and normal conscious cognition. In: Garcia-Rill, E. (Ed.), Arousal in Neurological and Psychiatric Diseases, pp. 1–24.

Trinder, J., Kleiman, J., Carrington, M., Smith, S., Breen, S., Tan, N., Kim, Y., 2001. Autonomic activity during human sleep as a function of time and sleep stage. J. Sleep Res. 10.

Tung, A., Szafran Martin, J., Bluhm, B., Mendelson Wallace, B., 2002. Sleep deprivation potentiates the onset and duration of loss of righting reflex induced by propofol and isoflurane. Anesthesiology 97, 906–911.

Uhlhaas, P.J., Pipa, G., Neuenschwander, S., Wibral, M., Singer, W., 2011. A new look at gamma? High- (>60 Hz) gamma-band activity in cortical networks: function, mechanisms and impairment. Prog. Biophys. Mol. Biol. 105, 14–28.

Vanderwolf, C.H., 1969. Hippocampal electrical activity and voluntary movement in the rat. Electroencephalogr. Clin. Neurophysiol. 26, 407–418.

Vanini, G., Watson, C.J., Lydic, R., Baghdoyan, H.A., 2008. Gamma-aminobutyric acid-mediated neurotransmission in the pontine reticular formation modulates hypnosis, immobility, and breathing during isoflurane anesthesia. Anesthesiology 109, 978–988.

Vanini, G., Nemanis, K., Baghdoyan, H.A., Lydic, R., 2014. GABAergic transmission in rat pontine reticular formation regulates the induction phase of anesthesia and modulates hyperalgesia caused by sleep deprivation. Eur. J. Neurosci. 40, 2264–2273.

Vanini, G., Bassana, M., Mast, M., Mondino, A., Cerda, I., Phyle, M., Chen, V., Colmenero, A.V., Hambrecht-Wiedbusch, V.S., Mashour, G.A., 2020. Activation of preoptic GABAergic or glutamatergic neurons modulates sleep-wake architecture, but not anesthetic state transitions. Curr. Biol. 30, 779–787.e774.

Vaughn, B.V., Quint, S.R., Messenheimer, J.A., Robertson, K.R., 1995. Heart period variability in sleep. Electroencephalogr. Clin. Neurophysiol. 94, 155–162.

Vijayan, S., Ching, S., Purdon, P.L., Brown, E.N., Kopell, N.J., 2013. Thalamocortical mechanisms for the anteriorization of alpha rhythms during propofol-induced unconsciousness. J. Neurosci. 33, 11070–11075.

Vorster, A.P., Krishnan, H.C., Cirelli, C., Lyons, L.C., 2014. Characterization of sleep in *Aplysia californica*. Sleep 37, 1453–1463.

Vyazovskiy, V.V., Palchykova, S., Achermann, P., Tobler, I., Deboer, T., 2017. Different effects of sleep deprivation and torpor on EEG slow-wave characteristics in Djungarian hamsters. Cerebr. Cortex 27, 950–961.

Walker, J.M., Glotzbach, S.F., Berger, R.J., Heller, H.C., 1977. Sleep and hibernation in ground squirrels (*Citellus spp.*): electrophysiological observations. Am. J. Physiol. 233, 213–221.

Wang, S., Steyn-Ross, M.L., Steyn-Ross, D., Wilson, M., Sleigh, J., 2014. EEG slow-wave coherence changes in propofol-induced general anesthesia: experiment and theory. Front. Syst. Neurosci. 8.

Wasilczuk, A.Z., Harrison, B.A., Kwasniewska, P., Ku, B., Kelz, M.B., McKinstry-Wu, A.R., Proekt, A., 2020. Resistance to state transitions in responsiveness is differentially modulated by different volatile anaesthetics in male mice. Br. J. Anaesth. 125, 308–320.

Winson, J., 1974. Patterns of hippocampal theta rhythm in the freely moving rat. Electroencephalogr. Clin. Neurophysiol. 36, 291–301.

Xiong, J., Kurz, A., Sessler, D.I., Plattner, O., Christensen, R., Dechert, M., Ikeda, T., 1996. Isoflurane produces marked and nonlinear decreases in the vasoconstriction and shivering thresholds. Anesthesiology 85, 240–245.

Zhang, X., Roy, R., Jensen, E., 2001. EEG complexity as a measure of depth of anesthesia for patients. IEEE Trans. Biomed. Eng. 48, 1424–1433.

Zoughi, T., Boostani, R., Deypir, M., 2012. A wavelet-based estimating depth of anesthesia. Eng. Appl. Artif. Intell. 25, 1710–1722.

CHAPTER 2

Polysomnography in humans and animal models: basic procedures and analysis

Pablo Torterolo[1,2], Joaquín Gonzalez[1], Santiago Castro-Zaballa[1], Matías Cavelli[1,3], Alejandra Mondino[1,4], Claudia Pascovich[1,5], Nicolás Rubido[6,7], Eric Murillo-Rodríguez[2,8] and Giancarlo Vanini[4]
[1]Laboratorio de Neurobiología del Sueño, Departamento de Fisiología, Facultad de Medicina, Universidad de la República, Montevideo, Uruguay; [2]Intercontinental Neuroscience Research Group, Mérida, Yucatán, México; [3]Department of Psychiatry, University of Wisconsin, Madison, WI, United States; [4]Department of Anesthesiology, University of Michigan, Ann Arbor, MI, United States; [5]Consciousness and Cognition Laboratory, Department of Psychology, University of Cambridge, Cambridge, United Kingdom; [6]Aberdeen Biomedical Imaging Centre, University of Aberdeen, Aberdeen, United Kingdom; [7]Instituto de Física, Facultad de Ciencias, Universidad de la República, Montevideo, Uruguay; [8]Laboratorio de Neurociencias Moleculares e Integrativas, Escuela de Medicina, División Ciencias de la Salud, Universidad Anáhuac Mayab, Mérida, Yucatán, Mexico

Introduction

Sleep is a critical physiological process. In most mammals and birds, two sleep states can be readily distinguished: rapid eye movement (REM) sleep and nonrapid eye movement (NREM) sleep [also called slow wave sleep (SWS)]. Together with the state of wakefulness (W), they constitute the wake-sleep cycle, one of the most visible circadian rhythms of the organism. Polysomnography (PSG) is the basic tool used to differentiate these behavioral states, both in clinical and human research settings, as well as in animal models.

Polysomnography

PSG is considered the gold standard method for studying sleep pathologies, and an essential tool for sleep research in both humans and animal models (Carskadon and Dement, 2017). It consists in the simultaneous recording of three biological signals: electroencephalogram (EEG), electromyogram (EMG), and eye movements [electrooculogram (EOG)]. Other bioelectrical signals can be also recorded according to the aim of the study (Keenan and Hirshkowitz, 2017).

Methodological Approaches for Sleep and Vigilance Research
ISBN 978-0-323-85235-7
https://doi.org/10.1016/B978-0-323-85235-7.00010-7

The EEG is produced by the summed electrical activities of populations of neurons, with a modest contribution from glial cells (Lopes da Silva, 2010). Pyramidal neurons of the cortex are the main contributors of the EEG signal, since they are arranged in palisades with the apical dendrites aligned perpendicularly to the cortical surface. The electrical fields generated by these neurons can be recorded by means of electrodes located from the cortical surface which is called electrocorticogram (ECoG) or intracranial EEG, or from the scalp (standard EEG). In the standard EEG, there is more distance between the source and the electrodes, and the signal is filtered out by the skull and scalp, which determines worse spatial resolution and reduction of the amplitude of the signal, mainly for oscillations higher than 30 Hz. On the contrary, oscillations up to 200 Hz can be easily recorded with the ECoG.

Several rhythmic oscillations can be observed in the EEG. These rhythms are generated in the thalamus and/or at cortical levels, and are modified according to the behavioral state (Torterolo et al., 2019). The EMG also changes abruptly across behavioral states, while the EOG tracks the eyes movements that characterize REM sleep.

Polysomnography in humans

Procedures

In standard EEG, the recording is obtained by placing electrodes on the scalp through a conductive gel. Individual electrodes or caps/nets with embedded electrodes are connected to a differential amplifier (Keenan and Hirshkowitz, 2017). Electrode locations and names are specified by the International 10–20 system for standard EEG clinical applications. However, a smaller number of electrodes may be used in PSG examination, which is incomplete for diagnosis of a neurological condition. Also, a smaller number of electrodes are typically used when recording term or premature neonates (Grigg-Damberger, 2016), which need to adapt the 10–20 system according to the neonate's head circumference. The number of electrodes may vary in research settings according to the goal of the study (up to 256 electrodes in high-density arrays).

The amplifiers increase (typically 1000–100,000 times or 60–100 dB of voltage gain) the voltage differences (ΔV) between the active and the referential electrode (or between two active electrodes in bipolar arrangements). The amplitude of the adult human standard EEG signal is about 10–100 μV.

Analog-to-digital sampling typically occurs at 256–512 Hz and 12 bits in clinical settings, but larger sampling rates are often used for research. Digital EEG signal is stored and can be filtered for display (and/or can be filtered before acquisition). Typical settings for the high-pass filter and a low-pass filter are 0.5–1 and 35–70 Hz, respectively. An additional notch (50 or 60 Hz) filter may be used to remove artifacts caused by electrical power lines.

For some conditions (usually for an evaluation for epilepsy surgery), the electrodes are placed on the brain surface by craniotomy to record the ECoG. In some cases, electrodes may be placed into deep brain structures such as the hippocampus. The procedure is the same as the standard EEG, but is typically recorded at higher sampling rates to observe the high-frequency components of the signals.

Since the EEG signal represents a ΔV between two electrodes, the EEG recording may be set in several ways or channels. This representation is known as montages. The montage could be bipolar, namely ΔV between two active electrodes; or referential, an active against a referential electrode. The reference could vary according to the objective of the study, but the referential electrode is usually placed in the midline position, both earlobes or mastoids. Also, an average of all the electrodes, Laplacian (Gordon and Rzempoluck, 2004) or computational such as reference electrode standardization technique or REST (Zheng et al., 2018), can be used as reference. For ECoG, it is also common to use the white matter as reference (Lachaux et al., 2003).

The EMG is recorded by bipolar electrodes on the skin above the muscle. The submental muscle is usually used, and the tibialis anterior can also be employed to record leg movements. The EOG is used to detect changes in eye movements, since moving the eyes creates changes in electrical potential (the eye behaves as a dipole). The surface electrodes are placed on the outer edge of each eye.

Other bioelectrical signals are often recorded. The respiratory effort is usually recorded by abdominal and thoracic piezoelectric bands that generate voltage changes with breathing movements. The ventilation can be also evaluated with respiratory inductance plethysmography by means of belts that capture the movement of chest and abdominal walls. The airflow measurement is carried out with a nasobuccal thermistor that detects temperature differences induced by breathing, and/or by a pressure transducer sensitive to air flow. In addition, signals such as oxygen saturation (by pulse oximetry), capnography (CO_2 sensor), electrocardiogram, snoring sensor, moving sensor are frequently recorded. Finally, a video-recording synchronized with the PSG is commonly used to track the behavior during sleep.

Simplified portable PSG equipment can also be used to perform the recordings at home, either for clinical exploration or research. Wireless systems technologies have been also developed (Markwald et al., 2016).

Sleep characteristics in humans

The EEG recording of W is marked by the presence of high frequency and low-voltage oscillations (cortical activation) (Carskadon and Dement, 2017). During relaxed W with eyes closed, a high-amplitude alpha (8–12 Hz) oscillation appears mainly in the occipital (visual) cortex.

At sleep onset, adults enter into NREM sleep. In humans, three NREM sleep phases are recognized: N1, N2, and N3, according to the depth of the state. N1 represents the transition from W to sleep and is characterized by the presence of relatively low-voltage, mixed frequency waves with a prominence of activity in the theta (4–7 Hz) range. N1 is a shallow sleep stage with low-arousal threshold. N2 is an unequivocal stage of sleep that is characterized by the presence of two types of EEG events: sleep spindles and K-complexes. The former is an event with a spindle shape of 0.5–2 s in duration and a frequency of 12–14 Hz that contrast with the lower frequency of the EEG background activity. K-complexes are often associated with sleep spindles, and consist of a brief negative sharp high-voltage peak (usually greater than 100 μV), followed by a slower positive complex and a final negative peak. The presence of low frequency (0.5–4 Hz, delta oscillations) of high-amplitude waves characterizes the EEG during N3. REM sleep (also called stage R) is a deep sleep stage even though the EEG has low-voltage mixed frequency activity that is similar to W or N1; hence, it is also called "paradoxical" sleep. Saw-tooth waves (theta activity, 4–7 Hz) are often observed in conjunction with bursts of REMs in the EOG. A decrease in the activity of the EMG caused by muscle atonia also characterizes REM sleep; usually small muscles twitches are also identified in the EMG. REM sleep occurs approximately 90 min after the onset of sleep, a parameter known as REM sleep latency (Carskadon and Dement, 2011). Dreams occur mainly during this sleep state.

Nighttime sleep in humans is characterized by the presence of four to five sleep cycles. They comprise the period between the onset of sleep until the end of the first episode of REM sleep or the period from the end of a REM sleep episode to the end of the subsequent episode. The average duration of sleep cycles is approximately 90 min in adults. Young adults spend 20%–28% of a night's sleep in REM sleep, 4%–5% in stage N1 sleep, 46%–50% in stage N2 sleep, and 20%–24% in stage N3. A shortening of sleep duration, which occurs in association with advancing age, is mainly related to the reduction in N3 sleep and REM sleep.

In newborns, W, immature NREM sleep known as quiet sleep (QS), immature REM sleep called active sleep (AS), and transitional or intermediate sleep (IS) are distinguished (Graven, 2006; Andre et al., 2010). The time spent in each stage of sleep differs throughout development. In the neonate, there is a predominance of AS. QS increases with gestational age and IS begins to disappear near term. The sleep-wake cycle in the neonatal stage also presents short sleep cycles (ultradian rhythm) (Mirmiran et al., 2003).

The results of the PSG are usually displayed in a hypnogram (see Fig. 2.1 for an example in animals). Thereafter, several parameters such as time, duration, and latencies of the different sleep phases are quantified.

Polysomnography in animal models

Procedures

Adapted PSG have been used to record different species. Rats, mice, and cats are the most utilized animal models for sleep research. Under general anesthesia, these animals are implanted with electrodes for PSG, usually, by means of stereotaxic surgery. Noninvasive PSG has been tried in dogs and equines, and is an option for veterinary medicine (Kis et al., 2014; Wohr et al., 2016; Gergely et al., 2020).

As the rat is one of the most common animal models employed to study sleep, we next describe the standard procedures to record and analyze the PSG in this model. In our experience, rats (Wistar, 250–300 g) are anesthetized by a mixture of ketamine-xylazine (90 mg/kg, 5 mg/kg i/p., respectively). The animals are positioned in a stereotaxic frame, and the skull is exposed. The brain electrical activity is recorded by means of stainless-steel screw electrodes (1.0 mm of diameter) placed on the skull with their tips touching the dura mater, i.e., ECoG. Usually, we use the montage illustrated in Fig. 2.1A. Six electrodes are located on the neocortex forming two anterior-posterior consecutive squares centered with respect to the midline, and the frontal square centered with respect to Bregma [all nearby neocortical electrodes are separated by the same distance, 5 mm, which is important for quantitative EEG (qEEG) analysis]. The electrodes are located bilaterally in primary motor cortex (M1: L $\pm$ 2.5 mm, AP + 2.5 mm), primary

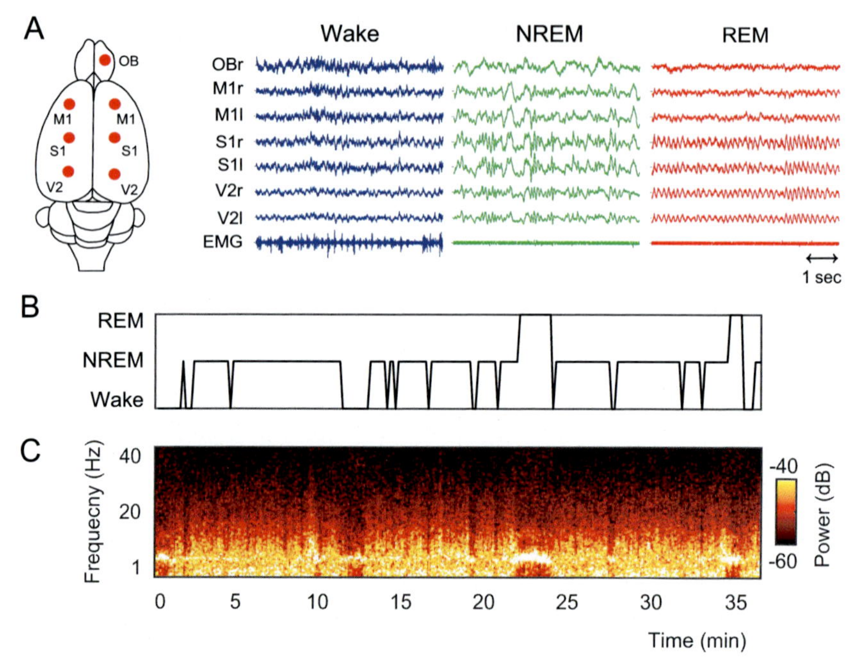

Figure 2.1 ***Polysomnographic recordings in the rat.*** (A) Schematic representation of the rat's brain along with the electrodes employed to record the intracranial electroencephalogram (referenced to the cerebellum). Representative traces are shown to the right of the panel for each one of the sleep-wake states. OB, olfactory bulb; M1, primary motor cortex; S1, primary somatosensory cortex; V2, secondary visual cortex; EMG, electromyogram; r and l, right and left hemispheres. (B) Hypnogram showing the sleep-wake states as a function of time. (C) Spectrogram showing the time-frequency representation of V2r cortex. *(Modified from Gonzalez, J., Cavelli, M., Mondino, A., Pascovich, C., Castro-Zaballa, S., Torterolo, P., Rubido, N., 2019. Decreased electrocortical temporal complexity distinguishes sleep from wakefulness. Sci. Rep. 9, 18457).*

somatosensory cortex (S1: L $\pm$ 2.5 mm, AP $-$ 2.5 mm) and secondary visual cortex (V2: L $\pm$ 2.5 mm, AP $-$ 7.5 mm, according to Paxinos and Watson atlas) (Paxinos and Watson, 2005). The other electrode is located over the right olfactory bulb (OB) (L: $+$1.25 mm, AP $+$ 7.5 mm). A reference electrode is positioned above the cerebellum. To record the EMG, a pair of electrodes are inserted into the neck muscles. In rodents, the recordings of the ECoG and EMG are enough to identify W, NREM, and REM sleep states (there is no need to record the EOG). The electrodes are soldered into a 12-pin socket and fixed onto the skull with acrylic cement. After the animals recover from surgery, they are adapted to the recording chamber for 1 week.

The recordings are performed through a rotating connector to allow tethered rats to move freely within the recording box (freely movement condition). Bioelectric signals are amplified (×1000), filtered (0.1–500 Hz), sampled (1024 Hz, 16 bits), and stored in a PC using an acquisition software.

As in humans, to EOG, ECG and respiratory activity can be included in the recording. Also, deep electrodes may be added. For example, in cats, it is common to implant bipolar electrodes into the hippocampus to record theta oscillations, or into the lateral geniculate nucleus to record the ponto-geniculo-occipital (PGO) waves (Torterolo et al., 2016).

Another approach for recording is known as head-fixed or semi-restricted condition. In this case, the animal must be adapted to be in the recording position for several weeks. Wireless recordings, for example, by means of telemetry, have been developed since the late 70s (Neuhaus and Borbely, 1978). New technologies are increasing the possibilities for wireless PSG as well for others bioelectrical recordings (Fan et al., 2011).

Data analysis

In the rat, W and sleep are defined by PSG as follows (Fig. 2.1A): (1) W, by the presence of low voltage fast waves in frontal cortex, a mixed theta (4.5–9 Hz) activity in occipital cortex and relatively high EMG activity. (2) Light sleep, by the occurrence of high voltage slow cortical waves interrupted by low voltage fast EEG activity. (3) SWS, by the occurrence of continuous high amplitude slow (0.5–4 Hz) frontal and occipital waves and sleep spindles (similar to humans' spindles) combined with a reduced EMG activity; light sleep and SWS are grouped as NREM sleep. (4) REM sleep, by the presence of low voltage fast frontal waves, a regular theta rhythm in the occipital cortex (the origin of this rhythm is the hippocampus), and a silent EMG except for occasional myoclonic twitching.

A transitional state from SWS to REM sleep is often identified in the rat, and called intermediate state (IS) (Gottesmann, 1996). This state is a mixture of sleep spindles and theta activity. In the cat, this state is also signaled by the PGO waves that are better recorded from the lateral geniculate nucleus (Callaway et al., 1987). Single and large PGOs characterize the IS. At the onset of REM sleep, the frequency of the PGOs increases; they appear in bursts and decrease in amplitude. In the rat, only the "P" component of the PGO waves can be recorded from the pons (Datta and Hobson, 2000).

Total time spent in W, LS, SWS, NREM (NREM = LS + SWS), and REM sleep, as well as the duration and the number of episodes are determined. NREM and REM sleep latencies (from the beginning of the recording) is also included in the analysis. As is shown in Fig. 2.1B and C, the hypnogram associated to the spectrogram (that shows the power spectrum of the recorded ECoG signal, see below) is used to represent the W and sleep profile throughout the recording time.

Quantitative electroencephalogram analysis

qEEG is the field concerned with the numerical analysis of EEG or ECoG data, and associated behavioral correlates. These include linear and nonlinear analysis of the electrical activity that can be done either for human or animal recordings. Next, we will briefly comment on some prototypical ECoG analysis from rats that we perform in our laboratory.

Spectral power

The power spectral density of the EEG provides the weight of the frequency components of the signal. Mathematically, for a signal $x(t)$, its power spectral density is defined as

$$P_x(f) = |x(f)|^2$$

where

$$x(f) = \int_{-\infty}^{+\infty} x(t) e^{2\pi i f t} dt$$

is the Fourier transform of x. In practice, it is calculated by means of the Fast Fourier transform (FFT). The modern generic FFT algorithm was developed by Cooley and Tukey in 1965 (Coolye and Tukey, 1965), which increases the computation speed of the Fourier analysis simplifying its complexity. However, the FFT is seldom employed alone. In contrast, it is common to estimate the power spectral density by means of Welch's algorithm, which significantly reduces noise in the spectrum estimate. This method relies upon estimating the power spectrum in short time windows, employing the FFT, and then averaging the results to obtain a better estimate (which reduces the nonstationary components). It should be noted that these short windows are usually multiplied by a windowing function (i.e., Hamming, Hanning or Slepian functions) to reduce the distortion produced by the windows edges. The result is a representation of the distribution of the energy or the amplitude of a signal in its different frequency components. The EEG power at a given frequency reflects the degree of local synchronization of the extracellular potential at that frequency (Buzsaki et al., 2012).

As it is shown in the spectrogram of Fig. 2.1C and in Fig. 2.2A, the analysis of the frequency content or power spectrum of the ECoG signal of the rat shows that in comparison with others behavioral states, the power of delta, theta, and sigma (sleep spindles) frequency bands during NREM sleep is high, while high frequency (>30 Hz) power values are low. During REM sleep, the power of high-frequency bands is intermediate between W and NREM sleep levels.

Another option to perform a spectral estimation is the wavelet transform (WT), in which any general function can be expressed as an infinite series of wavelets (Dhiman et al., 2013). Since WT allows the use of variable sized windows, it gives a more flexible time-frequency representation of a signal. To get a finer low-frequency resolution, long-time windows are used. In contrast, to get high-frequency information, short time windows are used.

Spectral coherence

Spectral coherence is a tool that is used to determine the degree of functional coupling between two cortical areas by means of the correlation between different frequency bands. In other words, it provides an objective

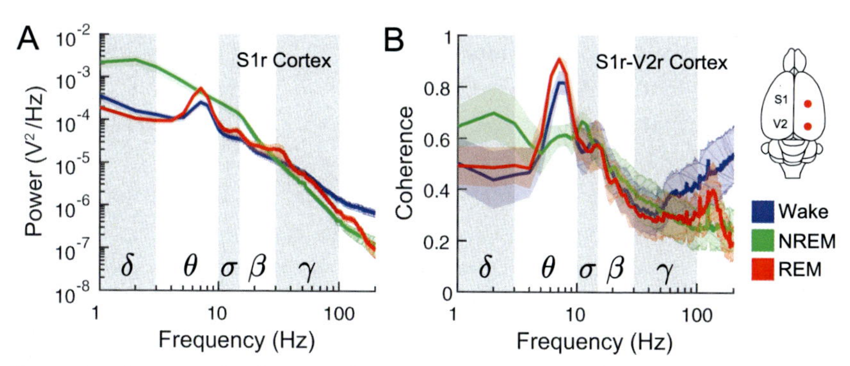

Figure 2.2 ***Power and coherence during wakefulness and sleep.*** (A) Power spectrum estimates from the right primary somatosensory cortex (S1r) during wakefulness, nonrapid eye movement and rapid eye movement sleep. Each trace shows the mean (solid line) +/− the standard error of the mean (shaded areas), $n = 6$. The frequency bands are shown with their corresponding Greek letters. (B) Coherence between S1r and right secondary visual cortex (V2r) during wakefulness, nonrapid eye movement and rapid eye movement sleep. Each trace shows the mean (solid line) +/− the standard error of the mean (shaded areas), $n = 6$.

index of functional interactions between different regions of the cerebral cortex. To be completely coherent, two waves must have a constant phase difference at a given frequency, and the relationship between the amplitudes at that frequency must be kept constant. This implies that two cortical areas that coordinate their electrical activity will present an increase in coherence between their electrical activities. It has therefore been proposed that the degree of coherence between the EEG of different cortices recorded simultaneously would reflect the strength of the functional interconnections (reentries) that occur between them (Bullock and McClune, 1989; Edelman and Tononi, 2000). The coherence is obtained from the cross spectral density (CSD, or the Fourier analysis of the cross-covariance function) between the two waves, normalized by the power of the spectral density of each wave. Therefore, the coherence between two waves *a* and *b*, at a given frequency *f*, is obtained as follows:

$$COH_{xy}(f) = \frac{\left|CSD_{xy}(f)\right|^2}{P_x(f)P_y(f)}$$

The coherence between two waveforms is a function of frequency and ranges from 0 for totally incoherent waveforms to 1 for maximal coherence. For two waveforms to be completely coherent at a particular frequency over a given time range, the phase shift between the waveforms must be constant and the amplitudes of the waves must have a constant ratio.

This function (magnitude-square cross coherence) evaluates both the amplitude and phase relationships. Importantly, the oscillations do not have to be fully synchronized to be consistent. These can be coupled between distant cortical areas with constant lag due to conduction and synaptic latencies, and still be coherent.

Fig. 2.2B shows an example where we analyze the intrahemispheric coherence between the right S1 and V2 channels across behavioral states. During W, there is a relatively large coherence in the theta band and in high-frequency components of the signal. On the contrary, during NREM sleep, coherence is high both in delta (slow oscillations) and sigma (spindle) bands. Interestingly, during REM sleep, there is a decrease in gamma coherence. High-gamma power (that reflect local synchronization) accompanied by minimal gamma coherence (that reflects long-range synchronization) is a trait that characterizes REM sleep, which is conserved in rodents, felines, and humans (Voss et al., 2009; Castro et al., 2013; Cavelli et al., 2015, 2017; Torterolo et al., 2016). A peak in what it is known as high-frequency oscillations (HFOs, about 120 Hz) is also observed (Cavelli et al., 2018).

Cross-frequency coupling

Apart from synchronization between areas at different frequency bands, the different frequency bands also interact within a single brain area in a phenomenon named cross-frequency coupling. This type of analysis thus allows to quantify whether fast oscillations (>30 Hz) "nest" in a certain phase of slower waves. One of the most studied forms of cross-frequency coupling is phase-amplitude coupling, which assesses whether the fast frequency envelope is modulated by the phase of a slower oscillation. This analysis is performed using the modulation index framework described by Tort (Tort et al., 2010). Briefly, the raw signal is filtered to obtain the slow frequency components, and then the phase time series is extracted from their analytical representation based on the Hilbert transform. In addition, the same raw signal is then also filtered to obtain the high-frequency components, and their amplitude time series are also obtained from their Hilbert analytical representation. Then, phase-amplitude distributions are computed between all slow-fast frequency combinations. Finally, the modulation index is obtained as $MI = (H\text{max} - H)/H\text{max}$, where Hmax is the maximum possible Shannon entropy for a given distribution [log(number of bins)], and H is the actual entropy of such distribution. For example, in Fig. 2.3, it is readily observed that the amplitude of high-frequency bands oscillations (120–160 Hz) is present in phase with the slow theta range oscillations (phase/amplitude coupling) during W and REM sleep (Cavelli et al., 2018; Gonzalez et al., 2020b). This is observable in both S1 and V2 cortex.

Complexity of the electrocorticogram signals

The complex nature of EEG or ECoG signals cannot be explained completely just by power and coherence analyses. In contrast, the field of nonlinear dynamics has developed measures and models that account for the complexity of the systems and their emerging interactions. A general approach to study time-signals is the characterization of their randomness or unpredictability of a signal.

One approach that has gained track over the last 20 years is the Ordinal Pattern (OP) analysis, which allows to encode any signal into OPs and approximate its Shannon entropy. The OPs are constructed by dividing a time-series $x(t)$ into nonoverlapping vectors (each of size D, commonly $D = 3,4$) and classifying each one according to the relative magnitude of its D elements. The classification is done by determining how many permutations are needed to order its elements increasingly; namely, an OP is

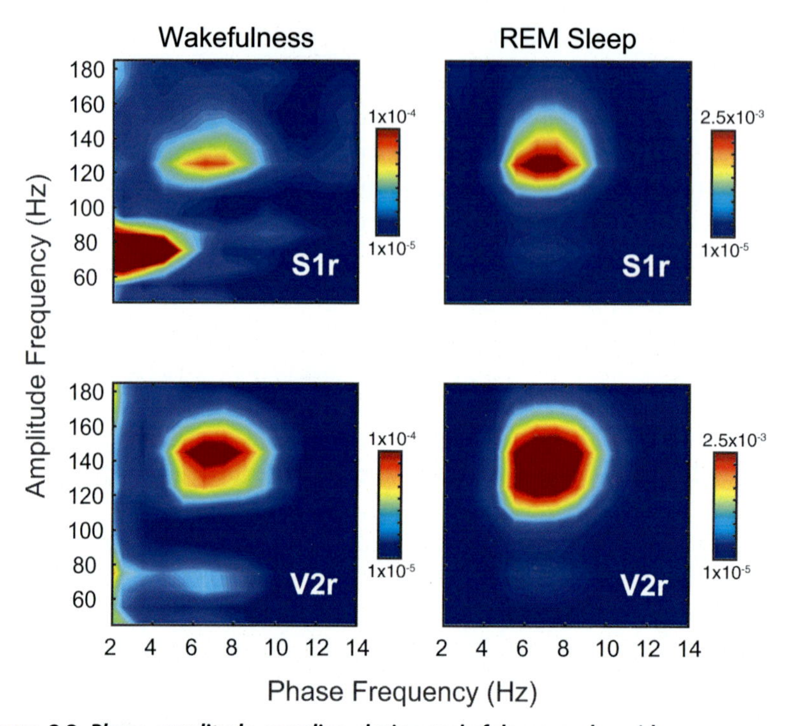

Figure 2.3 ***Phase amplitude coupling during wakefulness and rapid eye movement sleep.*** Comodulograms are shown for the S1r and V2r cortex. These plots indicate, through the color scale (modulation index), the amount of which the phase of the slow-frequency band (x-axis) modulates the amplitude of the high-frequency band (y-axis). Note that the modulation levels are one order of magnitude higher during rapid eye movement sleep (see the color calibration).

associated to represent the vector's permutations. For example, for $D = 2$, the time-series would be divided into vectors containing two consecutive values, such $\{x(t), x(t+1)\}$, These vectors have only two possible OPs for any time: either $x(t) > x(t+1)$ or $x(t) < x(t+1)$, which correspond to making 0 permutation or 1 permutation, respectively. It is worth noting that the number of possible permutations increases factorially with increasing vector length, i.e., for vectors of length D, there are $D!$ possible OPs. After that, the Shannon entropy

$$H(S) = -\sum_{x \in S} p(x) log[p(x)]$$

is computed from the OPs probability distribution, being S the alphabet containing each OP. This approximation is known as Permutation Entropy

(PeEn). In contrast to other methods, PeEn is a time-series complexity measure that is simple to implement, is robust to noise and short time-series, and works for arbitrary data sets. In particular, it has been shown that PeEn applied to EEG signals captures differences in waking and sleep states. Fig. 2.4 shows that the PeEN is larger during W and decreases during sleep (Gonzalez et al., 2019, 2020a). Importantly, the PeEn is highly dependent on the sampling frequencies employed, and can be therefore understood in terms of the ECoG's frequency content. Other metrics such as Lempel-Ziv are other options to evaluate the complexity of the signals.

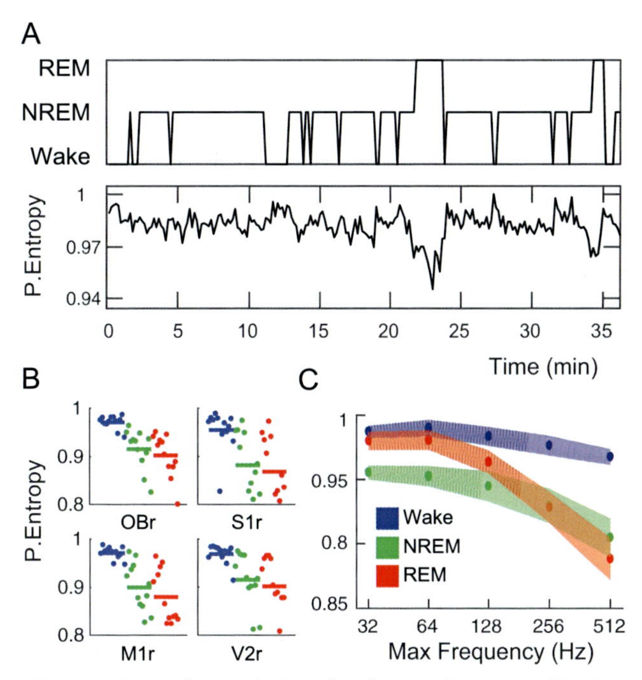

Figure 2.4 ***Temporal complexity during the sleep-wake states.*** (A) Top: Hypnogram showing the sleep-wake states as a function of time (same as in Fig. 2.1). Bottom: Temporal complexity as a function of time, estimated by Permutation Entropy. (B) PeEn values (for embedding dimension, $D = 3$) for 12 rats, differentiating each cortex electrode and sleep state (color code shown in panel C). Namely, each dot corresponds to the time-averaged PeEn value of each rat and cortex, where the horizontal bars are the population mean. (C) Average values of the seven cortical recording sites (shown in Fig. 2.1). Panel shows the averaged PeEn values as a function of the maximum frequency resolution (sampling frequency divided by 2, according to the Nyquist-Shannon criterion). Shaded areas in this panel depict the standard error of the mean for the PeEn, $n = 12$. *(Modified from Gonzalez, J., Cavelli, M., Mondino, A., Pascovich, C., Castro-Zaballa, S., Torterolo, P., Rubido, N., 2019. Decreased electrocortical temporal complexity distinguishes sleep from wakefulness. Sci. Rep. 9, 18457.)*

The entropy of a signal only gives information about the state of disorder of a single EEG channel. In our laboratory, we are also using a metric known as "mutual information" that assesses the amount of information shared between two cortical areas (Rubido et al., 2014; Bianco-Martinez et al., 2016; Garcia et al., 2020).

An important parameter to analyze is the sense of the information flow, i.e., "feedforward" (from primary to association cortex) versus "feedback" (from association to primary cortex), which has been seen to change across behavioral states. This could be done employing either "Granger Causality" or a time-lagged mutual information metric, such as Transfer Entropy (Imas et al., 2005). Both estimate the magnitude and direction of the temporal relationship between simultaneously recorded signals (Cekic et al., 2018) and are based on the statistical hypothesis that if A causes B, then the A's time series should be able to forecast or predict B.

Conclusion

In the present study, we have reviewed the main PSG procedures and data analysis for humans and animal models. Finally, we commented about important tools for the qEEG analysis that are widely used in research settings.

Acknowledgments

This study was supported by the "Programa de Desarrollo de las Ciencias Básicas, PEDECIBA" from Uruguay.

References

Andre, M., Lamblin, M.D., d'Allest, A.M., Curzi-Dascalova, L., Moussalli-Salefranque, F., T, S.N.T., Vecchierini-Blineau, M.F., Wallois, F., Walls-Esquivel, E., Plouin, P., 2010. Electroencephalography in premature and full-term infants. Developmental features and glossary. Neurophysiol. Clin. 40, 59–124.

Bianco-Martinez, E., Rubido, N., Antonopoulos Ch, G., Baptista, M.S., 2016. Successful network inference from time-series data using mutual information rate. Chaos 26, 043102.

Bullock, T.H., McClune, M.C., 1989. Lateral coherence of the electrocorticogram: a new measure of brain synchrony. Electroencephalogr. Clin. Neurophysiol. 73, 479–498.

Buzsaki, G., Anastassiou, C.A., Koch, C., 2012. The origin of extracellular fields and currents–EEG, ECoG, LFP and spikes. Nat. Rev. Neurosci. 13, 407–420.

Callaway, C.W., Lydic, R., Baghdoyan, H.A., Hobson, J.A., 1987. Pontogeniculooccipital waves: spontaneous visual system activity during rapid eye movement sleep. Cell. Mol. Neurobiol. 7, 105–149.

Carskadon, M.A., Dement, W., 2011. Normal human sleep: an overview. In: Kryger, M.H., Roth, T., Dement, W. (Eds.), Principles and Practices of Sleep Medicine. Elsevier-Saunders, Philadelphia, pp. 16–26.
Carskadon, M.A., Dement, W.C., 2017. Normal human sleep: an overview. In: Kryger, M.H., Roth, T., Dement, W.C. (Eds.), Principles and Practices of Sleep Medicine. Elsevier-Saunders, Philadelphia, pp. 15–24.
Castro, S., Falconi, A., Chase, M.H., Torterolo, P., 2013. Coherent neocortical 40-Hz oscillations are not present during REM sleep. Eur. J. Neurosci. 37, 1330–1339.
Cavelli, M., Castro, S., Schwarzkopf, N., Chase, M.H., Falconi, A., Torterolo, P., 2015. Coherent neocortical gamma oscillations decrease during REM sleep in the rat. Behav. Brain Res. 281, 318–325.
Cavelli, M., Castro-Zaballa, S., Mondino, A., Gonzalez, J., Falconi, A., Torterolo, P., 2017. Absence of EEG gamma coherence in a local activated neocortical state: a conserved trait of REM sleep. Trans. Brain Rhytmicity 2, 1–13.
Cavelli, M., Rojas-Libano, D., Schwarzkopf, N., Castro-Zaballa, S., Gonzalez, J., Mondino, A., Santana, N., Benedetto, L., Falconi, A., Torterolo, P., 2018. Power and coherence of cortical high-frequency oscillations during wakefulness and sleep. Eur. J. Neurosci. 48, 2728–2737.
Cekic, S., Grandjean, D., Renaud, O., 2018. Time, frequency, and time-varying Granger-causality measures in neuroscience. Stat. Med. 37, 1910–1931.
Coolye, J.W., Tukey, J.W., 1965. An algorithm for the machine calculation of complex Fourier series. Math. Comput. 19, 297–301.
Datta, S., Hobson, J.A., 2000. The rat as an experimental model for sleep neurophysiology. Behav. Neurosci. 114, 1239–1244.
Dhiman, R., Priyanka, Saini, J.S., 2013. Wavelet analysis of electrical signals from brain: the electroencephalogram. In: Singh, H., Awasthi, A.K., Mishra, R. (Eds.), Quality, Reliability, Security and Robustness in Heterogeneous Networks, pp. 283–289.
Edelman, G.M., Tononi, G., 2000. A Universe of Consciousness. Basic Books, New York.
Fan, D., Rich, D., Holtzman, T., Ruther, P., Dalley, J.W., Lopez, A., Rossi, M.A., Barter, J.W., Salas-Meza, D., Herwik, S., Holzhammer, T., Morizio, J., Yin, H.H., 2011. A wireless multi-channel recording system for freely behaving mice and rats. PLoS One 6, e22033.
Garcia, R.A., Marti, A.C., Cabeza, C., Rubido, N., 2020. Small-worldness favours network inference in synthetic neural networks. Sci. Rep. 10, 2296.
Gergely, A., Kiss, O., Reicher, V., Iotchev, I., Kovacs, E., Gombos, F., Benczur, A., Galambos, A., Topal, J., Kis, A., 2020. Reliability of family dogs' sleep structure scoring based on manual and automated sleep stage identification. Animals (Basel) 10.
Gonzalez, J., Cavelli, M., Mondino, A., Pascovich, C., Castro-Zaballa, S., Rubido, N., Torterolo, P., 2020a. Electrocortical temporal complexity during wakefulness and sleep: an updated account. Sleep Sci. 47–50.
Gonzalez, J., Cavelli, M., Mondino, A., Pascovich, C., Castro-Zaballa, S., Torterolo, P., Rubido, N., 2019. Decreased electrocortical temporal complexity distinguishes sleep from wakefulness. Sci. Rep. 9, 18457.
Gonzalez, J., Cavelli, M., Mondino, A., Rubido, N., Bl Tort, A., Torterolo, P., 2020b. Communication through coherence by means of cross-frequency coupling. Neuroscience 449, 157–164.
Gordon, R., Rzempoluck, E.J., 2004. Introduction to laplacian montages. Am. J. Electroneurodiagn. Technol. 44, 98–102.
Gottesmann, C., 1996. The transition from slow-wave sleep to paradoxical sleep: evolving facts and concepts of the neurophysiological processes underlying the intermediate stage of sleep. Neurosci. Biobehav. Rev. 20, 367–387.
Graven, S., 2006. Sleep and brain development. Clin. Perinatol. 33, 693–706 (vii).

Grigg-Damberger, M.M., 2016. The visual scoring of sleep in infants 0 to 2 Months of age. J. Clin. Sleep Med. 12, 429–445.
Imas, O.A., Ropella, K.M., Ward, B.D., Wood, J.D., Hudetz, A.G., 2005. Volatile anesthetics disrupt frontal-posterior recurrent information transfer at gamma frequencies in rat. Neurosci. Lett. 387, 145–150.
Keenan, S., Hirshkowitz, M., 2017. Sleep stage scoring. In: Kryger, M.H., Roth, T., Dement, W.C. (Eds.), Principles and Practices of Sleep Medicine. Elsevier-Saunders, Philadelphia, pp. 1567–1575.
Kis, A., Szakadat, S., Kovacs, E., Gacsi, M., Simor, P., Gombos, F., Topal, J., Miklosi, A., Bodizs, R., 2014. Development of a non-invasive polysomnography technique for dogs (*Canis familiaris*). Physiol. Behav. 130, 149–156.
Lachaux, J.P., Rudrauf, D., Kahane, P., 2003. Intracranial EEG and human brain mapping. J. Physiol. Paris 97, 613–628.
Lopes da Silva, F., 2010. EEG: origin and measurement. In: Mulert, C., Lemieuz, L. (Eds.), EEG-fMRI. Springer-Verlag, Berlin, pp. 19–38.
Markwald, R.R., Bessman, S.C., Reini, S.A., Drummond, S.P., 2016. Performance of a portable sleep monitoring device in individuals with high versus low sleep efficiency. J. Clin. Sleep Med. 12, 95–103.
Mirmiran, M., Maas, Y.G., Ariagno, R.L., 2003. Development of fetal and neonatal sleep and circadian rhythms. Sleep Med. Rev. 7, 321–334.
Neuhaus, H.U., Borbely, A.A., 1978. Sleep telemetry in the rat. II. Automatic identification and recording of vigilance states. Electroencephalogr. Clin. Neurophysiol. 44, 115–119.
Paxinos, G., Watson, C., 2005. The Rat Brain. Academic Press, New York.
Rubido, N., Martí, A.C., Bianco-Martínez, E., Grebogi, C., Baptista, M.S., Masoller, C., 2014. Exact detection of direct links in networks of interacting dynamical units. New J. Phys. 16, 093010.
Tort, A.B., Komorowski, R., Eichenbaum, H., Kopell, N., 2010. Measuring phase-amplitude coupling between neuronal oscillations of different frequencies. J. Neurophysiol. 104, 1195–1210.
Torterolo, P., Castro-Zaballa, S., Cavelli, M., Chase, M.H., Falconi, A., 2016. Neocortical 40 Hz oscillations during carbachol-induced rapid eye movement sleep and cataplexy. Eur. J. Neurosci. 43, 580–589.
Torterolo, P., Castro-Zaballa, S., Cavelli, M., Gonzalez, J., 2019. Arousal and normal conscious cognition. In: Garcia-Rill, E. (Ed.), Arousal in Neurological and Psychiatric Diseases, pp. 1–24.
Voss, U., Holzmann, R., Tuin, I., Hobson, J.A., 2009. Lucid dreaming: a state of consciousness with features of both waking and non-lucid dreaming. Sleep 32, 1191–1200.
Wohr, A., Kalus, M., Reese, S., Fuchs, C.S., Erhard, M., 2016. Equine sleep behaviour and physiology based on polysomnographic examinations. Equine Vet. J. 48, 9.
Zheng, G., Qi, X., Li, Y., Zhang, W., Yu, Y., 2018. A comparative study of standardized infinity reference and average reference for EEG of three typical brain states. Front. Neurosci. 12, 158.

CHAPTER 3

Electrophysiological studies and sleep-wake cycle*

Md Aftab Alam[1,2], Andrey Kostin[1], Md Noor Alam[1,3]
[1]Research Service (151A3), Veterans Affairs Greater Los Angeles Healthcare System, Sepulveda, CA, United States; [2]Department of Psychiatry, University of California, Los Angeles, CA, United States; [3]Department of Medicine, David Geffen School of Medicine, University of California, Los Angeles, CA, United States

Introduction

The experimental approaches and tools used for studying sleep- and wake-regulatory systems and the mechanisms underlying these complex processes have been continuously evolving with technological advances. In this regard, electrophysiological tools have been pivotal to understanding sleep-wake physiology and pathophysiology. Although questions like why do we sleep or what causes us to sleep have always fascinated the human mind, it was not until the invention of devices to record the encephalogram, and other electrical signals in the 1920s onwards that it became possible to assess the quantity and quality of sleep-wakefulness, the neural substrate involved, and its underlying mechanisms. It is the electrophysiological parameters including electroencephalogram (EEG), electromyogram (EMG), and electrooculogram that the mammalian sleep-wake cycle is objectively classified as waking (active and quiet), and two alternating and distinct stages of sleep, namely, nonrapid eye movement (nonREM) sleep (further classified into sub-stages roughly paralleling its depth as assessed by the EEG slow-wave activity) and REM sleep (Aserinsky and Kleitman, 1953; Brown et al., 2012; Alam, 2013).

The first scientific evidence that sleep is a central process came from the pathological findings of Von Economo during the early 20th century, indicating that the hypothalamic area of the brain controls sleep-wake function. He found that patients who suffered from the epidemic encephalitis lethargica and manifested insomnia had lesions or damages in the anterior hypothalamus, and those exhibiting excessive somnolence or

* This work was supported by the US Department of Veterans Affairs, Merit Award - BX005167.

Methodological Approaches for Sleep and Vigilance Research
ISBN 978-0-323-85235-7
https://doi.org/10.1016/B978-0-323-85235-7.00008-9

sleepiness had lesions in the posterior hypothalamus and midbrain (Brown et al., 2012; Alam, 2013; Saper and Fuller, 2017). These findings led to a series of experimental ablation, lesion, and stimulation studies that helped identify major sleep- and wake-regulatory anatomical regions in the brain. With technical advances, these studies have been followed by much-refined cell-specific lesion, stimulation, electrophysiological, pharmacological, optogenetic, chemogenetic, and molecular studies to precisely understand the neuronal groups, neurotransmitters and their receptors, neuronal circuitry, and synaptic interactions of the sleep-wake, and circadian regulatory neurons that orchestrate the timely onset and maintenance of sleep-wake function (Brown et al., 2012; Alam, 2013; Jones, 2013; Saper and Fuller, 2017; Scammell et al., 2017; Shiromani and Peever, 2017). Much evidence now supports that widespread structures in the mammalian central nervous system, including regions in the medulla, mesencephalon, basal forebrain, preoptic/hypothalamus, thalamus, and neocortex, have sleep- and wake-regulatory neural substrates and interact with each other and the circadian pacemaker to regulate sleep-wake function.

Electrophysiology, i.e., measuring changes in neurons' electrical properties and action potentials is one of the oldest tools used in sleep research. In fact, electrophysiological approaches have been pivotal to the functional characterization of individual neurons across the sleep-waking cycles and in understanding how collective signaling from those cell populations relates to the generation and maintenance of various stages of sleep-waking. Not surprisingly, a significant aspect of our current understanding of the neural control of sleep-wake regulatory mechanisms is based on electrophysiological studies. This is a broad area to be covered, given the number and diversified electrophysiological approaches that have been used in sleep-wake studies. This chapter briefly describes notable electrophysiological recording techniques or strategies that have been used to characterize neuronal activity profiles across sleep-wake cycles in freely behaving/restraint animals and comparing their relative strengths and limitations. We have also briefly described some notable findings of electrophysiological studies that have been central to our current understanding of sleep-wake physiology.

Electrophysiology and sleep-wake regulatory systems

One of the critical approaches of sleep-wake studies has been to assess the activity profiles of a significantly large number of neurons within a given region and determine how well it relates to the local circuitry's role in the sleep-wake function and integrates with the brain-wide sleep-wake neuronal network. The action potentials are fast events (microsecond timescale). Thus electrophysiological techniques are attractive tools for (a) a high-resolution characterization of neuronal activity across the dynamic sleep-wake stages and (b) studying the mechanisms by which synaptic inputs are transformed into action potential or neuronal behavior, which could be associated withxsleep-wake functions. Broadly, these electrophysiological techniques include (i) extracellular recording (electrode is placed outside the neuron); (ii) juxtacellular recording-labeling (electrode is in extracellular space but juxtaposition to the cell); (iii) patch-clamp recording (electrode is apposed to and form a tight seal with a patch of the membrane); and (iv) intracellular recording (electrode is inserted into the neuron). With technical advances, all of these techniques have improved significantly on both data acquisition and analysis fronts.

The best approach for characterizing a neuron's actual behavior across the sleep-wake cycle is to study them in freely behaving animals. Evidence suggests that sleep- and wake-active neurons are subject to different modulatory influences across spontaneous sleep-waking processes and during the waxing and waning of homeostatic sleep drive (Porkka-Heiskanen et al., 1997; Kostin et al., 2012; Alam et al., 2014; Saper and Fuller, 2017). These influences would mostly be absent in *in-vitro* preparations. Also, neurons in *in vitro* and anesthetized preparations may behave and respond differently to the same stimulus than in freely behaving conditions (Dean and Boulant, 1989; Alam et al., 1995a; Pack et al., 2001). Also, a neuronal recording done in anesthetized or head-fixed conditions might not sufficiently document movement- or active-waking related neurons, and thus the sampling may be biased.

In recent years, many variants of intracellular and patch-clamp techniques have been successfully used in anesthetized/head-fixed and a relatively small number of studies in freely behaving animals (Chorev et al., 2009; Lee et al., 2009; Cid and de la Prida, 2019). However, such studies have been limited in the sleep field due to the low yield and challenges faced with chronic recording. The mechanical difficulty of impaling a neuron and maintaining the delicate intracellular connection or patch seal makes it challenging to maintain stable recording chronically. The recording duration is typically short, and thus, in many cases, it may not be long enough to sufficiently document neuronal behavior across the sleep-wake cycle. Also, a much longer recording time is needed for examining neuronal behavior during waxing and waning of sleep pressure, which includes sleep deprivation (SD), or their responses to pharmacological manipulations.

In sleep research, extracellular unit recording and juxtacellular recording-labeling studies have been used for sleep-wake and pharmacological profiling of neurons at the system and network levels, whereas intracellular/patch-clamp studies have been used for examining their underlying synaptic mechanisms. This chapter is focused on describing extracellular and juxtacellular approaches used in sleep studies.

Extracellular recording: basic procedure

This is a tool to record action potentials generated by a neuron(s) by measuring the potential (voltage) difference between the tips of the recording electrode and a ground electrode placed into the extracellular space near the neurons of interest. The extracellular signals are picked up by a glass micropipette filled with an electrolyte solution or metallic microwires (e.g., steel, platinum, tungsten, nichrome, iridium) that are insulated (formver, teflon, polyimide are typical insulating material) except at the tip, or multichannel silicon or polymer-based probes.

The electrical signals recorded by the microelectrodes depend on their tip size and resistance. Electrodes with large tips have lower resistance and pick-up activities from several simultaneously active neurons nearby the electrode or record multiunit activity. On the other hand, smaller tips (5−10 μm) have higher resistance, which restricts the area from which potentials can be recorded and thus often pick up the activity of a single neuron in their close proximity. Electrodes of much higher dimensions (e.g., 50 μm) and low impedance are used for recording local field potentials that represent the aggregate activity of extracellular potentials from hundreds to thousands of neighboring neuronal populations (Mahmud et al., 2011). Depending upon the experimental need, different electrodes are used to record single unit, multiple units, and field potentials in various *in vivo* and *in vitro* models.

The extracellular signals picked up by the electrodes are relatively weak (amplitude in microvolts) and thus are amplified 100−1000X using high-gain amplifiers. The amplified and filtered signals from the amplifiers are displayed on an oscilloscope for visualization if desired, digitized using an analog-to-digital converter, and recorded on a computer using either open access or commercially available software developed for neural signal acquisition and processing. The recordings consist of a series of action potentials, depending upon the electrode size that are superimposed on the background noise.

In recent years, advances in electronics and computing have significantly increased the capabilities to isolate and simultaneously record a large number of neurons and the analyses of recorded signals. Nowadays, much advanced and automated systems for data acquisition, analysis, and interpretation are being used (e.g., Spikes-2, OmniPlex Neural Recording Data Acquisition System). Such programs reliably isolate a single unit from the background or differentiate and isolate multiple units from mixed signals based on series of electrophysiological criteria and statistical measures and can follow their discharge chronically across numerous sessions, even for days.

Depending upon the experimental need, many variants of extracellular recording techniques have been attempted and successfully used, broadly grouped into two categories: (a) single-site recording and (b) multisite recording.

Single-site extracellular unit recording and sleep-wake studies

The classic extracellular single-/multiunit recording technique has dominated since its inception and continues to be used in system neuroscience and is not new to the sleep field. In fact, the first such recording was conducted in the 1950s from the cortex in cat (Hubel, 1957). The next year, Strumwasser used relatively flexible stainless steel wires to record single-unit activity in unrestrained ground squirrels (Strumwasser, 1958). Earlier studies in the 1950s and 60s recognized that most of the neurons in the cortex and brainstem remain active across the sleep-wake cycle and discharge at a higher rate during waking and REM sleep than nonREM sleep (Saper et al., 2010; Brown et al., 2012; Jones, 2013).

Extracellular recordings make it possible to record simultaneously from several cells, and its recording stability allows monitoring of neuronal activity over extended periods. Therefore, this technique has been extensively used to comprehensively characterize and assess how the activity profiles of neurons in a given area relate to the onset and maintenance of spontaneous waking, nonREM sleep, and REM sleep, and to the waxing and waning of the homeostatic sleep pressure or other manipulations (Suntsova et al., 2002, 2007; Sakai, 2011; Alam et al., 2014, 2018).

A significant advantage of this technique is that it records real-time neurophysiological signaling and has much higher temporal resolution than the widely used immunohistochemical (IHC) marker of neuronal activation (c-fos protein expression), which requires ~30 min to express, and thus animals have to be maintained in the same behavioral state (forced waking or are sleep deprived for achieving consolidated sleep) typically for 60–90 min (Sherin et al., 1996; Gong et al., 2004; Gvilia et al., 2006). Also, the IHC technique requires more control and experimental groups. Still, because of the poor temporal resolution, it becomes difficult to establish if Fos + neurons are nonREM sleep-active or REM sleep-active or are associated with the initiation versus maintenance of nonREM sleep and or REM sleep. However, a significant limitation of extracellular unit recording is that, unlike IHC technique, relatively fewer neurons could be studied simultaneously (although newer generation of microelectrodes can simultaneously examine a large population of neurons), and neurotransmitter phenotypes of the recorded neurons cannot be ascertained.

Extracellular unit recording and analysis: basic procedure

While technical advances have helped improve data acquisition and analyses, the basic methodology for recording has not changed much. Briefly, a typical experimental preparation includes (i) implantation of a bundle of single, paired microwires or tetrodes (a popular approach given the extensive amount of data it generates and cost-effectiveness) into the area of interest for recording unit activity and (ii) implantation of EEG and EMG electrodes for the identification of sleep-wake stages. The microelectrodes could be static or driven using a microdrive.

Both microdrive and microelectrode technology have significantly evolved over the years (Fig. 3.1), enabling high throughput and reliable unit recording over an extended period (weeks) in freely behaving animals (Billard et al., 2018; Hong and Lieber, 2019). In recent years, microelectrodes have also been combined with optogenetic tools (optrode), which allows for the simultaneous optogenetic stimulation and recording of defined neuronal populations (Voigts et al., 2013; Xu et al., 2015; Zhong et al., 2019). Various versions of microelectrodes and microdrive systems are commercially available.

The recording is done after acclimatization of the animals with the recording setup. The microdrive is typically advanced in 5–30 μm steps, and microelectrodes are systematically scanned for the presence of spikes in amplified electrophysiological signals. The scanning is done during both waking and nonREM/REM sleep to avoid sampling bias and for ensuring detection of all neuronal types. Sorting of individual action potentials from filtered and amplified signals is done by customized or commercially available software (e.g., Spike-2, Cambridge Electronic Design, UK). Such

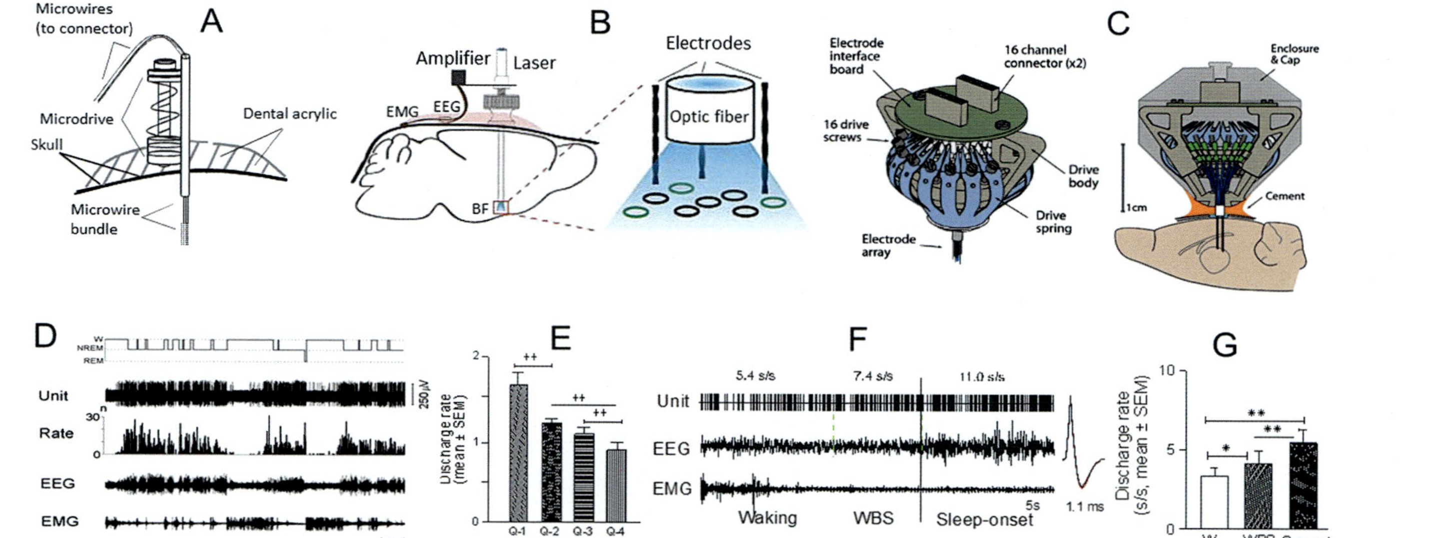

Figure 3.1 Various types of microdrives (A—C) and traces of recordings and the analysis done on neurons recorded across the sleep-wake cycle (D—G). (A), a simple mechanical microdrive that consists of one barrel. Typically, a bundle of 10—12 single or 5—6 paired microwires are passed through the barrel into the target (For details, see Alam et al., 1999; Suntsova 2002). (B), schematic of an optrode used for recording activity profiles of channelrhodopsin-2-tagged neurons in freely behaving mice. (C), an isometric view of the flexDrive (left) and cross-section of the drive and its placement on the mouse skull (right). This drive can be used to independently position 16 electrodes, including multiple fibers for optical control. (D), a 66 min continuous recording showing the discharge of a nonREM/REM sleep-active MnPO neuron across multiple sleep-waking cycles. (E), the difference in the firing rate of the same cell in different nonREM sleep quarters. Note that this sleep-active neuron's discharge rate was highest at sleep-onset and gradually declined with the progression of sleep or a decline in delta activity. (F), continuous recording showing discharge activity of individual sleep-active MnPO neuron during 15 s of waking, last 10 s of waking immediately preceding electroencephalogram (EEG) synchronization, and during 15 s of sleep onset. The spikes on the side represent averaged waveforms (with standard deviation in red, almost indistinguishable) of all the action potentials captured during the presented recording period. (G), mean discharge rates of all MnPO neurons recorded during the same time points as in (F). Note that MnPO sleep-active neurons as a group exhibit increased discharge before EEG synchronization and sleep-onset, indicating their role in sleep initiation. *(B) (Adopted from and for details see Xu, M., Chung, S., Zhang, S., Zhong, P., Ma, C., Chang, W.C., Weissbourd, B., Sakai, N., Luo, L., Nishino, S., Dan, Y., 2015. Basal forebrain circuit for sleep-wake control. Nat. Neurosci. 18, 1641—1647, (C) Voigts, J., Siegle, J.H., Pritchett, D.L., Moore, C.I., 2013. The flexDrive: an ultra-light implant for optical control and highly parallel chronic recording of neuronal ensembles in freely moving mice. Front. Syst. Neurosci. 7, 8, (D-E) Suntsova, N., Szymusiak, R., Alam, M.N., Guzman-Marin, R., McGinty, D., 2002. Sleep-waking discharge patterns of median preoptic nucleus neurons in rats. J. Physiol. 543, 665—677, and (F-G) Alam, M.A., Kostin, A., Siegel, J., McGinty, D., Szymusiak, R., Alam, M.N., 2018. Characteristics of sleep-active neurons in the medullary parafacial zone in rats. Sleep. 41, 1-13).*

software produces unit templates based on multiple parameters related to spike amplitude and shape for each distinct action potential present in the raw signal and sorts them throughout the recording session by matching them with the templates using various statistical measures. Spike shapes are based on a large number of analog to digital sampling points. All discharge activity of neurons is recorded across multiple sleep-wake cycles.

Typically, to identify sleep-wake state-related/-active neurons, the mean firing rates of neurons are calculated for each vigilance state from multiple (5–10) episodes of 30–300s artifact-free periods of active-waking, quiet-waking, nonREM sleep, and REM sleep, identified based on EEG and EMG parameters using standard criteria. Then the neurons are segregated by their nonREM sleep/wake, REM sleep/wake, and nonREM sleep/REM sleep discharge ratios, and a minimum of 20%–25% change between-state discharge rate criterion (Suntsova et al., 2002; Watson et al., 2005; Alam et al., 2018). Such standard has been validated by K-means cluster analysis followed by evaluating state-relatedness for each cluster member. Neurons exhibiting <25% change between states are considered as state-indifferent. To determine the neuronal participation in the onset or maintenance of sleep- or waking-state, changes in their discharge during state-transitions and during the early, middle, and late phases of waking, nonREM sleep, and REM sleep episodes are calculated (Fig. 3.1) (Suntsova et al., 2002; Sakai, 2011; Alam et al., 2018).

Total sleep-deprivation studies: To determine responses of sleep- and wake-active neurons to the waxing and waning of homeostatic sleep pressure, the activity profiles of these neurons are recorded during total SD and recovery sleep (RS). After isolating individual neurons, their baseline discharge activity is recorded across at least three sleep-wake cycles. Then, selected neurons with stable discharge patterns are recorded continuously during 2–3 h of SD followed by 2 h of a recovery period (Alam et al., 2014; Gvilia et al., 2006). During SD, animals are closely monitored both behaviorally as well as with reference to EEG changes. Each time the animal exhibits EEG synchronization for 5–10 s, it is gently awakened by tapping on the cage or slightly turning it. Each attempt to prevent the animal from entering into sleep, which progressively increases with SD, is recorded as an event. After SD, recording continues during a 2 h opportunity for an uninterrupted RS. Fig. 3.2 shows a representative recording session during which an isolated VLPO sleep-active neuron was recorded across the sleep-wake cycle during baseline and with the waxing and waning of sleep pressure.

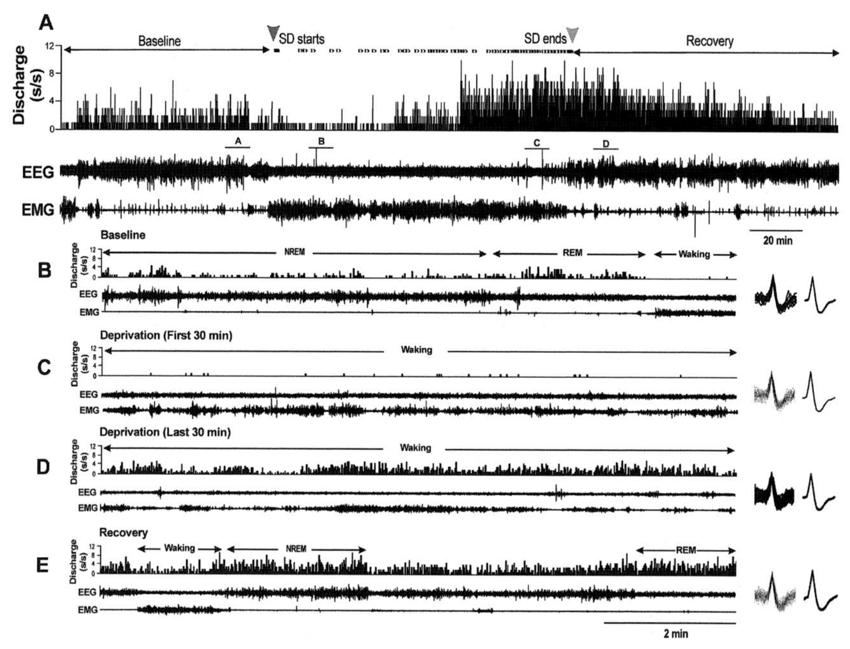

Figure 3.2 Responses of a MnPO sleep-active neuron across the waxing and waning of sleep drive. (A), from top to bottom, discharge rate histogram (spikes/s), cortical electroencephalogram and neck muscle EMG recordings during baseline, sleep deprivation (SD), and recovery sleep (RS). Red and gray arrowheads at the top indicate the start and end of SD, respectively. The *dots* between the *arrows* indicate times when the animal attempted to initiate sleep, and the experimenter had to intervene to maintain wakefulness. Note how the frequency of sleep attempts increases with increasing duration of SD. B–E: expansion of areas labeled in (A), showing recording during baseline (B), the first 30 min of SD (C), the last 30 min of SD (D), and early RS (E). The waveforms at the right of B–E are superimposed, and averaged action potentials recorded during the time shown in each figure, demonstrating the stability of unit recording across all three experimental conditions. *(Adopted from Alam, M.A., Kumar, S., McGinty, D., Alam, M.N., Szymusiak, R., 2014. Neuronal activity in the preoptic hypothalamus during sleep deprivation and recovery sleep. J. Neurophysiol. 111, 287–299).*

Selective REM sleep deprivation studies: In some studies, attempts have been made to examine how sleep- and wake-active neurons respond to selective REM SD. The experimental paradigm is the same as SD, except that animals are subjected to 2 h of selective REM-SD (Gvilia et al., 2006; Alam et al., 2014). During REM-SD, animals are closely monitored for electrophysiological and behavioral signs of REM sleep. Each time animal exhibits electrophysiological signs of REM sleep, i.e., EEG desynchronization and lower EMG tone after nonNREM sleep for 10 s; it is gently awakened by slightly touching the recording cable or turning the cage. Each attempt to prevent the animal from entering into REM sleep is recorded as an event. After REM-SD, animals are left undisturbed for 2 h of RS.

Extracellular unit recording-microdialysis

In some studies, the extracellular unit recording has been used in combination with microdialytic drug delivery adjacent to the recorded neurons for their pharmacological characterization across the sleep-wake state (Alam et al., 1999; Thakkar et al., 2003). In this case, the microdrive assembly consists of a guide cannula for the microdialysis probe. The microdialysis probe is implanted 0.2—0.5 mm adjacent to the microwires. The drug is typically microdialyzed for 5—15 min to determine its transient effects on neurons' spontaneous activity across the sleep-wake cycle without triggering strong behavioral changes. The neurons' baseline discharge activity is compared to the discharge during drug delivery and the wash-out period.

The advantage of this approach is that it involves a localized and better mode of drug delivery. Using this approach, effects of the multiple drugs on the activity profiles of neurons can be studied. Also, by analyzing the dialysate, the microenvironment of the recorded neurons can be simultaneously assessed. Finally, pharmacological responses of drugs on neuronal activity can be compared to changes occurring during spontaneous sleep-wake state changes. In some studies, adenosine analogs or nitric oxide donors were delivered adjacent to the neurons to imitate accumulations of those molecules during prolonged waking and access their effects on the discharge activity of neurons (Alam et al., 1999, 2014; Thakkar et al., 2003; Kostin et al., 2012). Fig. 3.3 shows a typical experimental session aimed at determining if VLPO sleep-active neurons via adenosine A_{2A} receptor contribute to homeostatic sleep response (Alam et al., 2014).

However, since the microdialysis probe provides a bathing medium for the entire extracellular environment of the recorded neurons, the site of drug action cannot be pinpointed. Perfused drugs could act (i) presynaptically on local interneurons that synapse directly on the recorded neurons, or (ii) presynaptically on the terminals of distant afferents to recorded neurons, and/or (iii) directly on the receptors located on the dendrites or cell body of the recorded neurons.

In some studies, the effects of local nonpharmacological manipulations, e.g., local warming and cooling, on the discharge activity of sleep- and

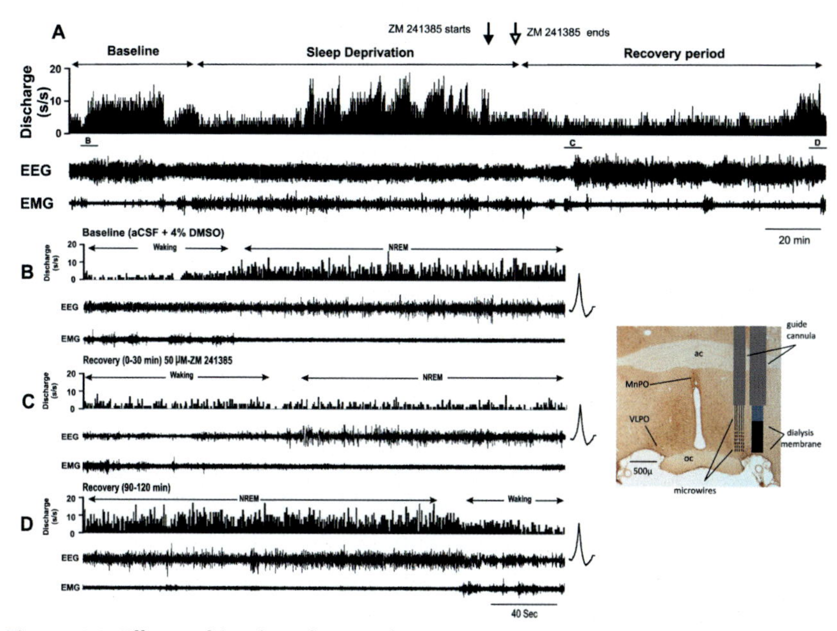

Figure 3.3 Effects of local perfusion of A_{2A} receptor antagonist, ZM-241385, on the discharge of a VLPO sleep-active neuron during sleep deprivation (SD) and recovery sleep (RS). (A), from top to bottom, discharge histogram (spikes/s), cortical electroencephalogram, and neck muscle EMG recordings during 40-min baseline waking and sleep, during 120 min of SD, and 120 min of recovery. Just prior to the end of the sleep deprivation period, perfusion was switched from control [artificial cerebrospinal fluid (aCSF) + 4% DMSO] to 50 μM ZM-241385, an adenosine A_{2A} antagonist or 10 min (arrows), then back to the vehicle for the remainder of the recording. (B), in the baseline condition, discharge of the neuron was strongly sleep-related. (A) the cell increased waking discharge, compared with baseline waking, during the second hour of SD (also see Fig. 3.2). Perfusion of ZM-241385 evoked suppression of waking discharge during the final 10 min of SD. (C), the cell exhibited marked suppression of nonREM sleep discharge during initial RS following ZM-24185 administration, compared with baseline nonREM discharge. (D), discharge suppression persisted throughout RS, eventually returning to baseline levels during the final 30 min of the RS period. The averaged action potentials demonstrate that the recording was stable across the experiment. (E), a schematic diagram showing the microelectrode-microdialysis assembly. These findings support that VLPO sleep-active neurons play a role in homeostatic sleep control via adenosine A_{2A}R. *(D) (Adopted from Alam, M.A., Kumar, S., McGinty, D., Alam, M.N., Szymusiak, R., 2014. Neuronal activity in the preoptic hypothalamus during sleep deprivation and recovery sleep. J. Neurophysiol. 111, 287–299).*

wake-active neurons have been examined (Alam et al., 1995a). Such studies helped establish that sleep-active preoptic area neurons are predominantly warm-sensitive, and therefore, their activation during sleep contributes to lowering the body temperature as well.

Extracellular unit recording-electrical stimulation

This is a classic approach, which has been used to examine the interactions between sleep and wake-regulatory neurons and stimulation-evoked sleep-wake changes to assess the role of specific brain structures in sleep-wake function. The advantage of electrical stimulation is its high-temporal resolution allowing the distinction of antidromic, monosynaptic, and polysynaptic responses. However, a significant limitation is the lack of selectivity for the local neurons or any specific neuronal group. Electrical stimulation may also excite axonal terminals and fibers of passage, making it difficult to ascertain the origin of the obtained responses.

Briefly, the area of interest is stimulated using monopolar or bipolar electrodes, and the response of single-pulse and or train stimulation on the activity of target neurons across sleep-wake stages is determined. The effective current spread around the electrode tips is calculated and stimulation parameters adjusted so that only the area of interest is sufficiently stimulated. The stimulation-evoked responses of the target area neurons are determined by various parameters, including the onset latency and duration of stimulation-induced effects from prestimulus time histograms. A typical recording procedure and experimental outcome of such a study are shown in Fig. 3.4 (For details, see Suntsova et al., 2007). This study used extracellular unit recording-electrical stimulation approach to examine how activation of sleep-regulatory MnPO neurons affects the discharge activity of wake-active putative hypocretin neurons in the perifornical area (PF-LHA).

Electrical stimulation has been very effectively used in the past to examine sleep-wake network interactions. However, in recent years, the development of genetically engineered mouse lines and retrograde viral vectors allows studying such interactions with much precision.

Multisite extracellular unit recording

A significant limitation of one-site unit recording studies is that its findings are limited to one particular area, while the onset and maintenance of sleep-wake function seemingly involve interactions among various sleep- and wake-regulatory neuronal groups localized in different brain regions. Such interactions can only be fully investigated by simultaneously recordings activity profiles of sleep- and wake-regulatory neurons from multiple regions.

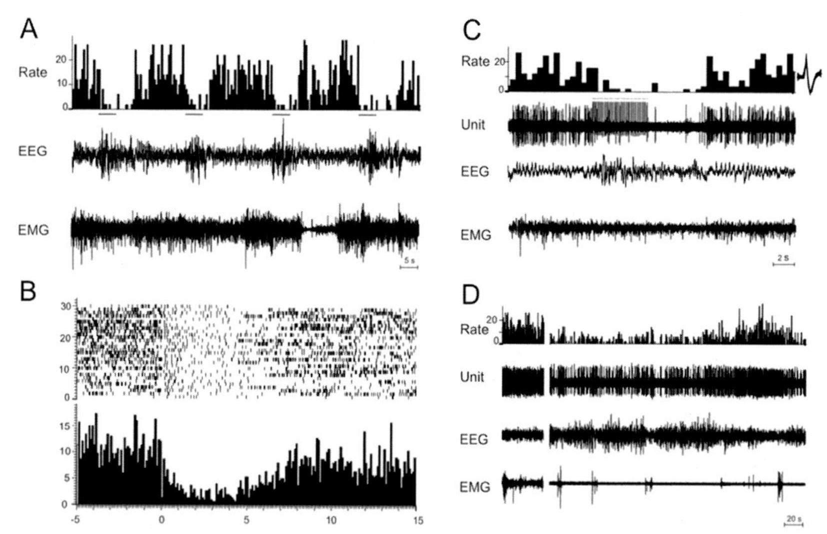

Figure 3.4 Effects of MnPO electrical stimulation on a wake/REM active neuron in the PF-LHA. Microwires were implanted into the PF-LHA to record the extracellular activity of neurons across the sleep-wake cycle, and a bipolar stimulating electrode was implanted into the MnPO. (A), changes in the discharge rate of an individual wake/REM active neuron in response to trains of stimuli (marked as *lines*, current intensity, 100 μA, interstimulus interval, 150 ms). (B), peristimulus and raster histograms. Point 0 corresponds to the beginning of the train. (C), expanded tracing from (A) with unit activity channel. (D), the sleep-waking discharge pattern of the neuron. In this study, most wake-REM and wake-active neurons exhibited discharge cessation/reduction in response to MnPO single-pulse and/or train stimulation during waking (Suntsova et al., 2007).

The fabrication of the multielectrode microdrive system to record simultaneously from multiple regions, especially in small rodents, has been challenging because of the proximity of target areas, drive mechanisms, size, and weight of the assembly. However, in recent years, with advances in miniaturization and three-dimensional printing, different variants of microdrives with multisite recording capabilities (e.g., Split-Drive, SystemDrive, Quad-Drive, microdrive arrays including tetrode bundle and options for optical fiber optogenetic manipulations of specific neuronal groups) have been fabricated and successfully used to simultaneously record electrophysiological activities from different brain regions (Lansink et al., 2007; Headley et al., 2015; Vinck et al., 2015; Billard et al., 2018; Ma et al., 2019).

A SystemDrive, which provides the ability to monitor and analyze neural activity from multiple and widely distributed cell groups simultaneously, has been successfully used in a recent study. In this study, using

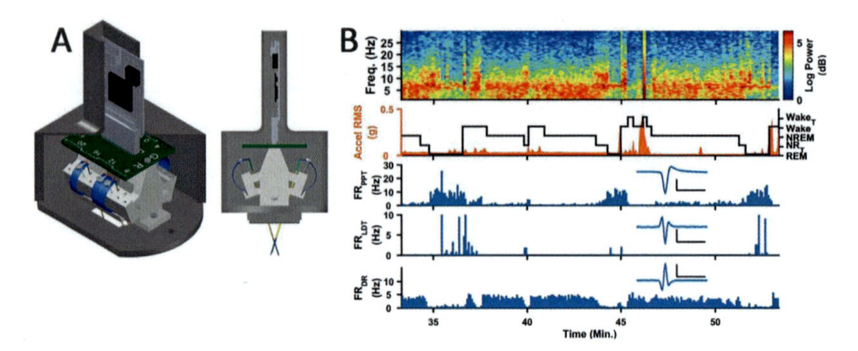

Figure 3.5 The microdrive system for multisite recording of extracellular discharge activity of neurons across sleep-wake cycle in rats. (A), Isometric (left) and side (right) three-dimensional—rendered computer-aided design model views of the complete microdrive system. Elements of the system include: the microdrive body (light gray structure), drive springs and screws, flexible drive axes, an electrode interface board (green printed circuit board) with a 36-position Omnetics connector, an Intan RHD2216 amplifier, and a head mount (dark gray base and cap). The cap of the head mount has been sliced to view the elements inside. (B) One hour continuous recording showing unit discharge from multiple brainstem cell groups across sleep-wake cycle that were recorded using this microdrive system. The traces represent from the top, power spectral density of one LFP channel; one channel of head acceleration (orange trace) with hypnogram overlaid (black trace); single-unit firing rates from three different sleep-wake regulatory structures, namely, PPT, LDT, and DR, with the inset representing the hour-long average waveforms. This study found that PPT and LDT neurons are REM-on while the neurons in DR are REM-off. *(Adopted from and for details, see Billard, M.W., Bahari, F., Kimbugwe, J., Alloway, K.D., Gluckman, B.J., 2018. The systemDrive: a multisite, multiregion microdrive with independent drive axis angling for chronic multimodal systems neuroscience recordings in freely behaving animals. eNeuro 5).*

SystemDrive, sleep-wake discharge profiles of neurons were simultaneously recorded from four sleep-wake regulatory areas, namely, VLPO in the preoptic region and pedunculopontine tegmental nucleus (PPT), laterodorsal tegmental nucleus (LDT), and dorsal raphe (DR) in the brainstem in rats (Fig. 3.5). The techniques used for multisite recording have been described in detail in another chapter.

Extracellular unit recording-optogenetics

In recent years, optrode, which allows for the simultaneous photostimulation and recording of channelrhodopsin2-tagged neurons, has been used to characterize the sleep-wake activity profiles of various neuronal groups in the sleep-wake neural network (Weber et al., 2015; Xu et al., 2015; Shiromani and Peever, 2017). Typically, such an optrode

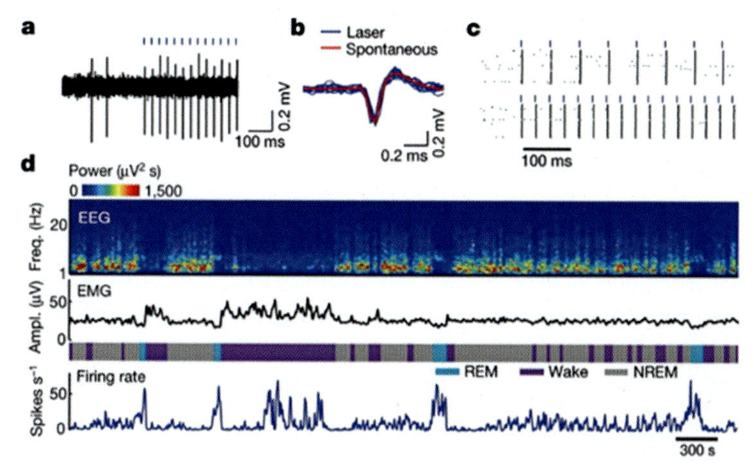

Figure 3.6 Firing rates of identified ventral medulla GABAergic neurons across sleep-wake states as recorded by an optrode. (A), recording showing spontaneous and laser-evoked spikes from a ventral medulla neuron. Blue ticks, laser pulses (30 Hz). (B), comparison between laser-evoked (blue) and averaged spontaneous (red) spike waveforms from this unit. (C), spike raster showing multiple trials of laser stimulation at 15 and 30 Hz d, firing rates of a representative ventral medulla GABAergic neuron across the sleep-wake cycle. *(Adopted from and for details, see Weber, F., Chung, S., Beier, K.T., Xu, M., Luo, L., Dan, Y., 2015. Control of REM sleep by ventral medulla GABAergic neurons. Nature 526, 435–438).*

consists of an optical fiber for optostimulation, surrounded by microwire electrodes twisted into stereotrodes or tetrodes for the extracellular unit recording of the stimulated neurons. A typical recording paradigm is shown in Fig. 3.6. In this study, optrode recordings from channelrhodopsin2-tagged ventral medulla GABAergic neurons showed that they were most active during REM sleep. During waking, these neurons were preferentially active during eating and grooming.

Extracellular unit recording: notable findings

Despite of its limitations, extracellular unit recording studies have yield a rich set of discoveries involving patterns of spiking activity in the sleep-wake neuronal network that underlie sleep-wake process, including: (i) that neurons exhibiting sleep-and wake-associated discharge profiles are distributed throughout the neuraxis and that it's the majority of the specific neuronal phenotype amongst the heterogeneous population that disposes a particular region to be sleep- or wake-active; (ii) that a large majority of the sleep-active neurons in various brain regions including those in the MnPO and VLPO are nonREM/REM-active and typically exhibit increasing discharge from waking > nonREM sleep transition > nonREM sleep > REM sleep; (iii) that sleep-active neurons in different brain regions may play complimentary and differential roles in sleep regulation. For example, sleep-active neurons in the VLPO and MnPO are more involved in homeostatic sleep regulation, whereas those in the parafacial zone are more involved in sleep maintenance (Suntsova et al., 2002; Alam et al., 2014, 2018); (iv) that sleep-active neurons in the POA are warm-sensitive (which explains a decline in body temperature during sleep) and that activation of these neurons inhibits multiple arousal systems (Alam et al., 1995a, 1995b; Methippara et al., 2003). These thermosensitive/sleep-active neurons have recently been identified as GABAergic and galaninergic (Kroeger et al., 2018); and (v) that the flip-flop switch model of sleep, which proposes mutually inhibitory interactions between sleep- and wake-active neurons, to a large extent was initially based on the extracellular discharge profiles of neurons in various sleep- and wake-promoting regions (Saper et al., 2010).

It is pertinent to note that in recent years, more and more genetically engineered mouse lines and viral vectors are becoming available. Optogenetic/chemogenetic and imaging techniques have also evolved significantly. Gradually, more of these approaches are being used in studying the activity profiles of specific neuronal phenotypes and their role in sleep-wake circuitry with greater precision (Shiromani and Peever, 2017). The findings of these refined approaches have generally been consistent with the results of the classic extracellular unit studies (Alam et al., 2014, 2018; Suntsova et al., 2007; Xu et al., 2015; Chung et al., 2017; Kroeger et al., 2018).

Juxtacellular recording-labeling

The juxtacellular recording-labeling technique has been used for anatomofunctional profiling of single neurons in several brain regions, including those involved in sleep-wake regulation and in many species (Pinault, 1996, 2005; Joshi and Hawken, 2006; Takahashi et al., 2006; Hassani et al., 2009; Katona et al., 2017; Cisse et al., 2020). It is less invasive than intracellular recording. This technique helps determine not only the extracellular discharge profiles of the neuron but also the recorded neuron can be stimulated and labeled with a tracer under direct electrophysiological control. The controlled cell stimulation can be used to validate the functional relation of the recorded neuron by its ability to initiate a physiological response (Houweling and Brecht, 2008; Diamantaki et al., 2016b).

Also, by combining juxtacellular recording-labeling with IHC, molecular, or genetic approaches (e.g., delivering plasmid DNA into the recorded neuron for its manipulation, optogenetic stimulation) not only the physiological profile of the recorded neuron in relation to particular behavior (e.g., sleep) rather the localization, morphology, anatomical connectivity, chemical, and molecular phenotyping (e.g., neurotransmitter content, receptor/protein mRNA, or ion channel) of the recorded neuron can also be characterized (Takahashi et al., 2006; Toney and Daws, 2006; Judkewitz et al., 2009; Pinault, 2011; Cisse et al., 2018; Cid and de la Prida, 2019). Therefore, the juxtacellular recording-labeling is a powerful tool that can help delineate the cellular, molecular, and network mechanisms of sleep-wake function under physiological and pathological conditions.

Basic procedure

Although developed in mid 1990s in anesthetized preparation, this technique has been used successfully in head-fixed and lately in freely behaving animals (Pinault, 1996; Herfst et al., 2012; Diamantaki et al., 2016a; Cid and de la Prida, 2019). For a detailed description of the equipment needed and the procedure of juxtacellular recording-labeling, please see earlier reviews (Pinault, 2011; Tang et al., 2014; Cid and de la Prida, 2019). The protocol for juxtacellular recording and labeling is essentially the same under anesthesia, head-fixed, and freely behaving conditions. Fig. 3.7 shows the assembled implant for juxtacellular recording in freely behaving rat and the procedure.

Briefly, the equipment required is similar to that of intracellular recording. It includes a stereotaxic apparatus, a micromanipulator, a conventional intracellular preamplifier, a signal conditioner, an oscilloscope, an analog-to-digital converter, a computer, and the software for data acquisition, storage, and off-line analyses. Typically, sharp micropipettes with fine capillaries (typical tip diameter, 0.5−1.5 μm; tip resistance, 10−25 MΩ) are used for recording extracellular signals. The micropipette is filled with 0.5M NaCl or 0.5M sodium acetate (some report suggests improved results with sodium acetate) typically containing 1%−3% of the tracer neurobiotin (biotin amide hydrochloride) or biocytin (biotinyl-L-lysine). Studies that may need a more extended survival period, using biotinylated dextran amines or neurobiotin-plus have been suggested (Cid and de la Prida, 2019).

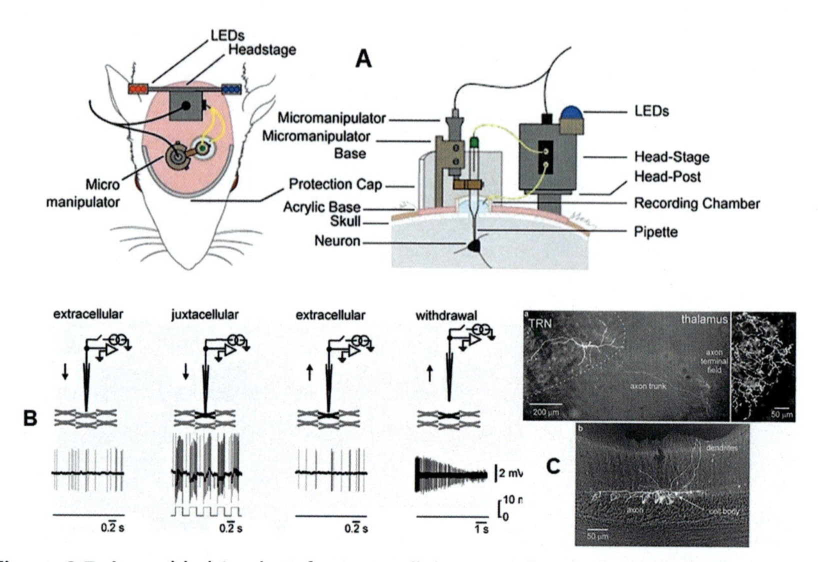

Figure 3.7 Assembled implant for juxtacellular recording in freely behaving animal and the recording/labeling procedure. (A), Schematic diagram showing the position of the individual implant components relative to the rat's head for obtaining juxtacellular-recordings in freely moving rats. (B) The four steps of the juxtacellular labeling. The tip of the micropipette presumably touches the neuron membrane without penetration during current injection and (C), juxtacellular labeling with biocytin of a projection (A) and local circuit (B) neurons. For details, please see Tang et al. (2014) and Pinault (2011). *(A) (Adopted from Tang, Q., Brecht, M., Burgalossi, A., 2014. Juxtacellular recording and morphological identification of single neurons in freely moving rats. Nat. Protoc. 9, 2369–2381 and (C) Pinault, D., 2011. The juxtacellular recording-labeling technique. In: Vertes, R.P., Stackman, R.W. (Eds.), Electrophysiological Recording Techniques, Neuromethods, vol. 54. Springer, New York, pp. 41–75).*

The basic procedure involves positioning animals in a stereotaxic frame for both surgery and subsequent recording, if under anesthetized preparation. EEG and EMG electrodes are implanted, and a small hole (1.2–2.0 mm in diameter) is drilled in the skull above the site of interest for the insertion of the micropipette. Although the removal of dura is not critical (the electrode can easily penetrate through it), it might increase the chances of obtaining stable juxtacellular recording as the micropipette enters the brain more cleanly. The filled micropipette is then inserted and advanced in small steps using a microdrive manipulator. After reaching the target area, the micropipette is advanced at a much slower pace (1–4 μm steps every 2–3 s), while looking for positive action potentials (or monitoring changes in electrode tip resistance to avoid bias toward firing neurons). The establishment of a juxtacellular configuration is signaled by a high signal-to-noise ratio of the spike signals and by ~two to three times increase in tip resistance (Tang et al., 2014; Cid and de la Prida, 2019).

Once a clear action potential is recognized, the neuron's discharge profiles are recorded across multiple sleep-wake cycles and during any other physiological or pharmacological manipulations, depending upon the experiment.

After functional characterization, juxtacellular labeling of the recorded neuron with the trace is initiated for subsequent morphological and IHC studies (Fig. 3.7). The tracer is ejected by passing 200–500 ms anodal current with a 50% duty cycle pulses through the recording electrode. While continuously monitoring the discharge of the recorded neuron, the intensity of current pulses is gradually increased (1–<10 nA, or 15–20 nA for few cycles for cells that resist entrainment) and adjusted to entrain the discharge of the recorded neuron to the timing of the current pulses. Such currents electroporate the cell membrane, creating small holes for the tracer to flow across the membrane and into the cytoplasm of the cell. In some cases, entrainment may require slight adjustments in microelectrode position so that it is in the juxtaposition of the neuron. The duration of entrainment needed for sufficient filling of the tracer depends on many factors, including the tracer's concentration, the filling solution used, the microelectrode's properties, and neuronal size or morphometry. A successful filling may take up to 20 min (Toney and Daws, 2006; Pinault, 2011; Cid and de la Prida, 2019).

Typically, only 1 cell per hemisphere per animal is labeled to avoid ambiguity, although if desired, a second penetration for recording and labeling another neuron can also be performed. After filling neuron with the tracer, pulse delivery is stopped. Typically after filling, the discharge rate increases, and spike shape changes, which gradually reverses with time. At this time, the neuronal discharge profile is re-examined to ensure that it is the same neuron and that it remains healthy following the labeling procedure. After labeling the cell, the micropipette is slowly retracted while monitoring cell firing.

In the end, the animals are perfused and the labeled neuron harvested for further ultrastructural, immune-histochemical, and molecular analyses. Typically animals are perfused after several hours, even days, to maximize diffusion/distribution of the tracer/dye for identifying *neuronal projections*. Passing fibers and or glia are rarely labeled.

Juxtacellular recording-labeling and sleep-wake studies

Many sleep labs have used the juxtacellular recording-labeling technique to determine the physiological and anatomical profiles of sleep- and wake-active neuronal groups, including cortical neurons (Averkin et al., 2016); cholinergic and GABAergic neurons in the basal forebrain (Manns et al., 2000; Hassani et al., 2009); orexin or hypocretin, melanin-concentrating hormone, and GABAergic neurons in the PF-LHA (Lee et al., 2005; Hassani et al., 2010); cholinergic and GABAergic neurons in the ponto-mesencephalic region (Cisse et al., 2018, 2020); GABAergic neurons in the hippocampus (Katona et al., 2017), and histamine neurons in the tuberomammillary nucleus (Takahashi et al., 2006). An advantage of the juxtacellular recording technique over conventional extracellular methods

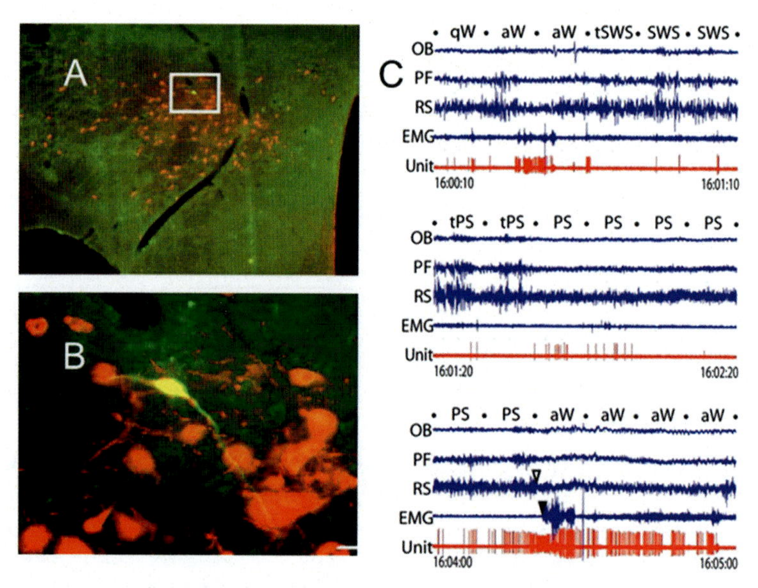

Figure 3.8 Juxtacellular labeling and recording of orexin/hypocretin neurons in the hypothalamus. (A), image showing red, Cy3-immunostained Orx neurons in the perifornical, lateral hypothalamic area. (B), magnified image of the box in (A) showing an Nb-labeled cell stained with green, Cy2, and red, Cy3, thus appearing yellow as a Nb+/Orx + cell. (C), three 1 min segments of the unit, electroencephalogram, and electromyogram activity during state transitions: from qW through aW, tSWS to SWS (top panels), from tPS to PS (middle panels), and from PS through aW (bottom panels). Note that the unit fired during aW and became quiet during SWS and PS, except for occasional spikes sometimes associated with twitches. *(Adapted from Lee, M.G., Hassani, O.K., Jones, B.E., 2005. Discharge of identified orexin/hypocretin neurons across the sleep-waking cycle. J. Neurosci. 25, 6716–6720, Copyright (2005), Society for Neuroscience).*

is that neurons can be identified irrespective of their spiking activity. This technique was the first to describe the sleep-wake activity profiles of identified orexin neurons (Fig. 3.8), a notable finding of that time. This technique's main limitations have been its low yield, problem maintaining stable chronic recording, and being very labor-intensive.

To date, this technique has been mostly used in anesthetized and head-fixed animals in the sleep field, where animal movement and/or brain pulsations are minimal. However, new generations of pipette-positioning devices/microdrives provide higher mechanical stability and longer recording time even in freely behaving animals (Tang et al., 2014; Cid and de la Prida, 2019). In recent years, juxtacellularly recorded neurons have been labeled with fluorescent dyes, DNA plasmid, and calcium or voltage-sensitive dyes for further functional, anatomical, and molecular characterization (Cid and de la Prida, 2019). These emerging approaches of combining juxtacellular recording-labeling in freely behaving conditions and other tools like optogenetics/chemogenetics that can address different functionalities of individual neurons are likely to provide a much detailed physiological, anatomical, and molecular footprint of sleep and wake-regulatory neurons. It remains to be seen that with these advances, juxtacellular recording-labeling remains a valuable tool in sleep studies.

Conclusions

This chapter has reviewed electrophysiological approaches that have been mainly used to characterize sleep-wake neural circuitries and their contributions to sleep-wake function. Neither extracellular unit recording nor juxtacellular recording-labeling is new. What is new is that with time, both of these techniques have gone through significant technological advances toward high-throughput and stable data acquisition, including miniaturization of devices, multisite recording, and analytical refinements. It has become possible to simultaneously record from multiple sites and perform juxtacellular recording/labeling in freely behaving animals. Also, with the availability of a growing number of transgenic mouse lines, these techniques will continue to be used to complement opto-/chemogenetic and imaging approaches to understand the molecular signature of the specific sleep- and wake-active neuronal groups and the mechanisms that underlie the sleep-wake process (Shiromani and Peever, 2017; Zhong et al., 2019; Cisse et al., 2020). We are witnessing more and more incorporation and reliance on those hybrid approaches in the sleep field that have certainly helped high-resolution monitoring of sleep-wake neuronal network activity. However, none of these techniques currently seem to simultaneously characterize both electrophysiological and anatomical profiles of a large number of neurons, which is needed for making system-level inferences about a higher brain function like sleep-wake regulation.

On the other hand, in recent years, optical imaging approaches have significantly improved in temporal and spatial resolution. The imaging approaches utilizing promotor-specific transgenic strategies and genetically encoded calcium and voltage indicators seem quite promising, given that (i) it is cell-specific and the activity profiles of a large population of a specific cell group can be simultaneously recorded for weeks; (ii) it is less invasive; and (iii) it can relate activity and anatomy (Antic et al., 2016). The voltage-sensitive dye imaging is becoming a convenient tool to study neuronal activity dynamics with spatiotemporal resolution seemingly at par to the electrophysiological recording. Also, in recent years, calcium imaging has been used to examine sleep-wake activity profiles of various cell groups in sleep-wake networks (Blanco-Centurion et al., 2019; Zhong et al., 2019; Ingiosi et al., 2020). The photobleaching of these probes remains a limiting factor for long-term recording.

In coming years, given the high-throughput, it seems likely that voltage-sensitive dye and calcium imaging will be used simultaneously or in combination with optogenetic/chemogenetic approaches for more detailed physiological, anatomical, and molecular characterization of distinct sleep- and wake-active neuronal groups and for addressing cellular and molecular questions about the working of sleep-wake regulatory systems. The adoption of these hybrid approaches has already started transforming the field of sleep research. With ongoing efforts to further improve these technologies, it seems probable that in the not-so-distant future, the extracellular unit recording (or potentially juxtacellular recording) as a standalone approach in basic research may become archaic which the new generation of sleep physiologists will find in old sleep studies.

References

Alam, M.A., Kostin, A., Siegel, J., McGinty, D., Szymusiak, R., Alam, M.N., 2018. Characteristics of sleep-active neurons in the medullary parafacial zone in rats. Sleep 41, 1–13.

Alam, M.A., Kumar, S., McGinty, D., Alam, M.N., Szymusiak, R., 2014. Neuronal activity in the preoptic hypothalamus during sleep deprivation and recovery sleep. J. Neurophysiol. 111, 287–299.

Alam, M.N., 2013. NREM sleep: anatomy and physiology. In: Encyclopaedia of Sleep. Elsevier, pp. 453–459.

Alam, M.N., McGinty, D., Szymusiak, R., 1995a. Neuronal discharge of preoptic/anterior hypothalamic thermosensitive neurons: relation to NREM sleep. Am. J. Physiol. 269, R1240–R1249.
Alam, M.N., Szymusiak, R., Gong, H., King, J., McGinty, D., 1999. Adenosinergic modulation of rat basal forebrain neurons during sleep and waking: neuronal recording with microdialysis. J. Physiol. 521 (Pt 3), 679–690.
Alam, N., Szymusiak, R., McGinty, D., 1995b. Local preoptic/anterior hypothalamic warming alters spontaneous and evoked neuronal activity in the magno-cellular basal forebrain. Brain Res. 696, 221–230.
Antic, S.D., Empson, R.M., Knopfel, T., 2016. Voltage imaging to understand connections and functions of neuronal circuits. J. Neurophysiol. 116, 135–152.
Aserinsky, E., Kleitman, N., 1953. Regularly occurring periods of eye motility, and concomitant phenomena, during sleep. Science 118, 273–274.
Averkin, R.G., Szemenyei, V., Borde, S., Tamas, G., 2016. Identified cellular correlates of neocortical ripple and high-gamma oscillations during spindles of natural sleep. Neuron 92, 916–928.
Billard, M.W., Bahari, F., Kimbugwe, J., Alloway, K.D., Gluckman, B.J., 2018. The systemDrive: a multisite, multiregion microdrive with independent drive Axis Angling for chronic multimodal systems neuroscience recordings in freely behaving animals. eNeuro 5.
Blanco-Centurion, C., Luo, S., Spergel, D.J., Vidal-Ortiz, A., Oprisan, S.A., Van den Pol, A.N., Liu, M., Shiromani, P.J., 2019. Dynamic network activation of hypothalamic MCH neurons in REM sleep and exploratory behavior. J. Neurosci. 39, 4986–4998.
Brown, R.E., Basheer, R., McKenna, J.T., Strecker, R.E., McCarley, R.W., 2012. Control of sleep and wakefulness. Physiol. Rev. 92, 1087–1187.
Chorev, E., Epsztein, J., Houweling, A.R., Lee, A.K., Brecht, M., 2009. Electrophysiological recordings from behaving animals—going beyond spikes. Curr. Opin. Neurobiol. 19, 513–519.
Chung, S., Weber, F., Zhong, P., Tan, C.L., Nguyen, T.N., Beier, K.T., Hormann, N., Chang, W.C., Zhang, Z., Do, J.P., Yao, S., Krashes, M.J., Tasic, B., Cetin, A., Zeng, H., Knight, Z.A., Luo, L., Dan, Y., 2017. Identification of preoptic sleep neurons using retrograde labelling and gene profiling. Nature 545, 477–481.
Cid, E., de la Prida, L.M., 2019. Methods for single-cell recording and labeling in vivo. J. Neurosci. Methods 325, 108354.
Cisse, Y., Ishibashi, M., Jost, J., Toossi, H., Mainville, L., Adamantidis, A., Leonard, C.S., Jones, B.E., 2020. Discharge and role of GABA pontomesencephalic neurons in cortical activity and sleep-wake states examined by optogenetics and juxtacellular recordings in mice. J. Neurosci. 40, 5970–5989.
Cisse, Y., Toossi, H., Ishibashi, M., Mainville, L., Leonard, C.S., Adamantidis, A., Jones, B.E., 2018. Discharge and role of acetylcholine pontomesencephalic neurons in cortical activity and sleep-wake states examined by optogenetics and juxtacellular recording in mice. eNeuro 5.
Dean, J.B., Boulant, J.A., 1989. In vitro localization of thermosensitive neurons in the rat diencephalon. Am. J. Physiol. 257, R57–R64.
Diamantaki, M., Frey, M., Berens, P., Preston-Ferrer, P., Burgalossi, A., 2016a. Sparse activity of identified dentate granule cells during spatial exploration. Elife 5.
Diamantaki, M., Frey, M., Preston-Ferrer, P., Burgalossi, A., 2016b. Priming spatial activity by single-cell stimulation in the dentate gyrus of freely moving rats. Curr. Biol. 26, 536–541.
Gong, H., McGinty, D., Guzman-Marin, R., Chew, K.T., Stewart, D., Szymusiak, R., 2004. Activation of c-fos in GABAergic neurones in the preoptic area during sleep and in response to sleep deprivation. J. Physiol. 556, 935–946.
Gvilia, I., Turner, A., McGinty, D., Szymusiak, R., 2006. Preoptic area neurons and the homeostatic regulation of rapid eye movement sleep. J. Neurosci. 26, 3037–3044.

Hassani, O.K., Henny, P., Lee, M.G., Jones, B.E., 2010. GABAergic neurons intermingled with orexin and MCH neurons in the lateral hypothalamus discharge maximally during sleep. Eur. J. Neurosci. 32, 448–457.

Hassani, O.K., Lee, M.G., Henny, P., Jones, B.E., 2009. Discharge profiles of identified GABAergic in comparison to cholinergic and putative glutamatergic basal forebrain neurons across the sleep-wake cycle. J. Neurosci. 29, 11828–11840.

Headley, D.B., DeLucca, M.V., Haufler, D., Pare, D., 2015. Incorporating 3D-printing technology in the design of head-caps and electrode drives for recording neurons in multiple brain regions. J. Neurophysiol. 113, 2721–2732.

Herfst, L., Burgalossi, A., Haskic, K., Tukker, J.J., Schmidt, M., Brecht, M., 2012. Friction-based stabilization of juxtacellular recordings in freely moving rats. J. Neurophysiol. 108, 697–707.

Hong, G., Lieber, C.M., 2019. Novel electrode technologies for neural recordings. Nat. Rev. Neurosci. 20, 330–345.

Houweling, A.R., Brecht, M., 2008. Behavioural report of single neuron stimulation in somatosensory cortex. Nature 451, 65–68.

Hubel, D.H., 1957. Tungsten microelectrode for recording from single units. Science 125, 549–550.

Ingiosi, A.M., Hayworth, C.R., Harvey, D.O., Singletary, K.G., Rempe, M.J., Wisor, J.P., Frank, M.G., 2020. A role for astroglial calcium in mammalian sleep and sleep regulation. Curr. Biol. 30, 4373–4383 e4377.

Jones, B., 2013. Electrophysiology of sleep-wake systems. In: Kushida, C. (Ed.), Encyclopedia of Sleep, vol. 1. Academic Press, Waltham, MA, pp. 477–484.

Joshi, S., Hawken, M.J., 2006. Loose-patch-juxtacellular recording in vivo—a method for functional characterization and labeling of neurons in macaque V1. J. Neurosci. Methods 156, 37–49.

Judkewitz, B., Rizzi, M., Kitamura, K., Hausser, M., 2009. Targeted single-cell electroporation of mammalian neurons in vivo. Nat. Protoc. 4, 862–869.

Katona, L., Micklem, B., Borhegyi, Z., Swiejkowski, D.A., Valenti, O., Viney, T.J., Kotzadimitriou, D., Klausberger, T., Somogyi, P., 2017. Behavior-dependent activity patterns of GABAergic long-range projecting neurons in the rat hippocampus. Hippocampus 27, 359–377.

Kostin, A., McGinty, D., Szymusiak, R., Alam, M.N., 2012. Mechanisms mediating effects of nitric oxide on perifornical lateral hypothalamic neurons. Neuroscience 220, 179–190.

Kroeger, D., Absi, G., Gagliardi, C., Bandaru, S.S., Madara, J.C., Ferrari, L.L., Arrigoni, E., Munzberg, H., Scammell, T.E., Saper, C.B., Vetrivelan, R., 2018. Galanin neurons in the ventrolateral preoptic area promote sleep and heat loss in mice. Nat. Commun. 9, 4129.

Lansink, C.S., Bakker, M., Buster, W., Lankelma, J., van der Blom, R., Westdorp, R., Joosten, R.N., McNaughton, B.L., Pennartz, C.M., 2007. A split microdrive for simultaneous multi-electrode recordings from two brain areas in awake small animals. J. Neurosci. Methods 162, 129–138.

Lee, A.K., Epsztein, J., Brecht, M., 2009. Head-anchored whole-cell recordings in freely moving rats. Nat. Protoc. 4, 385–392.

Lee, M.G., Hassani, O.K., Jones, B.E., 2005. Discharge of identified orexin/hypocretin neurons across the sleep-waking cycle. J. Neurosci. 25, 6716–6720.

Ma, J., Zhao, Z., Cui, S., Liu, F.Y., Yi, M., Wan, Y., 2019. A novel 3D-printed multi-drive system for synchronous electrophysiological recording in multiple brain regions. Front. Neurosci. 13, 1322.

Mahmud, M., Girardi, S., Maschietto, M., Pasqualotto, E., Vassanelli, S., 2011. An automated method to determine angular preferentiality using LFPs recorded from rat barrel

cortex by brain-chip interface under mechanical whisker stimulation. Annu. Int. Conf. IEEE Eng. Med. Biol. Soc. 2011, 2307–2310.
Manns, I.D., Alonso, A., Jones, B.E., 2000. Discharge properties of juxtacellularly labeled and immunohistochemically identified cholinergic basal forebrain neurons recorded in association with the electroencephalogram in anesthetized rats. J. Neurosci. 20, 1505–1518.
Methippara, M.M., Alam, M.N., Szymusiak, R., McGinty, D., 2003. Preoptic area warming inhibits wake-active neurons in the perifornical lateral hypothalamus. Brain Res. 960, 165–173.
Pack, C.C., Berezovskii, V.K., Born, R.T., 2001. Dynamic properties of neurons in cortical area MT in alert and anaesthetized macaque monkeys. Nature 414, 905–908.
Pinault, D., 1996. A novel single-cell staining procedure performed in vivo under electrophysiological control: morpho-functional features of juxtacellularly labeled thalamic cells and other central neurons with biocytin or neurobiotin. J. Neurosci. Methods 65, 113–136.
Pinault, D., 2005. A new stabilizing craniotomy-duratomy technique for single-cell anatomo-electrophysiological exploration of living intact brain networks. J. Neurosci. Methods 141, 231–242.
Pinault, D., 2011. The juxtacellular recording-labeling technique. In: Vertes, R.P., Stackman, R.W. (Eds.), Electrophysiological Recording Techniques, Neuromethods, vol. 54. Springer, New York, pp. 41–75.
Porkka-Heiskanen, T., Strecker, R.E., Thakkar, M., Bjorkum, A.A., Greene, R.W., McCarley, R.W., 1997. Adenosine: a mediator of the sleep-inducing effects of prolonged wakefulness. Science 276, 1265–1268.
Sakai, K., 2011. Sleep-waking discharge profiles of median preoptic and surrounding neurons in mice. Neuroscience 182, 144–161.
Saper, C.B., Fuller, P.M., 2017. Wake-sleep circuitry: an overview. Curr. Opin. Neurobiol. 44, 186–192.
Saper, C.B., Fuller, P.M., Pedersen, N.P., Lu, J., Scammell, T.E., 2010. Sleep state switching. Neuron 68, 1023–1042.
Scammell, T.E., Arrigoni, E., Lipton, J.O., 2017. Neural circuitry of wakefulness and sleep. Neuron 93, 747–765.
Sherin, J.E., Shiromani, P.J., McCarley, R.W., Saper, C.B., 1996. Activation of ventrolateral preoptic neurons during sleep. Science 271, 216–219.
Shiromani, P.J., Peever, J.H., 2017. New neuroscience tools that are identifying the sleep-wake circuit. Sleep 40.
Strumwasser, F., 1958. Long-term recording' from single neurons in brain of unrestrained mammals. Science 127, 469–470.
Suntsova, N., Guzman-Marin, R., Kumar, S., Alam, M.N., Szymusiak, R., McGinty, D., 2007. The median preoptic nucleus reciprocally modulates activity of arousal-related and sleep-related neurons in the perifornical lateral hypothalamus. J. Neurosci. 27, 1616–1630.
Suntsova, N., Szymusiak, R., Alam, M.N., Guzman-Marin, R., McGinty, D., 2002. Sleep-waking discharge patterns of median preoptic nucleus neurons in rats. J. Physiol. 543, 665–677.
Takahashi, K., Lin, J.S., Sakai, K., 2006. Neuronal activity of histaminergic tuberomammillary neurons during wake-sleep states in the mouse. J. Neurosci. 26, 10292–10298.
Tang, Q., Brecht, M., Burgalossi, A., 2014. Juxtacellular recording and morphological identification of single neurons in freely moving rats. Nat. Protoc. 9, 2369–2381.

Thakkar, M.M., Delgiacco, R.A., Strecker, R.E., McCarley, R.W., 2003. Adenosinergic inhibition of basal forebrain wakefulness-active neurons: a simultaneous unit recording and microdialysis study in freely behaving cats. Neuroscience 122, 1107–1113.

Toney, G.M., Daws, L.C., 2006. Juxtacellular labeling and chemical phenotyping of extracellularly recorded neurons in vivo. Methods Mol. Biol. 337, 127–137.

Vinck, M., Bos, J.J., Van Mourik-Donga, L.A., Oplaat, K.T., Klein, G.A., Jackson, J.C., Gentet, L.J., Pennartz, C.M., 2015. Cell-type and state-dependent synchronization among rodent somatosensory, visual, perirhinal cortex, and Hippocampus CA1. Front. Syst. Neurosci. 9, 187.

Voigts, J., Siegle, J.H., Pritchett, D.L., Moore, C.I., 2013. The flexDrive: an ultra-light implant for optical control and highly parallel chronic recording of neuronal ensembles in freely moving mice. Front. Syst. Neurosci. 7, 8.

Watson, P.A., Ellwood-Yen, K., King, J.C., Wongvipat, J., Lebeau, M.M., Sawyers, C.L., 2005. Context-dependent hormone-refractory progression revealed through characterization of a novel murine prostate cancer cell line. Cancer Res. 65, 11565–11571.

Weber, F., Chung, S., Beier, K.T., Xu, M., Luo, L., Dan, Y., 2015. Control of REM sleep by ventral medulla GABAergic neurons. Nature 526, 435–438.

Xu, M., Chung, S., Zhang, S., Zhong, P., Ma, C., Chang, W.C., Weissbourd, B., Sakai, N., Luo, L., Nishino, S., Dan, Y., 2015. Basal forebrain circuit for sleep-wake control. Nat. Neurosci. 18, 1641–1647.

Zhong, P., Zhang, Z., Barger, Z., Ma, C., Liu, D., Ding, X., Dan, Y., 2019. Control of non-REM sleep by midbrain neurotensinergic neurons. Neuron 104, 795–809 e796.

CHAPTER 4

Physiologic systems dynamics, coupling and network interactions across the sleep-wake cycle

Plamen Ch. Ivanov[1,2,3], Ronny P. Bartsch[4]
[1]Keck Laboratory for Network Physiology, Department of Physics, Boston University, Boston, MA, United States; [2]Division of Sleep Medicine, Brigham and Women's Hospital, Harvard Medical School, Boston, MA, United States; [3]Institute of Solid State Physics, Bulgarian Academy of Sciences, Sofia, Bulgaria; [4]Department of Physics, Bar-Ilan University, Ramat Gan, Israel

Physiologic dynamics across sleep stages and circadian phases

Integrated physiologic systems under neural regulation, such as the cardiac and respiratory system, exhibit complex dynamics characterized by continuous "erratic" fluctuations. Investigations in recent years have shown that these fluctuations are not simply noise due to external perturbations, and instead carry important information related to the underlying mechanisms of physiologic control, and how these mechanisms are modulated as a function of physiological states. Different levels of activation of the sympathetic and parasympathetic branch of the central nervous system during sleep and wake state, changes in sympathovagal balance with sleep-stage transitions and circadian rhythms affect the regulation of physiologic systems and their output signals across the sleep-wake cycle. Measures of physiologic variability provide insight into key aspects of sleep regulation under health and disease.

The normal electrical activity of the heart is usually described as a regular sinus rhythm. However, cardiac interbeat intervals fluctuate in an irregular manner in healthy subjects even at rest and during sleep (Fig. 4.1). In recent years, the intriguing statistical properties of interbeat interval sequences have attracted the attention of researchers from different fields. Analysis of heartbeat fluctuations focused initially on short-time oscillations associated with breathing, blood pressure, and neuroautonomic control (Kitney and Rompelman, 1980; Akselrod et al., 1981). Studies of longer heartbeat records revealed 1/f-like behavior in the power spectrum of

Methodological Approaches for Sleep and Vigilance Research
ISBN 978-0-323-85235-7
https://doi.org/10.1016/B978-0-323-85235-7.00006-5

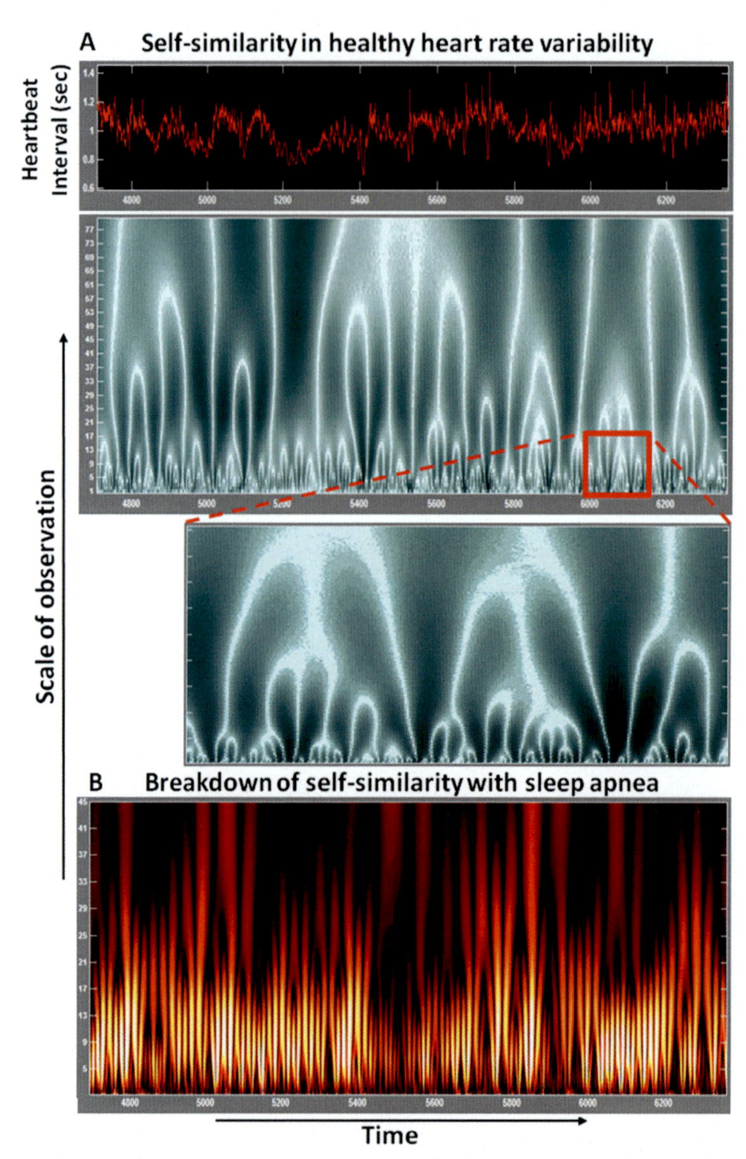

Figure 4.1 Complex temporal organization of physiologic variability. Seemingly random heartbeat interval fluctuations derived from a 30-min electrocardiography recording of a healthy subject during sleep (A, top panel) reveal a hierarchical structure across scales characterized by arches (A, middle panel) that bifurcate in a self-similar manner at smaller and smaller scales, demonstrating presence of scale-invariance (scaling) in temporal organization. Brighter colors correspond to large heartbeat interval fluctuations and white tracks (arches) show evolution of the fluctuations in time (horizontal axis: $\approx$ 1700 heartbeats) and across different scales of observation (vertical axis: from $\approx$ 5 s to 10 min, with large scales at the top). The plot shows a hierarchical self-similar structure formed by the white arches—a magnification of the central portion of the middle panel in A with 200 heartbeats on the

heartbeat intervals indicating the presence of a complex temporal organization over a broad range of frequencies and time scales (Kobayashi and Musha, 1982). Recent analyses of very long time series (up to 24 h) show that under healthy conditions, interbeat interval increments exhibit power-law anticorrelations (Peng et al., 1993, 1995; Ivanov et al., 2002a; Ivanov, 2003) and follow a functional form for their distributions that remains invariant at different time scales of observation and is universal across subjects (Ivanov et al, 1996, 1998c; Kiyono et al., 2004). Furthermore, heartbeat fluctuations exhibit turbulence-like nonlinear dynamics characterized by a broad multifractal spectrum (Ivanov et al., 1999a, 2001). The robustness of these dynamical features across time scales (i.e., scaling behavior) is indicative of the basic mechanisms regulating physiologic dynamics and their fluctuations through coupled feedback loops operating on different time scales (Ivanov et al., 1998a; Ashkenazy et al., 2002; Ivanov and Lo, 2002; Kantelhardt et al., 2003a). Scaling behavior has been observed in the dynamics of various physiological systems (Hausdorff et al., 2001; Lo et al., 2002, 2004; Suki et al., 2003; Hu et al., 2004a; Rostig et al., 2005; Bartsch et al., 2007; Ivanov et al., 2009; Dvir et al., 2018), and the derived scaling measures are reliable markers of diagnosis and prognosis, reflecting alterations in physiological regulation in response to aging and pathologic perturbations (Lipsitz et al., 1990; Kaplan et al., 1991; Iyengar et al., 1996; Bernaola-Galván et al., 2001; Goldberger et al., 2002; Schmitt and Ivanov, 2007; Schumann et al., 2010).

Sleep-wake cycles and the endogenous circadian rhythms are associated with periodic changes in key physiological processes (Moelgaard et al., 1991; Malik and Camm, 1995; Berne and Levy, 2001). In addition to the known periodic rhythms with a characteristic time scale, the endogenous

horizontal axis and smaller scales of observation from $\approx$ 5 s to 1.5 min on the vertical axis shows similar branching patterns. Such complex self-similar organization in heartbeat variability results from nonlinear feedback interactions between multiple sympathetic and parasympathetic inputs that underlie cardiac control during sleep and operate over a broad range of time scales (Ivanov et al., 1998a). For sleep disorders such as sleep apnea, this self-similarity breaks down to a repetitive pattern at small and intermediate scales of observation (panel B), indicating that physiologic variability can provide important insights into the mechanisms of physiologic regulation during sleep. *(Adapted from Ivanov, P.C., Amaral, L.A.N., Goldberger, A.L., Havlin, S., Rosenblum, M.G., Stanley, H.E., Struzik, Z., 2001. From $1/f$ noise to multifractal cascades in heartbeat dynamics. Chaos 11, 641–652.)*

mechanisms of sleep and circadian regulation influence the dynamics of physiological systems across multiple time scales, leading to systematic changes in their scaling properties. Establishing association between distinct physiologic states and complexity measures of physiologic dynamics is essential to elucidate the underlying neuroautonomic mechanisms of physiological regulation.

Sleep-wake differences in the distributions of heartbeat fluctuations

Typically, differences in cardiac dynamics during sleep and wake are reflected in the average and standard deviation of heartbeat intervals (Fig. 4.2) (Moelgaard et al., 1991). Novel methods designed to quantify probability distributions of physiologic fluctuations embedded in nonstationary signals [cumulative variation amplitude analysis (CVAA) (Ivanov et al., 1996)], reveal that despite individual differences the distribution of the amplitudes of the cumulative variations in heartbeat intervals are characterized by a homogenous function that is universal for all subjects and remains unchanged (invariant) with the time scale of analysis from seconds to hours (Ivanov et al, 1996, 1998b, 1998c; Kiyono et al., 2004). Such robust scaling behavior is reminiscent of a wide class of well-studied physical systems exhibiting universal scaling properties at criticality (Stanley, 1971; Vicsek, 1992; Stanley et al., 2000), and indicates presence of a multicomponent mechanism of cardiac regulation with feedbacks operating on multiple time scales (Ivanov et al., 1998a; Kantelhardt et al., 2003a). The collapse of the individual distributions for all healthy subjects after rescaling their individual parameter is indicative of a universal structure, in the sense that there is a closed mathematical scaling form describing in a unified quantitative way the cardiac dynamics of healthy subjects over a broad range of time scales (Ivanov et al., 1998b).

This universal behavior across subjects and time scales holds for both sleep and wake phase, however, with slower decay in the tail for the sleep-state compared to the wake-state (Fig. 4.3) (Ivanov et al., 1998b, 1998c). Counterintuitively, the slower decaying tail of the distribution of heartbeat fluctuations during sleep indicates higher probability of larger variations in comparison with the daytime dynamics during wake state (Fig. 4.3). Modeling approaches to heart rate dynamics based on stochastic feedback mechanisms suggest that the marked change in the distribution of the cumulative heartbeat variations we observe during transition from wake to

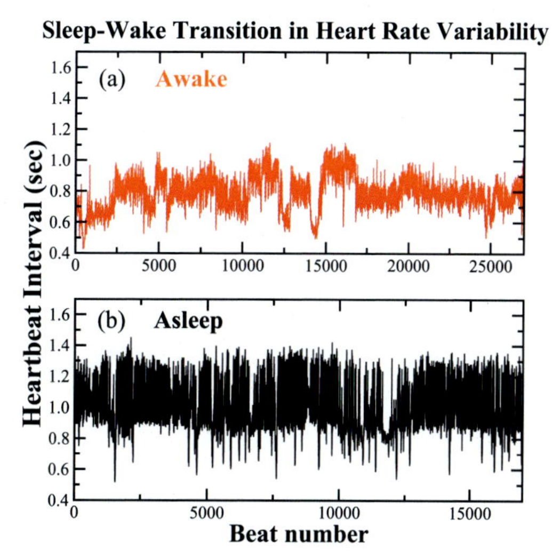

Figure 4.2 Changes in the degree of sympathetic and parasympathetic activation during wake and sleep lead to transitions in physiologic variability. In cardiac dynamics, the transition from wake to sleep state is typically associated with decrease in the average heart rate (increased average interbeat interval). However, other key aspects of heart rate variability also change with sleep-wake transitions. (A) Higher sympathetic tone during wake is reflected in smaller amplitude of beat-to-beat increments and pronounced trends embedded in heartbeat variability. (B) In contrast, dominant parasympathetic tone during sleep leads to significantly higher amplitude of heartbeat interval increments and reduced low-frequency trends. *(Adapted from Ivanov, P.C., Bunde, A., Amaral, L.A.N., Havlin, S., Fritsch-Yelle, J., Baevsky, R.M., Stanley, H.E., Goldberger, A.L., 1999b. Sleep-wake differences in scaling behavior of the human heartbeat: analysis of terrestrial and long-term space flight data. Eur. Lett. 48, 594–600. https://doi.org/10.1209/epl/i1999-00525-0.)*

sleep state is due to the relative decrease of sympathetic tone inputs to cardiac dynamics in relation to parasympathetic tone during sleep (Ivanov et al., 1998a).

Sleep-stage stratification in the standard deviation of heartbeat intervals and their fluctuations

Healthy sleep consists of cycles of approximately one to two hours duration. Each cycle is characterized by a sequence of sleep stages usually starting with light sleep, followed by deep sleep and rapid eye movement (REM) sleep (Carskadon and Dement, 1994). While the specific functions of the different sleep stages are not yet well understood, deep sleep is considered to be essential for physical rest, while light and REM sleep are important for

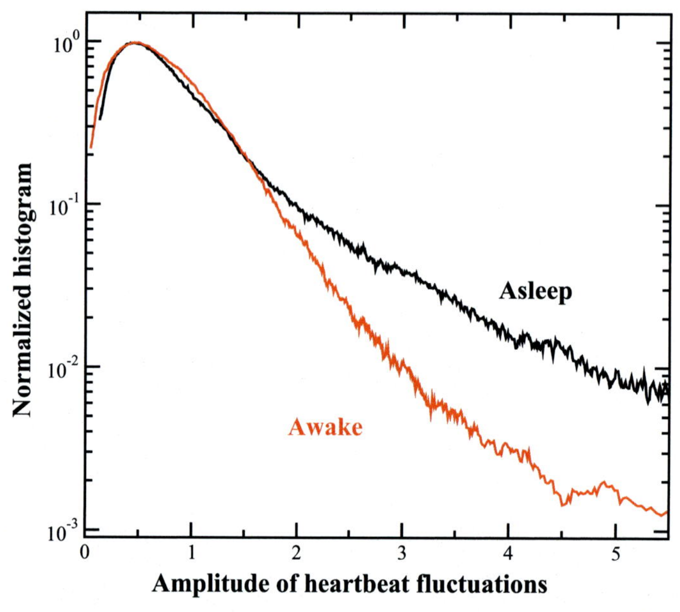

Figure 4.3 Probability distribution of the amplitude of heartbeat fluctuations obtained by cumulative variation amplitude analysis (Ivanov et al., 1996) reveal universal functional form for healthy subjects that remain invariant at different time scales of analysis. Markedly different functional forms characterize heartbeat fluctuations during wake and sleep. In contrast to the belief that heart rate variability is higher during daily activity compared to a more metronome-like and less varied heartbeat during sleep, normalized distributions of the amplitude of heartbeat interval increments under healthy condition show significantly higher probability (more than 1 decade on the vertical axis) for large amplitude of fluctuations during sleep, as indicated by the slower-decaying tail of the distribution during sleep compared to wake. Curves represent data collapse of normalized probability distributions from 18 young healthy subjects (mean age 34 years) based on 6 h continuous electrocardiography recordings for sleep as well as wake state. *(Adapted from Ivanov, P.C., Rosenblum, M.G., Peng, C.-K., Mietus, J., Havlin, S., Stanley, H.E., Goldberger, A.L., 1996. Scaling behaviour of heartbeat intervals obtained by wavelet-based time-series analysis. Nature 383, 323–327 and Ivanov, P.C., Rosenblum, M.G., Peng, C.-K., Mietus, J., Havlin, S., Stanley, H.E., Goldberger, A.L., 1998c. Scaling and universality in heart rate variability distributions. Phys. A 249, 587–593. https://doi.org/10.1016/S0378-4371(97)00522-0.)*

memory consolidation (Carskadon and Dement, 1994; Stickgold et al., 2001; Stickgold, 2005). It is known that changes in the physiological processes are associated with wake or sleep state and with different sleep stages. Thus, investigations have focused on how temporal characteristics of cardiac dynamics are affected by changes in physiologic regulation during wake, sleep, and different sleep stages.

Recent studies have reported reduced heart rate variability (HRV) (Corino et al., 2006) and alterations of the fractal and nonlinear properties of heartbeat fluctuations with healthy aging (Lipsitz and Goldberger, 1992; Umetani et al., 1998), and these changes in cardiac dynamics have been associated with higher cardiac risk in elderly (Moelgaard et al., 1991). Sleep dynamics have also been found to change with advanced age, e.g., elderly subjects typically have more fragmented sleep with frequent arousals and reduced duration of deep sleep (Bliwise, 1993). Sympathetic nerve activity measurements as well as spectral analysis of HRV across sleep stages show dominant parasympathetic control during non-REM sleep, and activation of the sympathetic nervous system during REM and wake (Somers et al., 1993; Baharav et al., 1995) leading to different morphology in the heartbeat interval time series across sleep stages (Fig. 4.4). The intricate mechanism of interaction between sleep regulation and cardiac control, and whether this interaction declines with age, however, remains not well understood.

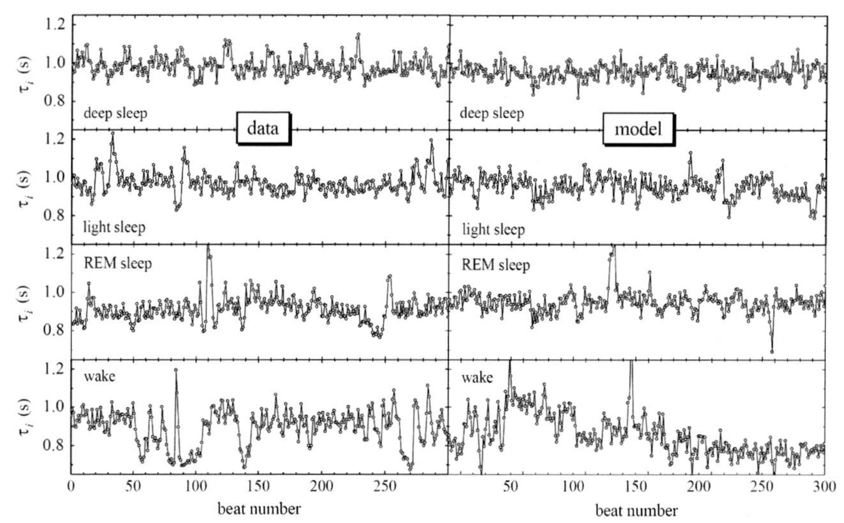

Figure 4.4 On the left: representative recordings of 300 time intervals between successive heartbeats for a healthy subject during deep sleep, light sleep, rapid eye movement sleep, and an intermediate wake state (from top to bottom). On the right: corresponding series simulated by a stochastic feedback model that takes into account changes in sympathetic and parasympathetic inputs during different sleep stages and reproduces dynamic characteristics of the original data. *(Adapted from Kantelhardt, J.W., Havlin, S., Ivanov, P.C., 2003a. Modeling transient correlations in heartbeat dynamics during sleep. Eur. Lett. 62, 147—153; Ivanov, P.C., Amaral, L.A.N., Goldberger, A.L., Stanley, H.E., 1998a. Stochastic feedback and the regulation of biological rhythms. EPL Europhys. Lett. 43, 363. doi:10.1209/epl/i1998-00366-3.)*

To test how sleep dynamics affect temporal scaling and nonlinear organization of heartbeat fluctuations across time scales, we compare heartbeat dynamics of healthy young and healthy elderly subjects for different sleep stages. One possible hypothesis is that there will be a well-pronounced stratification pattern in the values of static (e.g., mean, standard deviation) and dynamic (scaling and nonlinear) measures of heartbeat dynamics across sleep stages, and that this stratification will be present in both healthy young and healthy elderly subjects. An alternative hypothesis is that while there are significant differences in cardiac control with sleep stage transitions for young subjects, the static and dynamic features of heartbeat fluctuations in healthy elderly subjects remain unchanged across sleep stages due to age-related loss of cardiac variability and decline in neuroautonomic responsiveness to changes in sleep regulation. Testing these hypotheses addresses the question of how sleep regulation influences cardiac dynamics, and whether this influence changes with healthy aging, as sleep disorders have been associated with increased cardiovascular risk (Tsuji et al., 1994; Grote et al., 1999; Mohsenin, 2001). Probing for changes in the complex temporal organization of heartbeat fluctuations during different sleep stages can thus help elucidate the mechanisms through which alterations of sleep regulation in elderly may contribute to cardiac risk. This approach is in line with earlier studies where combinations of HRV measures as outlined in the HRV task force 1996 (Electrophysiology Task Force of the European Society of Cardiology the North American Society of Pacing, 1996) were utilized to increase their power in stratifying age- and disease-related cardiac risk (Hallstrom et al., 2004).

To test whether HRV changes according to sleep-stage transitions, and whether this variability is reduced in healthy elderly subjects for certain sleep stages, we first estimate for each subject and for each sleep stage the standard deviation of the RR interbeat intervals σ_{RR} (SDNN) and the standard deviation of the increments in the consecutive interbeat intervals $\sigma_{\Delta RR}$ (RMSSD). For healthy young subjects ($n = 13$, group mean age 33 years), we find that the group average σ_{RR} is highest during wake, lower for REM and light sleep, and lowest during deep sleep (Fig. 4.5A), with statistically significant difference between sleep stages. Notably, healthy elderly subjects ($n = 24$, group mean age 78 years) exhibit very similar stratification in σ_{RR} across sleep stages (Fig. 4.5A). These observations demonstrate that the variability in cardiac dynamics in elderly subjects exhibits a remarkably similar response to sleep-stage transitions and stratification pattern across sleep stages as the one observed for young subjects.

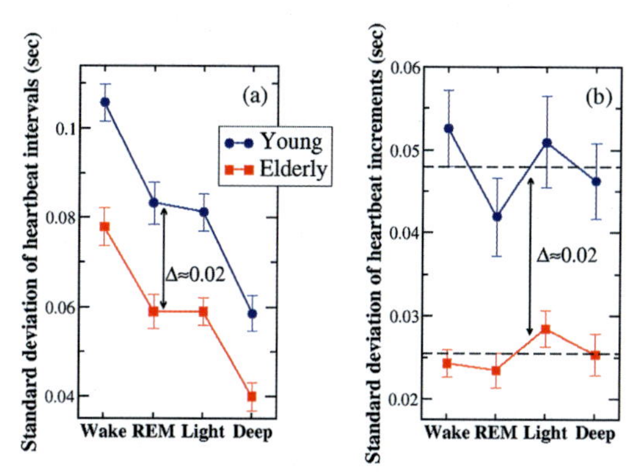

Figure 4.5 Changes in sympathovagal balance associated with sleep-stage transitions lead to stratification in basic measures of physiologic variability. (A) A well pronounced "chair-like" stratification pattern in the standard deviation of heartbeat intervals σ_{RR} (SDNN) reflects a gradual decrease of sympathetic tone with transitions from wake to REM and light sleep to deep sleep in both healthy young and elderly subjects. MANOVA-tests show that σ_{RR} changes significantly with age and also with sleep stage (P-value $< .0001$). (B) In contrast, the standard deviation of heartbeat interval increments $\sigma_{\Delta RR}$ (RMSSD) does not significantly change (within error bars) with transitions across sleep stages. Both measures in (A) and (B) exhibit the same vertical shift of $\Delta \approx 0.02$ s due to suppressed parasympathetic (vagal) tone in elderly subjects. MANOVA-test shows that $\sigma_{\Delta RR}$ changes significantly (P-value $< .0001$) with age and but not with sleep stage (P-value $= .23$). These observations indicate that the standard deviation of heartbeat increments $\sigma_{\Delta RR}$ (RMSSD) is a measure of parasympathetic tone during sleep, while the standard deviation of heartbeat intervals σ_{RR} (SDNN) is a measure of both sympathetic and parasympathetic tone, the balance of which gradually changes from wake to sleep phase and across sleep stages. *(Adapted from Schmitt, D.T., Stein, P.K., Ivanov, P.C., 2009. Stratification pattern of static and scale-invariant dynamic measures of heartbeat fluctuations across sleep stages in young and elderly. IEEE Trans. Biomed. Eng. 56, 1564–1573. https://doi.org/10.1109/TBME.2009.2014819.)*

However, the elderly group average values for σ_{RR} are significantly lower for all sleep stages compared to the young group. The observed identical shift in σ_{RR} for all sleep stages (Fig. 4.5A) indicates significantly reduced HRV in elderly subjects – a finding also observed in other studies (Tsuji et al., 1994; Iyengar et al., 1996; Crasset et al., 2001; Brandenberger et al., 2003). Nevertheless, the response of cardiac dynamics to sleep-stage transitions in healthy elderly subjects remains the same as in young subjects, as evident from the similar stratification patterns in σ_{RR} for both groups, indicating a robust sleep-regulatory mechanism that does not break down with aging.

In contrast to σ_{RR}, another standard static measure $\sigma_{\Delta RR}$ (RMSSD), which quantifies the variability in consecutive beat-to-beat increments, does not exhibit sleep-stage stratification—behavior observed for both young and elderly subjects (Fig. 4.5B). Remarkably, there is a significant reduction in the values of $\sigma_{\Delta RR}$ for elderly subjects compared to the young group with the same shift of $\Delta \approx 0.02s$ as for the σ_{RR}.

We note that for both young and elderly groups, the difference in the group average standard deviation between wake and deep sleep of 0.04 s is approximately twice larger than the vertical shift Δ we find between young and elderly. These results indicate that the effect of sleep regulation on HRV, as measured by σ_{RR}, is stronger than the effect of aging (average age difference of 45 years between the young and elderly groups). The pronounced stratification pattern in σ_{RR} across sleep stages relates to reduction in sympathetic tone during light and deep sleep compared to wake and REM (Somers et al., 1993; Baharav et al., 1995). As the sympathetic activity is represented by the low-frequency range in the heart rate power spectrum (Baharav et al., 1995; Bonnet and Arand, 1997), bursts of sympathetic tone and parasympathetic withdrawal lead to increased nonstationarity (pronounced trends) in interbeat interval time series leading to higher values of σ_{RR} during wake. With gradual decrease of sympathetic tone during light and deep sleep, the degree of nonstationary trends also decreases, and therefore, σ_{RR} is reduced. This is also observed for the group of healthy elderly subjects, however, with lower values of σ_{RR} for all sleep stages due to reduced parasympathetic tone in elderly.

There is a strong positive correlation between σ_{RR} and low-frequency HRV, as well as between $\sigma_{\Delta RR}$ and high-frequency variability (Otzenberger et al., 1998). As $\sigma_{\Delta RR}$ represents high-frequency parasympathetic inputs and filters out low-frequency sympathetic inputs, the fact that there is no significant reduction in $\sigma_{\Delta RR}$ between wake, REM, light and deep sleep indicates that parasympathetic tone does not significantly change. This, in turn, shows that the statistically significant drop in σ_{RR} between REM and deep sleep in both healthy young and elderly subjects is due to a significant reduction of sympathetic activity. On the other hand, the drop of $\Delta \approx 0.02s$ in σ_{RR} between young and elderly subjects must be caused by suppression of parasympathetic tone in elderly, since there is a similar drop of $\Delta \approx 0.02s$ in $\sigma_{\Delta RR}$, which is a measure sensitive only to parasympathetic tone. While both σ_{RR} and $\sigma_{\Delta RR}$ are static measures, they reflect fundamentally different aspects of cardiac control. The standard deviation of heartbeat intervals σ_{RR} captures both sympathetic and parasympathetic activity, and thus changes across sleep stages as well as with advanced age. In contrast, the complementary measure $\sigma_{\Delta RR}$ (standard deviation of the increments in consecutive heartbeat intervals) is sensitive only to changes in the parasympathetic activity, and thus, changes only with age.

Sleep-wake differences in the correlations of heartbeat fluctuations

The empirical observations of significant sleep-wake differences in the probability distributions of the amplitudes of the fluctuations in the heartbeat intervals (Fig. 4.3) as well as in static measures of heartbeat variability (Fig. 4.5), raise the question whether changes in physiologic regulation and in neuroautonomic cardiac control during sleep and wake state lead to characteristic differences in the temporal correlations of cardiac dynamics. Studies of continuous heartbeat interval time series recorded from healthy subjects during 6 h of sleep (midnight to 6 a.m.) and diurnal wake activity (noon to 6 p.m.) have utilized the detrended fluctuation analysis (DFA) method (Peng et al., 1994) to detect and quantify correlations in heartbeat fluctuations over a range of time scales from seconds to hours. One of the advantages of the DFA method is that it can reliably quantify scale-invariant structure in the temporal organization of time series in the presence of linear and higher-order polynomial trends, data artifacts, and nonstationarity (Hu et al., 2001; Chen et al., 2002, 2005; Xu et al., 2005, 2011; Ma et al., 2010). These analyses have uncovered long-range power-law correlations over 2 decades of time scales, indicating a robust scale-invariant structure in heartbeat fluctuations ranging from seconds to hours—a long-term memory mechanism that correlates a heartbeat interval at a given moment in time with intervals thousands of heartbeats later. During wake, these temporal power-law correlations are characterized by a scaling exponent $\alpha_W = 1.05$ and with a significantly lower exponent $\alpha_S = 0.85$ during sleep (Ivanov et al., 1999b). These findings indicate that with transition from wake to sleep, dominant parasympathetic tone leads to a complete reorganization in the temporal correlations of heartbeat fluctuations and to alteration of underlying memory process that occurs over a broad range of time scales. This effect of sleep regulation is observed also under pathologic conditions such as heart failure (Ivanov et al., 1999b), however, with much higher values for the correlation exponent α during both wake with $\alpha_W = 1.2$ and sleep with $\alpha_S = 0.95$ due to the increased level of sympathetic tone in heart failure (Fig. 4.6).

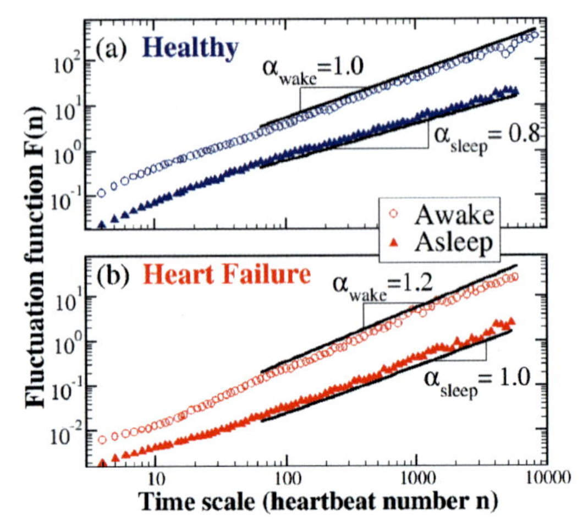

Figure 4.6 Sleep-wake transitions modulate physiologic dynamics on scales from a few seconds to many hours. Competing sympathetic and parasympathetic inputs lead to complex self-similar organization in heartbeat fluctuations (Fig. 4.1) that is quantified by temporal correlations (vertical axis) over a broad range of time scales (horizontal axis) characterized by the correlation exponent α (slope of the curve)—for white noise $\alpha = 0.5$; for Brownian motion (integrated white noise) $\alpha = 1.5$; $\alpha = 1.0$ indicates strong long-range correlations and $\alpha < 0.5$ indicates anticorrelations. Scaling curves F(n) in the plot represent long-range power-law correlations (straight line on log-log plot) quantified by the DFA method (Peng et al., 1994) and indicate that temporal organization of heartbeat fluctuations over a range of time scales dramatically changes with transition from wake to sleep as evidenced by significant reduction in the correlation exponent α. Such scale-invariant temporal organization (scaling behavior) is indicative of a long memory mechanism that correlates heartbeats at a given moment in time with heartbeats that occur thousands of seconds later. Notably, sleep regulation impacts this memory mechanism in cardiac dynamics across time scales for both (A) healthy subjects and (B) subjects with congestive heart failure with similar reduction in the correlation exponent α related to sympathetic withdrawal during sleep that affects intermediate and large time scales. *(Adapted from Ivanov, P.C., Bunde, A., Amaral, L.A.N., Havlin, S., Fritsch-Yelle, J., Baevsky, R.M., Stanley, H.E., Goldberger, A.L., 1999b. Sleep-wake differences in scaling behavior of the human heartbeat: analysis of terrestrial and long-term space flight data. Eur. Lett. 48, 594–600. https://doi.org/10.1209/epl/i1999-00525-0, Schmitt, D.T., Ivanov, P.C., 2007. Fractal scale-invariant and nonlinear properties of cardiac dynamics remain stable with advanced age: a new mechanistic picture of cardiac control in healthy elderly. Am. J. Physiol. Regul. Integr. Comp. Physiol. 293(5):R1923-R1937. https://doi.org/10.1152/ajpregu.00372.2007, and Ivanov, P.C., 2007. Scale-invariant aspects of cardiac dynamics — Observing sleep stages and circadian phases. IEEE Eng. Med. Biol. 26(6):33-37. https://doi.org/10.1109/MEMB.2007.907093.)*

Empirical analyses show that for all individuals, heartbeat dynamics during sleep are characterized by a smaller exponent (Fig. 4.6), suggesting stronger anticorrelations in heartbeat fluctuations during sleep compared with wake state (note that for correlated processes characterized by correlation exponent $\alpha < 1.5$, the fluctuations are anticorrelated with $\alpha < 0.5$). The findings of stronger anticorrelations (Ivanov et al., 1999b), as well as higher probability for larger amplitude of heartbeat fluctuations during sleep (Ivanov et al, 1996, 1998b, 1998c), suggest that the observed dynamical characteristics are related to intrinsic mechanisms of neuroautonomic control during sleep and wake phases, and support a reassessment of the sleep as an active dynamical state. We note that the average difference in the scaling exponents ($\alpha_W - \alpha_S = 0.2$ for both healthy and heart failure groups) is comparable with the scaling difference between health and disease indicating a surprisingly strong impact of sleep regulation on sleep dynamics. We also note that the scaling exponents for the heart failure group during sleep are close to the exponents observed for the healthy group during wake (Ivanov et al., 1999b). Since heart failure occurs when the cardiac output is not adequate to meet the metabolic demands of the body, one would anticipate that the manifestations of heart failure would be most severe during physical stress when metabolic demands are greatest, and least severe when metabolic demands are minimal, i.e., during rest or sleep. The observed scaling behavior in temporal correlations of heartbeat fluctuations is consistent with these physiological considerations: the correlation exponents for heart failure subjects is closer to normal during sleep and minimal physical activity. Of related interest, recent studies indicate that sudden death in individuals with underlying heart disease is most likely to occur in the hours just after awakening (Muller et al., 1985; Peters Robert et al., 1994; Behrens et al., 1997), raising the question of how circadian rhythms influence the embedded temporal organization and associated scaling behavior in cardiac dynamics (Hu et al., 2004b; Ivanov et al., 2007). Sleep-wake transitions in long-term correlations are also present in other physiologic systems, including respiratory dynamics of interbreath intervals, where such transitions in temporal organization of physiological fluctuations are also a function of age (Ashkenazy et al., 2002; Kantelhardt et al., 2003b; Suki et al., 2003; Rostig et al., 2005; Schumann et al., 2010).

The transition from wake to sleep phase is associated with significant reduction in sympathetic tone as reflected by changes in static measures (Fig. 4.5), in the temporal organization of physiological dynamics (Fig. 4.6) as well as in the pronounced difference in the temporal correlations and scaling behavior of heartbeat fluctuations during different sleep stages (Fig. 4.7), with significantly different values for the correlation exponent α. Specifically, a clear sleep-stage stratification was reported for the correlation exponent α for both healthy young and elderly subjects, with highest value for wake state, followed by a lower value for REM sleep and even lower values for light and deep sleep (Fig. 4.7), in accordance with decline in sympathetic tone. In summary, these investigations demonstrate (1) a significant change in the temporal organization of heartbeat fluctuations in response to transitions between the wake and the sleep state and across sleep stages (Ivanov et al., 1999b; Bunde et al., 2000; Kantelhardt et al., 2002, 2003a; Dvir et al., 2002) and (2) that heartbeat correlations during wake and sleep and within each sleep stage remain robust, and do not significantly change with healthy aging (Schmitt and Ivanov, 2007; Schmitt et al., 2009).

Scale-invariant fractal processes with identical self-similar temporal organization over time scales characterized by long-range power-law correlations, as quantified by the DFA scaling exponent α (Figs. 4.6 and 4.7), may exhibit very different nonlinear (multifractal) properties (Stanley et al., 1999; Ivanov et al., 2002b). It has been demonstrated that nonlinear properties of physiologic dynamics are encoded in the magnitude of their fluctuations (Ashkenazy et al., 2001). It has been found that information contained in the temporal organization of the magnitude and the sign time series, derived from the fluctuations in the output signal of a physiological system, is independent and complementary to the correlation properties of the original signal, and reflect different aspects of the underlying mechanisms of regulation (Ashkenazy et al., 2000, 2003; Ivanov et al., 2003; Gómez-Extremera et al., 2016). For cardiac dynamics of healthy subjects during daily activity routine, it was shown that heartbeat intervals exhibit power-law correlations at intermediate and large time scales characterized by a scaling exponent $\alpha \approx 1$ (Peng et al., 1995), while the time series of the magnitude of increments in consecutive heartbeat intervals is characterized by $\alpha_{mag} \approx 1.75$ (Ashkenazy et al., 2000, 2001). Furthermore, while temporal correlations in heartbeat intervals reflect linear properties of cardiac dynamics, the temporal structure in magnitude of heartbeat increments relates to nonlinear properties encoded in the Fourier phases (Theiler et al., 1992; Ashkenazy et al., 2003). For certain pathologic conditions, such as congestive heart failure, which are associated with loss of nonlinearity in cardiac dynamics (Lefebvre et al., 1993; Poon and Merrill, 1997; Ivanov et al., 2001), previous studies have reported breakdown of the nonlinear multifractal spectrum (Ivanov et al., 1999a) and reduced scaling

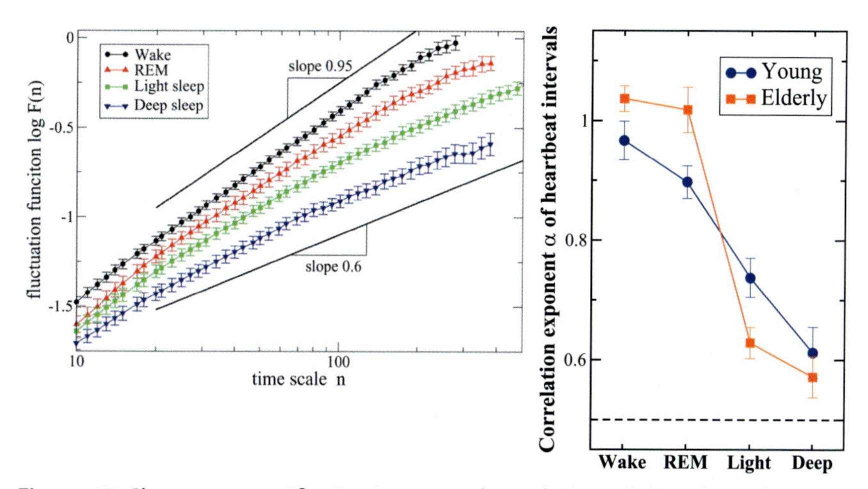

Figure 4.7 Sleep-stage stratification in temporal correlations of physiologic dynamics. Left panel: Log-log plots of the group average fluctuation functions F(n) versus time scale n measured in heartbeats for a group of young subjects using detrended fluctuation analysis analysis of heartbeat interval time series recorded during different sleep stages. Scaling curves F(n) for all sleep stages indicate presence of long-range power-law correlations characterized by scaling exponent α (slope of the curve) over a broad range of time scales from few seconds to hundreds of heartbeats indicating robust scale-invariant temporal organization. The correlation exponent varies from $\alpha \approx 1$ for wake (indicating strong correlations) to $\alpha \approx 0.6$ for deep sleep (weak correlations), suggesting change in the temporal organization of heartbeat fluctuations with transitions across sleep stages. Error bars represent the standard error. The scaling curves F(n) are vertically offset for clarity. Right panel: With transitions from wake to rapid eye movement (REM), light and deep sleep the correlation exponent α of heartbeat intervals significantly decreases for both healthy young and elderly subjects. Student's t-test for young and elderly subjects shows statistically significant difference between the group average exponent α_W for wake and α_D for deep sleep with P-value $= 2 \times 10^{-7}$ (young group) and P-value $= 10^{-17}$ (elderly), as well as between α_{REM} for REM sleep and α_L for light sleep with P-value $= 9 \times 10^{-4}$ (young) and P-value $= 6.5 \times 10^{-12}$ (elderly). Significant change in the heartbeat correlation exponent α across sleep stages (MANOVA test P-value $< .0001$) but an insignificant change with age (MANOVA test P-value $= .93$), indicating an impact of sleep regulation on cardiac dynamics that exceeds effects of healthy aging (Schmitt and Ivanov, 2007). Dominant sympathetic tone during wake and REM is associated with stronger heartbeat correlations, while dominant parasympathetic tone during light and deep sleep leads to weaker correlations with lower exponent α. Such pronounced sleep-stage stratification in the degree of heartbeat interval correlations indicate a strong coupling between sleep and cardiac regulation. Uncorrelated white noise behavior, i.e., absence of long-term memory is characterized by $\alpha = 0.5$ (horizontal dashed line). A similar sleep-stage stratification pattern is observed also for respiration (Kantelhardt et al., 2003b; Suki et al., 2003; Rostig et al., 2005; Schumann et al., 2010). *(Adapted from Schmitt, D.T., Stein, P.K., Ivanov, P.C., 2009. Stratification pattern of static and scale-invariant dynamic measures of heartbeat fluctuations across sleep stages in young and elderly. IEEE Trans. Biomed. Eng. 56, 1564–1573. https://doi.org/10.1109/TBME.2009.2014819.)*

exponent of $\alpha_{mag} \approx 1.5$ (Ashkenazy et al., 2001) related to loss of nonlinearity. Positive correlations in the magnitude series ($\alpha_{mag} > 1.5$) are a reliable marker of long-term nonlinear properties related to the width of the multifractal spectrum (Ashkenazy et al., 2003). Thus, the magnitude and sign analysis (MSA) (Ashkenazy et al., 2000, 2001) is a complementary method to the DFA because it can distinguish physiologic signals with identical long-range power-law correlations but different nonlinear properties and different temporal organization for the magnitude and sign series of physiologic fluctuations.

Empirical studies have utilized the MSA method to investigate how linear and nonlinear properties of heartbeat dynamics change during different stages of sleep. Quantifying correlations in the sign and magnitude time series derived from heartbeat increments in recordings of interbeat intervals from healthy subjects during sleep, these studies reported that the sign series exhibits anticorrelated behavior at short time scales, characterized by a correlation exponent α_{sign} with smallest value for deep sleep, larger for light sleep, and largest value for REM sleep (Fig. 4.8). The magnitude series, on the other hand, exhibits uncorrelated behavior for deep sleep with $\alpha_{mag} \approx 1.5$ (random walk), while long-range correlations with increasing exponent α_{mag} are present during light and REM sleep with largest exponent $\alpha_{mag} \approx 1.75$ during wake (Fig. 4.8) (Kantelhardt et al., 2002; Ivanov, 2007).

The observations of positive power-law correlations for the magnitude of the heartbeat increments during wake and REM sleep and decline of these correlations during light and deep sleep (Fig. 4.8) indicate a dramatic change in the degree of nonlinearity in cardiac dynamics during different sleep stages—from highly nonlinear dynamics during wake and REM to linear behavior during deep sleep. Notably such decrease in the nonlinearity of cardiac dynamics is paralleled by increase in the degree of anticorrelations in the sign of heartbeat fluctuations—i.e., decreasing value of the correlation exponent α_{sign} (Fig. 4.8)—with transition from wake to REM, light, and deep sleep. These findings demonstrate a remarkably complex influence of sleep regulation on cardiac dynamics. Increase in sympathetic tone from deep sleep to light sleep, REM and wake is associated with gradual increase in the degree of long-range correlations (Figs. 4.6 and 4.7) as well as in increase in nonlinearity ($\alpha_{mag} > 1.5$, Fig. 4.8) in heartbeat interval time series. In contrast, increasing contribution of the parasympathetic tone during light sleep and deep sleep leads to stronger anticorrelations in the sign of heartbeat increments. Similar sleep-stage stratification pattern in

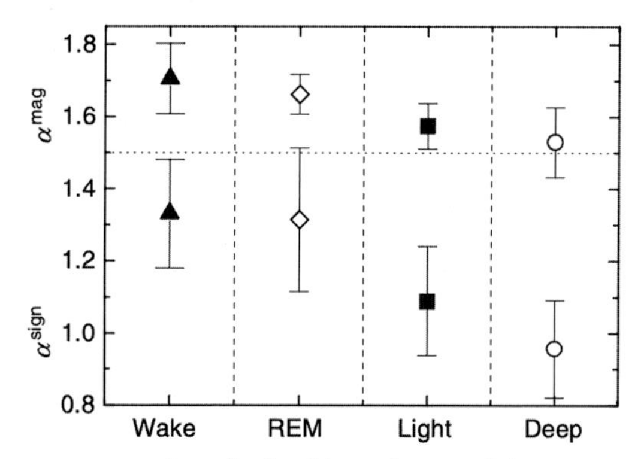

Figure 4.8 Group average values for healthy subjects of the exponents α_{mag} for the integrated magnitude series and α_{sign} for the integrated sign series obtained from consecutive increments in heartbeat intervals during wake state, rapid eye movement (REM) sleep, light and deep sleep. Increasing values of α_{mag} from deep sleep to wake indicate increasing degree of nonlinearity in cardiac dynamics in response to increase in sympathetic tone. In contrast, increasing values of α_{sign} indicate decline in anti-correlations of heartbeat fluctuations with transition from deep to light sleep, REM and wake, associated with decline in the relative contribution of parasympathetic activity. *(Adapted from Kantelhardt JW, Ashkenazy Y, Ivanov PC, Bunde A, Havlin S, Penzel T, Peter J-H, Stanley HE. Characterization of sleep stages by correlations in the magnitude and sign of heartbeat increments. Physical Review E 2002;65(5):051908(6); Ivanov PC. Scale-invariant aspects of cardiac dynamics - Observing sleep stages and circadian phases. IEEE Engineering in Medicine and Biology Magazine 2007; 26(6):33-37.)*

nonlinear and linear characteristics of heartbeat fluctuations is also observed for healthy elderly subjects (Schmitt et al., 2009). A physiologically motivated stochastic model has been developed to account for the complex modulation across time scales in correlations and nonlinear features of heartbeat dynamics during sleep and wake and across sleep-stage transitions (Ivanov et al., 1998a; Kantelhardt et al., 2002, 2003a).

The sleep-wake changes in the scaling characteristics we observe possibly indicate different regimes of intrinsic neuroautonomic regulation of cardiac dynamics, which may switch on and off in accordance with circadian rhythms. These findings raise the intriguing possibility that the transition between the sleep and wake phases is a period of potentially increased neuroautonomic instability because it requires a transition from strongly to weakly anticorrelated regulation of the heart, i.e., a phase transition from one fractal state of self-organization over a range of time scales to another. This hypothesis triggered further investigations on the potential influence of the circadian rhythms on cardiac vulnerability, as outlined in the following subjection. These transitions in the scale-invariant temporal organization of the heartbeat fluctuations between sleep and wake state have been further investigated and confirmed by recent studies (Kiyono et al., 2005). We also note that similar transitions have been observed in the correlation scaling properties of heartbeat dynamics during rest and exercise (Karasik et al., 2002; Martinis et al., 2004).

Temporal correlations in heartbeat dynamics change with circadian phase

Epidemiological studies have reported a robust day/night pattern in the incidence of adverse cardiovascular events with a peak approximately at 10 a.m (Muller et al., 1985). This peak has been traditionally attributed to day/night patterns in behaviors including activity levels that affect cardiovascular variables, such as autonomic balance, blood pressure, and platelet aggregability, in vulnerable individuals (Shea et al., 2007). However, endogenous influences from the circadian pacemaker [suprachiasmatic nuclei of the hypothalamus (SCN)], independent from external behavioral effects, may also contribute to this daily pattern of adverse cardiovascular events. Indeed, the endogenous circadian pacemaker influences key physiologic functions, such as body temperature and heart rate, and is normally synchronized with the sleep/wake cycle. These circadian influences could occur via hormonal effects, direct neuronal links between the SCN and the sympathetic system (Buijs et al., 2003) and through circadian modulation of the sympathovagal balance (Hilton et al., 2000). Testing this hypothesis, recent studies (Hu et al., 2004b) demonstrated that dynamical scale-invariant features of heartbeat fluctuations related to underlying mechanisms of cardiac control (Kobayashi and Musha, 1982; Peng et al., 1995; Goldberger, 1996; Ivanov et al., 1996, 1998a, 2001), exhibit a significant endogenous circadian rhythm, independent from extrinsic scheduled behaviors and the sleep/wake cycle. These dynamical features of heartbeat fluctuations move closer to the features observed under pathologic conditions (Peng et al., 1995; Goldberger, 1996; Ho et al., 1997) at the endogenous circadian phase corresponding to approximately 10 a.m (Hu et al., 2004b).

To separate internal circadian factors from behavior-related factors, these studies investigated heartbeat dynamics in healthy young subjects recorded throughout a 10-day protocol in which the sleep-wake cycle and scheduled activity were desynchronized from the endogenous circadian cycle. Subjects' sleep-wake behavior cycles were adjusted to 28 h. This 28-h recurring sleep/wake schedule was repeated for seven cycles in the absence of bright light, so that the body clock oscillates at its inherent rate. Subjects were asked to repeat the same activity schedule in all seven wake periods so that, statistically, the same behaviors occur at each circadian phase throughout all seven 28-h cycles. Since the endogenous circadian pacemaker is normally synchronized with the sleep-wake cycle, and sleep and wake states have different effects on the temporal correlations and nonlinear characteristics of cardiac dynamics (Figs. 4.6—4.8) (Ivanov et al., 1998c, 1999b; Bunde et al., 2000; Kantelhardt et al., 2002), heartbeat data were analyzed for the wake and sleep-opportunity periods separately across circadian phases (Fig. 4.9). Thus, averaging the results of the analyses according to the circadian phase yields effects caused only by the endogenous circadian rhythms independent of behavioral factors due to the forced desynchrony protocol scheduled activities and behaviors are equally distributed across all circadian phases.

Analysis of interbeat interval data separated into 1-h windows along the circadian cycle show that the power-law scaling exponent α characterizing the temporal correlations in heartbeat dynamics exhibits a significant circadian rhythm, with a sharp peak at the circadian phase corresponding to approximately 10 a.m (Fig. 4.9), coinciding with the window of cardiac vulnerability reported in epidemiological clinical studies (Muller et al., 1985). We find that this peak in the value of the scaling exponent is independent of the scheduled behaviors and occurs during both sleep and awake periods scheduled across different circadian phases (Hu et al., 2004b). Since cardiac dynamics under pathologic conditions such as congestive heart failure are associated with a larger value of the scaling exponent α (Fig. 4.6), these findings suggest that endogenous circadian-mediated influences on cardiac control are involved in cardiac vulnerability by shifting the scaling exponent α closer to the values observed for heart failure subjects. Furthermore, these investigations demonstrate that the peak in the correlation exponent α at approximately 10 a.m. (corresponding to 90 degrees circadian phase) is not related to the circadian-mediated influence on the mean activity levels, leading to changes in the average heart rate that displays a very different circadian rhythm with a peak in the window 5—9 p.m. (about 180—240 degrees circadian phase) (Fig. 4.9D) (Ivanov et al., 2007). In addition, long-term correlations in motor activity and the associated scaling exponent α do not show a circadian rhythm indicating that circadian modulation of cardiac risk is not influenced by endogenous circadian modulation in motor activity (Ivanov et al., 2007).

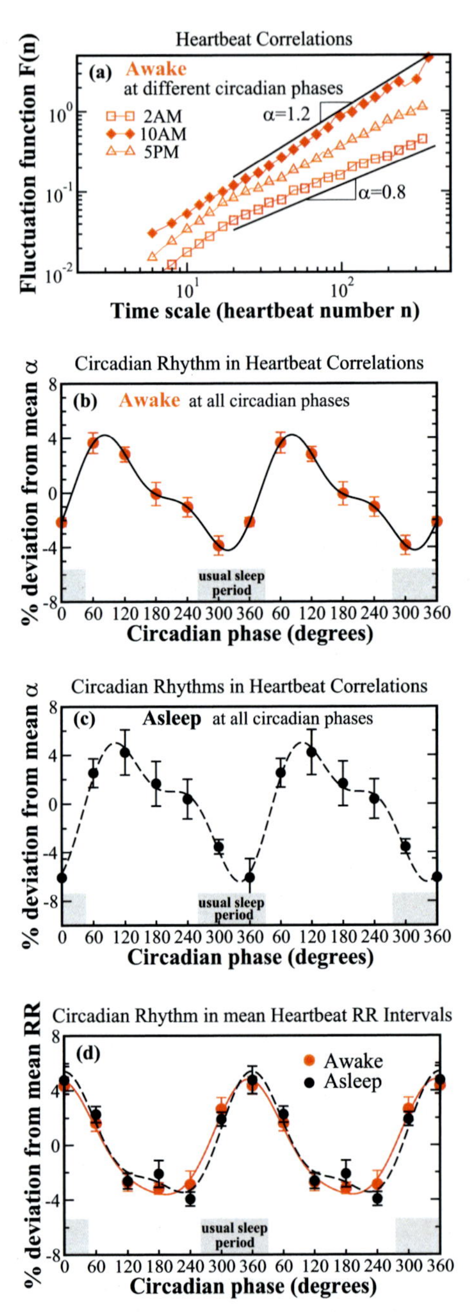

Figure 4.9 Circadian influence on physiologic dynamics. (A) Reorganization of temporal correlations in heartbeat intervals (vertical axis) over a broad range of time scales (horizontal axis) at different circadian phases. The correlation exponent α, obtained using the DFA method, gradually changes from $\alpha = 1.2$ in the morning hours (elevated sympathetic tone) to $\alpha = 1.0$ during daytime, and $\alpha = 0.8$ during the habitual sleep period (decreased sympathetic tone). Scaling curves F(n) show results for one

In summary, key static, dynamic, scale-invariant linear and nonlinear features of heartbeat dynamics, associated with sympathovagal control and reflecting various aspects of the underlying mechanisms of cardiac regulation, change significantly with transition from wake to sleep state, across sleep stages and circadian phases under both healthy and pathologic conditions. These findings indicate that sleep-wake and circadian cycles do not simply modulate basic physiologic functions by generating rhythms with a fixed periodicity, but also influence the neural regulation of fundamental physiologic systems, such as the cardiovascular and the respiratory system, simultaneously over a broad range of time scales. The empirical observations indicate that the neural pathways and systems involved in sleep and circadian regulation play an important role in the emergence and modulation of scale-invariant and nonlinear patterns in the temporal organization of physiologic dynamics.

representative subject at different circadian phases while awake. Because the subject was awake across all circadian phases while controlling for the daily behavior pattern, this change in the temporal organization of heartbeat fluctuations across time scales indicates strong endogenous circadian influence on cardiac dynamics that is mediated through the suprachiasmatic nucleus in the brain and the sympathetic innervation of the heart. Circadian rhythms in the exponent α obtained from healthy subjects during a 7-day forced desynchrony protocol where sleep/wake behavior cycles were adjusted to 28 h and where subjects were awake (B) or asleep (C) across all circadian phases. Peaks at 60–90 degrees circadian phase (9–11 a.m.) indicate strong heartbeat correlations regardless of whether or not the subjects were awake or asleep. The decrease in α during the usual sleep period (shaded regions) is accompanied by an increase in the mean heartbeat RR interval (decreased heart rate), shown in (D). Decreased heart rate and decreased α during the usual sleep period (Figs. 4.6 and 4.7) are consistent with a relative increase of parasympathetic tone in cardiac neuroautonomic regulation. However, this is not a simple relationship across the entire circadian cycle, because the circadian peaks in α at 60–90 degrees circadian phase during both sleep and wakefulness are not accompanied by a peak in the heart rate (D), which is located at $\approx$ 180 degrees. This important dissociation between mean heart rate and the exponent α is notable because α provides a unique insight into cardiac dynamic regulation by quantifying self-similar structures in heartbeat fluctuations over a range of time scales (Fig. 4.1), which cannot be accounted for by the traditional concept of homeostatic equilibrium. A circadian phase of 0 degree corresponds to the diurnal minimum in core body temperature. *(Adapted from Hu, K., Ivanov, P.C., Hilton, M.F., Chen, Z., Ayers, R.T., Stanley, H.E., Shea, S.A., 2004b. Endogenous circadian rhythm in an index of cardiac vulnerability independent of changes in behavior. Proc. Natl. Acad. Sci. U S A 101, 18223–18227. https://doi.org/10.1073/pnas.0408243101 and Ivanov, P.C., Hu, K., Hilton, M.F., Shea, S.A., Stanley, H.E., 2007. Endogenous circadian rhythm in human motor activity uncoupled from circadian influences on cardiac dynamics. Proc. Natl. Acad. Sci. U S A 104, 20702–20707. https://doi.org/10.1073/pnas.0709957104.)*

Physiological systems interactions during wake and sleep

The human organism is an integrated network where complex physiological systems, each with its own regulatory mechanisms, continuously interact to support physiological functions and to generate distinct physiological states. These interactions are mediated through intrinsic feedback coupling mechanisms that are affected by sleep. Cardiorespiratory coupling, a prime example of interaction between key physiologic systems, is strongly influenced by sleep regulation and exhibits phase transitions across sleep stages. Furthermore, sleep affects the entire network of interactions among diverse physiologic systems, where different sleep stages are characterized by specific physiologic network connectivity and topology, demonstrating a robust interplay between physiological systems interactions, network dynamics, and physiologic function.

In clinical studies, the interaction between two key physiological systems under neural regulation, the cardiac and the respiratory system, is traditionally identified through the respiratory sinus arrhythmia (RSA), which accounts for the periodic variation of the heart rate within a breathing cycle (Angelone and Coulter, 1964; Song and Lehrer, 2003). Recent advances in the field of nonlinear dynamics and statistical physics have led to the development of advanced phase-synchronization approaches, tailored to detect coupling between nonlinear systems with nonstationary output signals (Rosenblum et al., 1996; Pikovsky et al., 2001; Xu et al., 2006; Chen et al., 2006), that have identified a previously unknown form of cardiorespiratory coupling—cardiorespiratory phase synchronization (CRPS) (Schäfer et al., 1998). In the context of cardiorespiratory coupling, phase-synchronization is defined as a consistent occurrence of heartbeats at the same relative phases within consecutive breathing cycles, despite continuous fluctuations in the duration of heartbeat intervals and respiratory cycles (Fig. 4.10). Earlier studies have focused on short data segments to identify certain degrees of CRPS in healthy young subjects (Schäfer et al., 1998; Prokhorov et al., 2003) and in newborn infants (Mrowka et al., 2000). A more recent study has identified an intriguing effect of maternal-fetal heart rate synchronization (Leeuwen et al., 2009). However, both cardiac and respiratory dynamics exhibit transient changes in the long-term correlations associated with different physiologic states and conditions (Figs. 4.7 and 4.9), raising the hypothesis that CRPS may also be affected by such changes in relation to underlying mechanisms of physiologic control during wake and sleep state and different

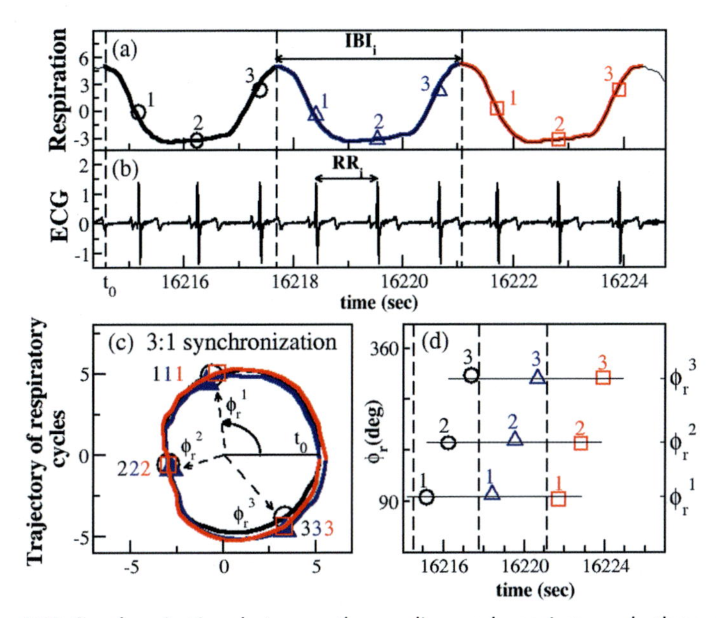

Figure 4.10 Synchronization between the cardiac and respiratory rhythms during sleep. Coupling feedback mechanisms between the cardiac and the respiratory system lead to continuous adjustment of (A) interbreath IBI and (B) heartbeat RR intervals. Heartbeats occur at the same relative phase (marked by symbols) within consecutive breathing cycles (shown in different color)—i.e., their respective frequencies and phases "lock" at a particular ratio. (C) 3:1 synchronization between breathing and heartbeat intervals, where for each breathing cycle (represented by a close-to-circular trajectory) there are three heartbeats that occur at the same relative phases $\phi_r(t)$ within consecutive breathing cycles (symbols collapse). (D) Phase synchrogram: segments of horizontal lines indicate cardiorespiratory synchronization. Different symbols represent heartbeats in different breathing cycles as in (A) and (C); vertical dashed lines show the beginning of each breathing cycle. *(Adapted from Bartsch, R.P., Schumann, A.Y., Kantelhardt, J.W., Penzel, T., Ivanov, P.C., 2012. Phase transitions in physiologic coupling. Proc. Natl. Acad. Sci. U S A 109, 10181–10186. https://doi.org/10.1073/pnas.1204568109.)*

sleep stages. Moreover, a fundamental question of relevance to basic physiology is whether RSA and CRPS represent different aspects of cardiorespiratory coupling, and how these two forms of physiologic interaction are influenced by physiologic states.

To gain insight into the mechanism of cardiorespiratory coupling, recent studies investigate CRPS during (i) transitions from one physiological state to another and (ii) under different physiologic conditions. Because sleep stages are well-defined physiologic states and are associated with distinct mechanisms of autonomic control (Schmitt et al., 2009;

Schumann et al., 2010), these studies have addressed the question whether cardiorespiratory coupling is characterized by a specific degree of CRPS during each sleep stage.

Nonlinear oscillatory systems characterized by nonidentical intrinsic frequencies and highly irregular signal output can synchronize even when their coupling is weak—i.e., their respective frequencies and phases "lock" at a particular ratio (Rosenblum et al., 1996; Pikovsky et al., 2001) (Fig. 4.10). Despite the significant difference in the periodicity of the cardiac and respiratory rhythms represented by the heartbeat and interbreath intervals, and the complex variability in these rhythms (Fig. 4.11A–E), episodes of regular cardiorespiratory phase relationship emerge. The key feature of this phase relation is that sequences of heartbeats consistently occur at the same respiratory phases for consecutive breathing cycles (Fig. 4.10C). This leads to segments of horizontal lines in the phase synchrogram over consolidated periods of time (Figs. 4.10D and 4.11C and F). Such phase synchronization of heartbeats and respiration is a manifestation of the temporal organization and adjustment of the cardiac and respiratory rhythms due to their underlying coupling. Because the variability as well as the scale-invariant and nonlinear features of the cardiac and respiratory systems change with physiological state (e.g., wake/sleep and different sleep stages (Ivanov et al., 1999b; Bunde et al., 2000; Kantelhardt et al., 2003a, 2003b; Schumann et al., 2010)) and with physiologic conditions (e.g., disease and age (Brandenberger et al., 2003; Schumann et al., 2010)), quantifying CRPS across physiologic states and conditions can provide insight into how physiologic regulation affects cardiorespiratory coupling.

Analyses of data obtained from healthy subjects in different age groups during sleep show a significant change in the degree of CRPS for different sleep stages: low degree of phase synchronization during REM sleep, higher during wake state, and much stronger synchronization coupling during light sleep and deep sleep, indicating a dramatic $\approx 400\%$ increase in the degree of cardiorespiratory coupling with transition from REM to deep sleep (Fig. 4.12A). These results show a remarkable sensitivity of the cardiorespiratory coupling in response to sleep-wake and sleep-stage transitions.

Aging is traditionally associated with the process of decline of physiologic function and reduction of physiologic complexity (Goldberger et al., 1990; Kaplan et al., 1991; Bliwise, 1993; Arking, 2006) that result from changes in the underlying control mechanisms and regulatory networks of

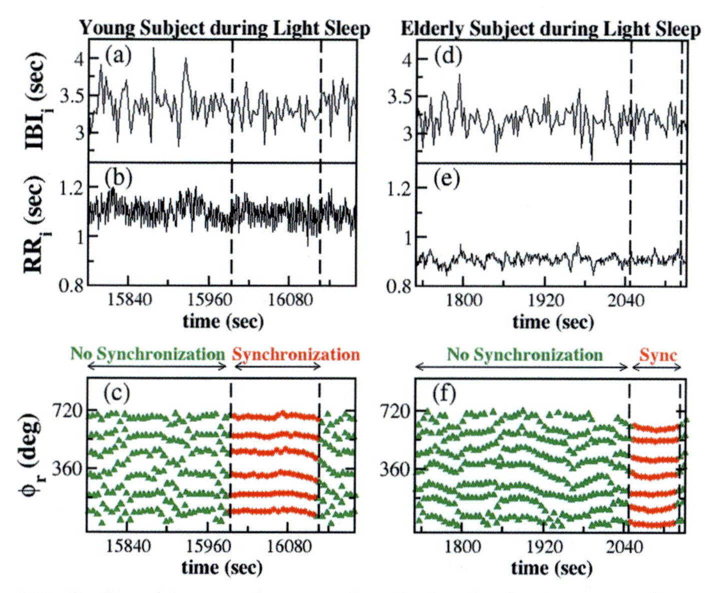

Figure 4.11 Cardiorespiratory phase synchronization in the presence of continuous fluctuations in both interbreath IBI and heartbeat RR intervals in both young (A, B) and elderly (D, E) subjects is a manifestation of the temporal organization and adjustment of the cardiac and respiratory rhythms due to their nonlinear coupling. Segments in red color in (C) and (F) correspond to periods of cardiorespiratory synchronization. An episode of 3:1 synchronization is shown for the young subject, i.e., six heartbeats within two breathing cycles are consistently placed at the same respiratory phases ϕ_r over many consecutive breathing cycles (6 horizontal red segments within 2×360 degrees); a segment of 7:2 synchronization, i.e., seven heartbeats are synchronized with each 2 breathing cycles for the elderly subject over an extended period of time (7 horizontal red segments within 2×360 degrees). Notably, young subjects exhibit longer periods of cardiorespiratory synchronization despite significantly higher interbreath and heartbeat variability compared to the elderly subject, indicating stronger coupling (longer red segments in (C) compared to (F)). *(Adapted from Bartsch, R.P., Schumann, A.Y., Kantelhardt, J.W., Penzel, T., Ivanov, P.C., 2012. Phase transitions in physiologic coupling. Proc. Natl. Acad. Sci. U S A 109, 10181–10186. https://doi.org/10.1073/pnas.1204568109.)*

neural and metabolic pathways that interact through coupled cascades of nonlinear feedback loops over a range of timescales (Schmitt and Ivanov, 2007). Although healthy aging is accompanied by a reduction in the variability of heart rate and respiratory rate (Tsuji et al., 1994) leading to decreased responsiveness and loss of sensitivity to external and internal stimuli (Bernaola-Galván et al., 2001), recent studies have shown that

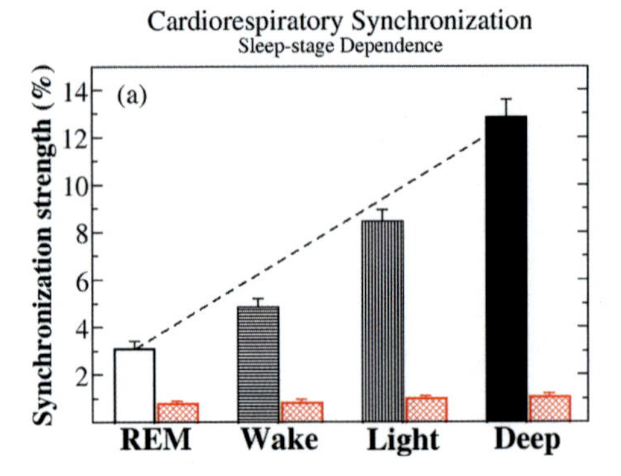

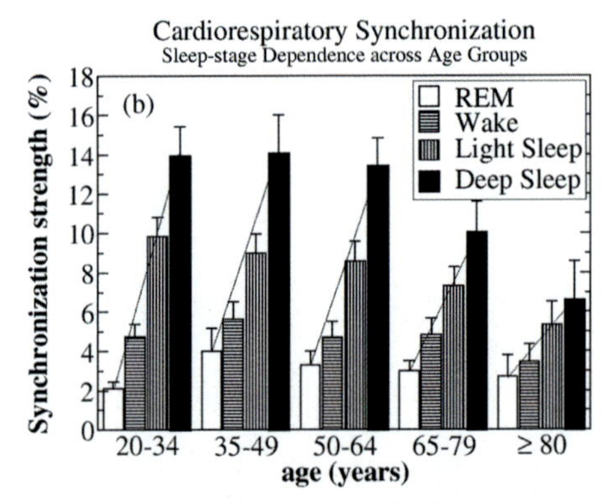

Figure 4.12 Phase transition in physiologic coupling in response to changes in wake and sleep regulation. (A) A ≈ 400% increase in cardiorespiratory phase synchronization (CRPS) during deep sleep and light sleep compared to REM and wake, indicate a significant modulation in cardiorespiratory coupling due to sleep. With increase of parasympathetic tone during light and deep sleep, the strength of cardiorespiratory coupling, as represented by the degree of synchronization, also increases despite higher amplitude in heartbeat and respiratory variability (Figs. 4.2 and 4.3). Examining changes in CRPS between the sleep stages for each subject in a database of 190 subjects, one-way ANOVA test with repeated measures (comparisons between all sleep stages) confirms statistically significant difference with P-value $< 10^{-3}$ (all pairwise multiple comparison procedures by Tukey's test yield $P < .05$). Statistical significance of the results for each sleep stage is further demonstrated by a comparison to a surrogate test (red bars), where breathing cycles and heartbeat intervals were randomized, thus eliminating their coupling. (B) Sleep-stage stratification pattern in CRPS is stable across all age groups. Suppression of parasympathetic tone in elderly subjects and increased sympathetic tone leads to reduction in CRPS across the entire sleep period including all sleep stages—significant ≈ 40% reduction in CRPS; Mann–Whitney rank sum test $P = .021$. Age-related reduction in CRPS is most pronounced during light and deep sleep with ≈ 50% when comparing the youngest with the oldest group—an effect that is independent of the decline in total light and deep sleep duration with aging. In contrast, the n:m synchronization during REM and wake does not significantly change with age. These intriguing results indicate that the decline in CRPS with age is primarily mediated through the neuroautonomic mechanisms regulating deep sleep and light sleep. *(Adapted from Bartsch, R.P., Schumann, A.Y., Kantelhardt, J.W., Penzel, T., Ivanov, P.C., 2012. Phase transitions in physiologic coupling. Proc. Natl. Acad. Sci. U S A 109, 10181–10186. https://doi.org/10.1073/pnas.1204568109.)*

certain scaling and nonlinear properties remain intact (Schmitt and Ivanov, 2007; Schmitt et al., 2009; Schumann et al., 2010). These observations indicate that, although the coupled feedback loops controlling physiologic dynamics across different timescales may still be present in healthy elderly subjects (hence preserving the scaling features), the reduction in physiologic responsiveness to external and internal stimuli suggests a reduced coupling strength of these feedback interactions with advanced age. Thus, one can hypothesize that the strength of CRPS as a measure of cardiorespiratory coupling decreases with age. Analyses of subjects from different age groups show a gradual and significant decrease in the degree of CRPS across age groups (Fig. 4.12B). Taking into account all n:m synchronization ratios, there is significant $\approx 40\%$ decline in synchronization when comparing young versus elderly (>80 years old) subjects. Remarkably, the decline in synchronization is not monotonous with age: there is weak, not statistically significant decline for the middle-age groups (35–49 and 50–64 years old) and a large drop in CRPS for the elderly groups (65–79 and ≥ 80 years old). These results indicate that the strength of cardiorespiratory coupling is significantly affected by aging.

This reduction in CRPS with aging (Fig. 4.12B) cannot be attributed to changes in sleep architecture that are traditionally associated with marked decline of deep sleep duration in elderly (Bliwise, 1993; Schumann et al., 2010) because phase synchronization analyses are based on the percentage of synchronization within a given sleep stage, irrespective of age-related changes in sleep-stage total duration. Because total light sleep duration increases with aging (Bliwise, 1993; Schumann et al., 2010), our observation of decreased synchronization in elderly during light sleep (Fig. 4.12B) indicates that the impact of aging on cardiorespiratory coupling is not a consequence of age-related changes in sleep-stage duration and sleep architecture. Finally, these observations demonstrate that, despite the significant reduction in CRPS with aging (Fig. 4.12B), the consistent stratification in the degree of CRPS across sleep stages is preserved for all age groups (Fig. 4.12B) and persists for different n:m ratios (Bartsch et al., 2012).

Coexisting forms of physiologic coupling

Nonlinear coupling and interactions between organ systems influence their output dynamics and facilitate coordination of their function, leading to another level of complexity. Integrated organ systems can communicate through several independent mechanisms of interaction, which operate at different time scales and are represented by different forms of coupling that simultaneously coexist. Neuroautonomic regulation of physiologic systems changes with transition from one physiologic state to another, leading to transitions in scaling and nonlinear features of their output dynamics (Ivanov et al., 1999b, 2007; Kantelhardt et al., 2002; Karasik et al., 2002; Schumann et al., 2010). Correspondingly, physiologic coupling also undergoes transitions to facilitate and optimize organ interactions during different physiologic states. Specifically, at any given moment pairs of systems interact through complementary forms of coupling that exhibit different strength and different stratification patterns across physiologic states (Bartsch et al., 2014; Bartsch and Ivanov, 2014).

One form of cardiorespiratory interaction that is traditionally utilized in clinical studies is the RSA (Angelone and Coulter, 1964; Song and Lehrer, 2003). RSA is quantified by the amplitude of heart rate modulation due to inspiration and expiration within each respiratory cycle (Fig. 4.13). The amplitude of RSA exhibits a pronounced and statistically significant stratification pattern across different sleep stages: it is lowest during wake periods and gradually increases during REM, light sleep and deep sleep. In particular, periods of non-REM sleep (light sleep and deep sleep) are characterized by $\approx$ 40% higher RSA amplitude compared to REM and wake (Fig. 4.14), indicating a strong dependence of this form of coupling on the underlying neuroautonomic control during sleep. With transition from wake to REM, light and deep sleep, the sympathetic tone of autonomic control decreases while the level of parasympathetic activity remains unchanged (Schmitt et al., 2009). This yields a relative dominance of parasympathetic tone in cardiac and respiratory regulation, which in turn leads to lower respiratory rates in non-REM sleep associated with higher RSA amplitude. Indeed, experimental studies in healthy subjects show a decrease in the RSA amplitude with increasing respiratory frequency under paced respiration (Angelone and Coulter, 1964; Song and Lehrer, 2003). However, while the respiratory frequency decreases just with $\approx$ 10% during transitions from wake to light and deep sleep, there is a highly nonlinear response in the RSA amplitude which increases with $\approx$ 40% for these transitions. In contrast to RSA, CRPS is weaker during REM as compared to wake, while during light and deep sleep there is a dramatic increase in the degree of CRPS of $\approx$ 400% compared to REM (Fig. 4.14). Notably, CRPS exhibits a factor of 10 stronger response to sleep-stage transitions compared to RSA.

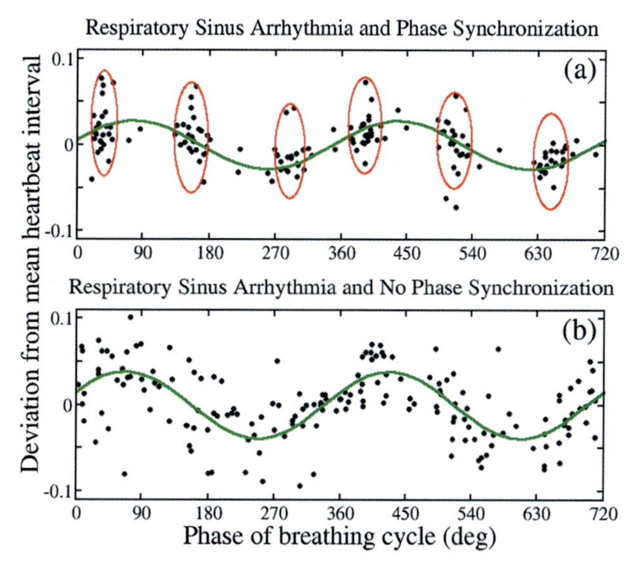

Figure 4.13 Cardiorespiratory phase synchronization (CRPS) and respiratory sinus arrhythmia (RSA) represent different aspects of cardiorespiratory coupling. (A) Whereas RSA leads to periodic modulation of the heart rate within each breathing cycle (highlighted by a sinusoid line fitted to the data points) associated with increase in heart rate during inspiration and decrease during expiration (Angelone and Coulter, 1964; Song and Lehrer, 2003), phase synchronization leads to clustering of heartbeats at certain phases ϕ_r of the breathing cycle (highlighted by red *ovals*). Shown are consecutive heartbeats over a period of 200 s. The horizontal axis indicates the phases ϕ_r of the breathing cycle where heartbeats occur, and the vertical axis indicates the deviation of each heartbeat interval from the mean interval. Heartbeats are plotted over pairs of consecutive breathing cycles (0–720 degrees) to better visualize rhythmicity. Data are selected from a subject during deep sleep. (B) For the same subject as in (A), heartbeats from another period of 200 s also during deep sleep are plotted over pairs of consecutive breathing cycles. Data show well-pronounced RSA with a similar amplitude as in (A), however, heartbeats are homogeneously distributed across all phases of the respiratory cycles, indicating absence of synchronization and reduced cardiorespiratory coupling. *(Adapted from Bartsch PR, Ivanov PC. Coexisting forms of coupling and phase-transitions in physiological networks. Commun. Comput. Inf. Sci. 2014; 438: 270-287.)*

Recent investigations demonstrate that distinct forms of cardiorespiratory interaction represent the dynamics at different time scales (Bartsch and Ivanov, 2014). Furthermore, different forms of cardiorespiratory coupling exhibit different sleep-wake and sleep-stage stratification patterns (Bartsch and Ivanov, 2014), indicating that these forms of coupling are independent and represent different aspects of physiologic interaction that are affected in

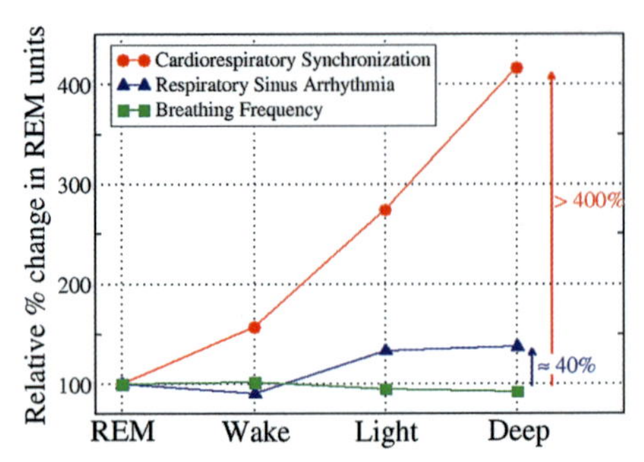

Figure 4.14 Transitions in cardiorespiratory phase synchronization, respiratory sinus arrhythmia (RSA), and average breathing frequency across sleep stages. Average values for each sleep stage are normalized on the corresponding values during rapid eye movement (REM) sleep. While with transition from wake and REM to light and deep sleep, the average breathing frequency decreases by $\approx$ 10%, the two forms of cardiorespiratory coupling—RSA and phase synchronization increase. The response of cardiorespiratory synchronization to the relative decrease in sympathetic tone during light and deep sleep, however, is by a factor of 10 higher than the response in RSA, indicating that sleep regulation affects very differently these two aspects of cardio-respiratory coupling. *(Adapted from Bartsch RP, Liu KKL, Ma QDY, Ivanov PC. Three independent forms of cardio-respiratory coupling: transitions across sleep stages. Comput. Cardiol. 2014; 41:781-784.)*

a different way by changes in neuroautonomic control across physiologic states (Fig. 4.14). Empirical observations in subjects where the strength of RSA remains constant during a given physiologic state, while CRPS may be present or not in separate time windows of the same physiologic state (Fig. 4.13), indicate that different forms of cardiorespiratory coupling are independent and may serve different physiologic functions (Bartsch et al., 2012). In particular, while the RSA amplitude increases nonlinearly with decreasing breathing frequencies (increase in parasympathetic tone), the degree of CRPS remains unchanged (Bartsch et al., 2012), demonstrating the dissociation between these two forms of cardiorespiratory coupling.

Moreover, these independent forms of cardiorespiratory coupling are modulated differently and are not of constant strength but are rather transient and intermittent with "on" and "off" periods even within a given physiologic state. Indeed, CRPS is usually observed in relatively short epochs of 20–45 s, rarely exceeding 100 s duration (Bartsch and Ivanov, 2014). Even for the same subject within the same sleep stage, there are time periods when CRPS is not present (Fig. 4.13), indicating that "on" and "off" switching of this form of cardiorespiratory interaction is not always triggered by transitions across physiological states.

Network interactions among physiological systems

Sleep affects the entire network of interactions among diverse physiological systems (Fig. 4.15). To understand how different organ systems interact as a network, a new time delay stability (TDS) method (Bashan et al., 2012; Bartsch et al., 2015) was developed to probe interactions between the brain, cardiac, respiratory, and other physiological systems, and to quantify how network interactions change across physiological states. Because brain dynamics are characterized by EEG signals with different brain rhythms (spectral frequencies) dominant at different scalp locations and during different physiological states, the TDS method is tailored to study how bursts in EEG frequency bands from certain brain areas are coupled with corresponding bursts in the heart and respiratory rate, and in the dynamics of other physiological systems.

Recent investigations have shown that the network of brain, respiratory, cardiac, muscle tone, chin, and eye movement interactions exhibits different topology during different sleep stages. Specifically, the physiologic network is characterized by high connectivity during wake and light sleep, lower during REM, and lowest connectivity during deep sleep (Fig. 4.16A). A similar sleep-stage stratification pattern is also observed for the network links strength (Fig. 4.16B). Such transitions in network structure and in the strength of network links indicate strong relation between network topology and physiologic function. Traditionally, differences between sleep stages are attributed to modulation in the sympathovagal balance with dominant sympathetic tone during wake and REM (Otzenberger et al., 1998): spectral, scale-invariant, and nonlinear characteristics of the dynamics of individual physiologic systems indicate higher degree of temporal correlations and nonlinearity during wake and REM compared to non-REM (light and deep sleep), where physiologic dynamics during non-REM exhibit weaker correlations and loss of nonlinearity (Bunde et al., 2000; Kantelhardt et al., 2002; Schumann et al., 2010) (Figs. 4.3, 4.7 and 4.8). In contrast, the network of physiologic -interactions among organ systems shows a completely different picture: network characteristics during light sleep are much closer to those during wake and very different from deep sleep. Furthermore, not only network connectivity but also the overall strength of physiologic interactions is significantly higher during wake and light sleep, intermediate during REM, and much lower during deep sleep (Bashan et al., 2012; Bartsch et al., 2015; Liu et al., 2015b). Similar stratification pattern in network structure and link strength was found for the dynamic

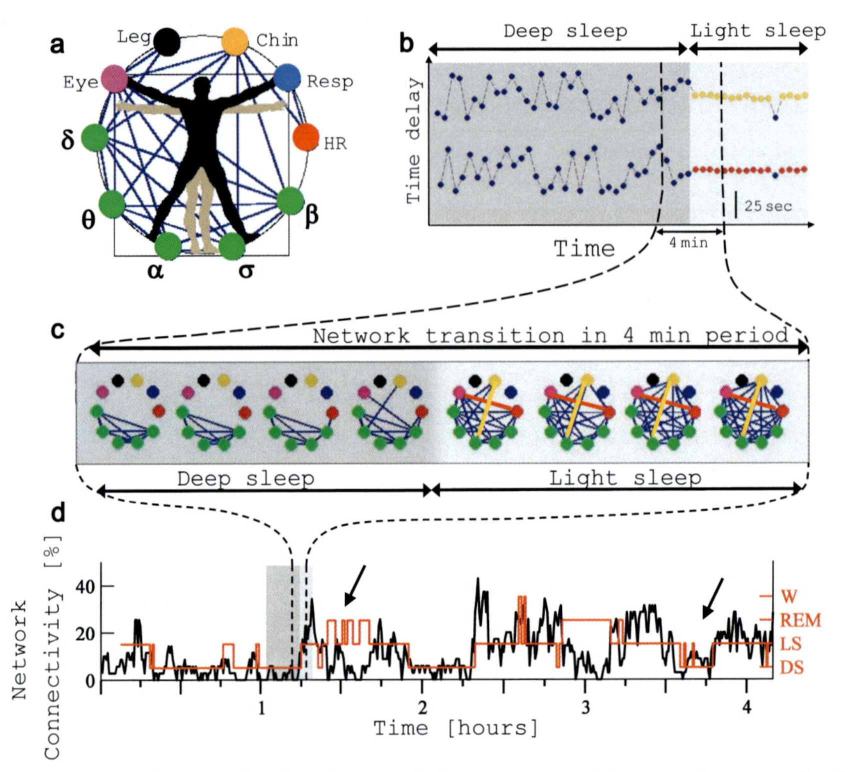

Figure 4.15 Influence of wake, sleep, and sleep-stage transitions on the network of interactions between physiological systems. (A) Dynamical network of physiologic interactions where 10 network nodes represent six physiological systems—brain activity (EEG waves: δ, θ, α, σ and β), cardiac (HR), respiratory (Resp), chin muscle tone, leg and eye movements. (B) Transitions in physiologic interactions across sleep stages. The time delay between two pairs of signals, (top) α-brain waves and chin muscle tone, and (bottom) HR and eye movement, quantifies their physiological interaction: highly irregular behavior (blue *dots*) during deep sleep is followed by a period of stable (almost constant) time delay during light sleep indicating a period of stable physiologic interaction (marked by red *dots* for the HR-eye and orange *dots* for the α-chin interaction). (C) Sleep-stage transitions are associated with changes in network structure: network evolution represented by snapshots over 30 s windows during a 4 min period at the transition from deep sleep (dark gray segment) to light sleep (light gray segment) shown in (B). During deep sleep, the network consists mainly of brain-brain links. With transition to light sleep, links between other physiological systems (network nodes) emerge and the network becomes highly connected. The stable α-chin and HR-eye interactions during light sleep in (B) are shown in (C) by an orange and a red network link, respectively. (D) Physiological network connectivity for one subject during 4 h of sleep calculated in 30 s windows as the fraction (%) of present links out of all possible network links. Red line marks the sleep stages. Low connectivity is consistently observed during deep sleep (0:30—1:15 h and 1:50—2:20 h) and rapid eye movement sleep (1:30—1:45 h and 2:50—3:10 h), while transitions to light sleep and wake are associated with a significant increase in physiologic network connectivity. *(Adapted from Bashan, A., Bartsch, R.P., Kantelhardt, J.W., Havlin, S., Ivanov, P.C., 2012. Network physiology reveals relations between network topology and physiological function. Nat. Commun. 3, 702.)*

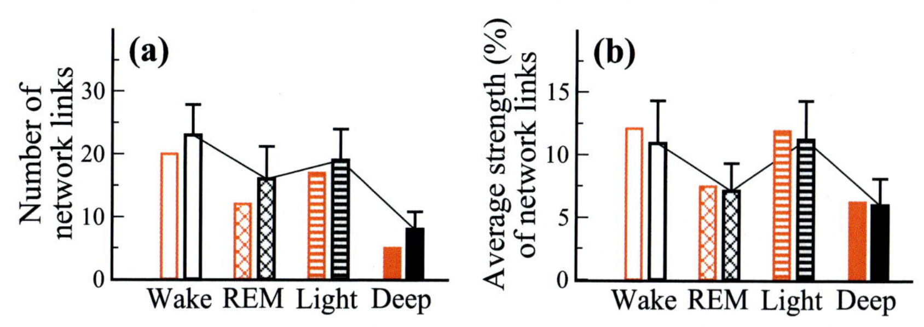

Figure 4.16 Transitions in neuroautonomic regulation associated with different sleep stages lead to transitions in structural and dynamical characteristics of the network of interactions between physiological systems. Both (A) network connectivity (number of network links) and (B) network link strength (average strength of interactions between all systems in the network) exhibit a pronounced sleep-stage stratification pattern with less and weaker links during deep sleep and rapid eye movement compared to wake and light sleep. Data from an individual subject are represented by red bars and group averaged data by black bars. Network connectivity and link strength are obtained from the interactions between six physiological systems—brain EEG activity, heart rate, respiration, chin muscle tone, leg and eye movement. *(Adapted from Bashan, A., Bartsch, R.P., Kantelhardt, J.W., Havlin, S., Ivanov, P.C., 2012. Network physiology reveals relations between network topology and physiological function. Nat. Commun. 3, 702.)*

interactions of bursts in brain rhythms across cortical locations (Liu et al., 2015a). Moreover, the functional form of coupling profiles between cortical rhythms as well as between the brain and other organ systems change with transition from wake to sleep and across sleep stages (Lin et al., 2016, 2020), indicating that an "alphabet" of coupling profiles uniquely represent cross-communications among physiologic systems for each physiologic state. These empirical observations indicate that while sleep-stage related modulation in sympathovagal balance plays a key role in regulating individual physiological systems (Figs. 4.5–4.7), it does not fully account for the stratification of physiologic network topology and dynamics across sleep stages, showing that the proposed network approach captures principally new information.

Physiologic network reorganization with transitions from wake to sleep and across sleep stages occurs globally at the level of the entire network as well as at each network node (physiological system) (Fig. 4.17), indicating a hierarchical organization and modular structure of physiologic network interactions that responds to modulation in autonomic function during

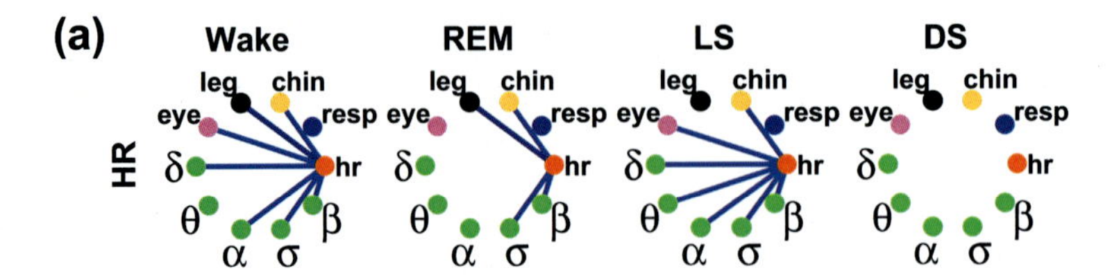

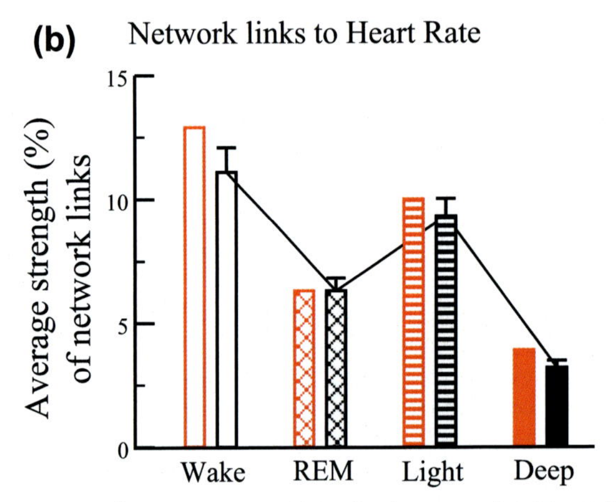

Figure 4.17 Sleep-stage transitions lead to complex hierarchical reorganization in the structure and dynamics of the network of physiologic interactions. Transitions in connectivity and average link strength in the entire network (shown in Fig. 4.17) are associated with corresponding increase or reduction in the number of links and links strength for each network node (i.e., physiological system): (A) the number of links to a specific network node (heart) and (B) the average strength of the links connecting the heart to the rest of the network exhibit a sleep-stage stratification pattern with lower connectivity and weaker links during deep sleep and rapid eye movement, and much higher connectivity with stronger links during light sleep and wake. Data from an individual subject are represented by red bars and group averaged data by black bars in panel (B). The absence of a link between heart rate and respiration in the physiological network shown in (A) does not indicate absence of cardiorespiratory coupling but rather that this coupling as represented by time delay stability (TDS) is rarely stable for periods longer than 3–4 min that are considered in this analysis of physiologic interactions. Network connectivity and link strength are obtained from the interactions between the heart and five other physiological systems (brain EEG δ, θ, α, σ and β activity, respiration, chin muscle tone, leg and eye movement) using the TDS method (Bashan et al., 2012; Bartsch et al., 2015). Note that the TDS measure reflects a different aspect of cardiorespiratory coupling compared to cardiorespiratory phase synchronization and respiratory sinus arrhythmia (Figs. 4.13 and 4.14), where the response to changes in sleep regulation across sleep stages leads to markedly different stratification patterns. *(Adapted from Ivanov, P.C., Bartsch, R.P., 2014. In: D'Agostino, G., Scala, A. (Eds.), Network Physiology: Mapping Interactions between Networks of Physiologic Networks. Springer International Publishing, pp. 203–222.)*

sleep stages. This demonstrates a robust interplay between network structure and physiologic function, and the necessity to measure not only the dynamics of individual physiological systems (cardiac, respiratory, brain, temperature, muscle tone, eye movements, etc.) but also their coupling and network interactions to comprehensively study and quantify the impact of sleep regulation on physiologic function in health and disease (J Randall Moorman et al., 2016). The presented in this chapter modern advances in understanding physiologic regulation underline the necessity to channel future efforts in basic research and clinical practice within the new conceptual framework and interdisciplinary field of Network Physiology (Ivanov and Bartsch, 2014; Ivanov et al., 2016, 2017).

Acknowledgments

We acknowledge support from the W. M. Keck Foundation, National Institutes of Health (NIH Grant 1R01- HL098437), and the US-Israel Binational Science Foundation (BSF Grant 2012219).

References

Akselrod, S., Gordon, D., Ubel, F.A., Shannon, D.C., Berger, A.C., Cohen, R.J., 1981. Power spectrum analysis of heart rate fluctuation: a quantitative probe of beat-to-beat cardiovascular control. Science 213, 220–222. https://doi.org/10.1126/science.6166045.

Angelone, A., Coulter, N.A., 1964. Respiratory sinus arrhythmia: a frequency dependent phenomenon. J. Appl. Physiol. 19, 479–482.

Arking, R., 2006. Biology of Aging: Observations and Principles, 3 edition. Oxford University Press, USA.

Ashkenazy, Y., Hausdorff, J.M., Ivanov, P.C., Stanley, H.E., 2002. A stochastic model of human gait dynamics. Phys. A 316, 662–670. https://doi.org/10.1016/S0378-4371(02)01453-X.

Ashkenazy, Y., Havlin, S., Ivanov, P.C., Peng, C.-K., Schulte-Frohlinde, V., Stanley, H.E., 2003. Magnitude and sign scaling in power-law correlated time series. Phys. A 323, 19–41. https://doi.org/10.1016/S0378-4371(03)00008-6.

Ashkenazy, Y., Ivanov, P.C., Havlin, S., Peng, C.-K., Goldberger, A.L., Stanley, H.E., 2001. Magnitude and sign correlations in heartbeat fluctuations. Phys. Rev. Lett. 86, 1900–1903. https://doi.org/10.1103/PhysRevLett.86.1900.

Ashkenazy, Y., Ivanov, P.C., Havlin, S., Peng, C.-K., Yamamoto, Y., Goldberger, A.L., Stanley, H.E., 2000. Decomposition of heartbeat time series: scaling analysis of the sign sequence. Comput. Cardiol. 27, 139–142. https://doi.org/10.1109/CIC.2000.898475.

Baharav, A., Kotagal, S., Gibbons, V., Rubin, B.K., Pratt, G., Karin, J., Akselrod, S., 1995. Fluctuations in autonomic nervous activity during sleep displayed by power spectrum analysis of heart rate variability. Neurology 45, 1183–1187.

Bartsch, R., Plotnik, M., Kantelhardt, J.W., Havlin, S., Giladi, N., Hausdorff, J.M., 2007. Fluctuation and synchronization of gait intervals and gait force profiles distinguish stages of Parkinson's disease. Phys. A 383, 455–465. https://doi.org/10.1016/j.physa.2007.04.120.

Bartsch, R.P., Ivanov, P.C., 2014. Coexisting forms of coupling and phase-transitions in physiological networks. Nonlinear Dyn. Electron. Syst. 438, 270–287.
Bartsch, R.P., Liu, K.K.L., Bashan, A., Ivanov, P.C., 2015. Network physiology: how organ systems dynamically interact. PLoS One 10, e0142143. https://doi.org/10.1371/journal.pone.0142143.
Bartsch, R.P., Liu, K.K.L., Ma, Q.D.Y., Ivanov, P.C., 2014. Three independent forms of cardio-respiratory coupling: transitions across sleep stages. Comput. Cardiol. 41, 781–784.
Bartsch, R.P., Schumann, A.Y., Kantelhardt, J.W., Penzel, T., Ivanov, P.C., 2012. Phase transitions in physiologic coupling. Proc. Natl. Acad. Sci. U S A 109, 10181–10186. https://doi.org/10.1073/pnas.1204568109.
Bashan, A., Bartsch, R.P., Kantelhardt, J.W., Havlin, S., Ivanov, P.C., 2012. Network physiology reveals relations between network topology and physiological function. Nat. Commun. 3, 702.
Behrens, S., Ney, G., Gross Fisher, S., Fletcher, R.D., Franz, M.R., Singh, S.N., 1997. Effects of amiodarone on the circadiam pattern of sudden cardiac death (department of veterans affairs Congestive Heart Failure-Survival Trial of Antiarrhythmic Therapy). Am. J. Cardiol. 80, 45–48. https://doi.org/10.1016/S0002-9149(97)00281-6.
Bernaola-Galván, P., Ivanov, P.C., Amaral, L.A.N., Stanley, H.E., 2001. Scale invariance in the nonstationarity of human heart rate. Phys. Rev. Lett. 87, 168105. https://doi.org/10.1103/PhysRevLett.87.168105.
Berne, R.M., Levy, M.N., 2001. Cardiovascular Physiology. Mosby, St. Louis, MO.
Bliwise, D.L., 1993. Sleep in normal aging and dementia. Sleep 16, 40–81.
Bonnet, M., Arand, D., 1997. Heart rate variability: sleep stage, time of night, and arousal influences. Electroencephalogr. Clin. Neurophysiol. 102, 390–396.
Brandenberger, G., Viola, A.U., Ehrhart, J., Charloux, A., Geny, B., Piquard, F., Simon, C., 2003. Age-related changes in cardiac autonomic control during sleep. J. Sleep Res. 12, 173–180.
Buijs, R.M., la Fleur, S.E., Wortel, J., Heyningen, C.V., Zuiddam, L., Mettenleiter, T.C., Kalsbeek, A., Nagai, K., Niijima, A., 2003. The suprachiasmatic nucleus balances sympathetic and parasympathetic output to peripheral organs through separate pre-autonomic neurons. J. Comp. Neurol. 464, 36–48. https://doi.org/10.1002/cne.10765.
Bunde, A., Havlin, S., Kantelhardt, J.W., Penzel, T., Peter, J.-H., Voigt, K., 2000. Correlated and uncorrelated regions in heart-rate fluctuations during sleep. Phys. Rev. Lett. 85, 3736–3739. https://doi.org/10.1103/PhysRevLett.85.3736.
Carskadon, M.A., Dement, W.C., 1994. In: Kryger, M.H., Roth, T., Dement, W.C. (Eds.), Principles and Practice of Sleep Medicine. WB Saunders Company, Philadelphia, PA, pp. 16–25.
Chen, Z., Hu, K., Carpena, P., Bernaola-Galvan, P., Stanley, H.E., Ivanov, P.C., 2005. Effect of nonlinear filters on detrended fluctuation analysis. Phys. Rev. E 71, 011104. https://doi.org/10.1103/PhysRevE.71.011104.
Chen, Z., Hu, K., Stanley, H.E., Novak, V., Ivanov, P.C., 2006. Cross-correlation of instantaneous phase increments in pressure-flow fluctuations: applications to cerebral autoregulation. Phys. Rev. E 73, 031915.
Chen, Z., Ivanov, P.C., Hu, K., Stanley, H.E., 2002. Effect of nonstationarities on detrended fluctuation analysis. Phys. Rev. E 65, 041107. https://doi.org/10.1103/PhysRevE.65.041107.
Corino, V.D.A., Matteuccib, M., Cravelloc, L., Ferraric, E., Ferrarid, A.A., Mainardi, L.T., 2006. Long-term heart rate variability as a predictor of patient age. Comput. Methods Progr. Biomed. 82, 248–257. https://doi.org/10.1016/j.cmpb.2006.04.005.

Crasset, V., Mezzetti, S., Antoine, M., Linkowski, P., Degaute, J.P., Borne, P. van de, 2001. Effects of aging and cardiac denervation on heart rate variability during sleep. Circulation 103, 84–88.

Dvir, H., Elbaz, I., Havlin, S., Appelbaum, L., Ivanov, P.C., Bartsch, R.P., 2018. Neuronal noise as an origin of sleep arousals and its role in sudden infant death syndrome. Sci. Adv. 4, eaar6277. https://doi.org/10.1126/sciadv.aar6277.

Dvir, I., Adler, Y., Freimark, D., Lavie, P., 2002. Evidence for fractal correlation properties in variations of peripheral arterial tone during REM sleep. Am. J. Physiol. Heart Circ. Physiol. 283, H434–H439. https://doi.org/10.1152/ajpheart.00336.2001.

Electrophysiology Task Force of the European Society of Cardiology the North American Society of Pacing, 1996. Heart rate variability. Circulation 93, 1043–1065. https://doi.org/10.1161/01.CIR.93.5.1043.

Goldberger, A.L., 1996. Non-linear dynamics for clinicians: chaos theory, fractals, and complexity at the bedside. Lancet 347, 1312.

Goldberger, A.L., Amaral, L.A.N., Hausdorff, J.M., Ivanov, P.C., Peng, C.-K., Stanley, H.E., 2002. Fractal dynamics in physiology: alterations with disease and aging. Proc. Natl. Acad. Sci. U S A 99, 2466–2472.

Goldberger, A.L., Rigney, D.R., West, B.J., 1990. Chaos and fractals in human physiology. Sci. Am. 262, 42–49.

Gómez-Extremera, M., Carpena, P., Ivanov, P.C., Bernaola-Galván, P.A., 2016. Magnitude and sign of long-range correlated time series: decomposition and surrogate signal generation. Phys. Rev. E 93, 042201. https://doi.org/10.1103/PhysRevE.93.042201.

Grote, L., Ploch, T., Heitmann, J., Knaack, L., Penzel, T., Peter, J.H., 1999. Sleep-related breathing disorder is an independent risk factor for systemic hypertension. Am. J. Respir. Crit. Care Med. 160, 1875–1882.

Hallstrom, A.P., Stein, P.K., Schneider, R., Hodges, M., Schmidt, G., Ulm, K., 2004. Structural relationships between measures based on heart beat intervals: potential for improved risk assessment. IEEE Trans. Biomed. Eng. 51, 1414–1420. https://doi.org/10.1109/TBME.2004.828049.

Hausdorff, M., Ashkenazy, Y., Peng, C.-K., Ivanov, P.C., Stanley, H.E., Goldberger, A.L., 2001. When human walking becomes random walking: fractal analysis and modeling of gait rhythm fluctuations. Phys. A 302, 138–147.

Hilton, M.F., Umali, M.U., Czeisler, C.A., Wyatt, J.K., Shea, S.A., 2000. Endogenous circadian control of the human autonomic nervous system. Comput. Cardiol. 27, 197–200.

Ho, K.K.L., Moody, G.B., Peng, C.-K., Mietus, J.E., Larson, M.G., Levy, D., Goldberger, A.L., 1997. Predicting survival in heart failure cases and controls using fully automated methods for deriving nonlinear and conventional indices of heart rate dynamics. Circulation 96, 842–848.

Hu, K., Ivanov, P.C., Chen, Z., Carpena, P., Stanley, H.E., 2001. Effect of trends on detrended fluctuation analysis. Phys. Rev. E 64, 011114. https://doi.org/10.1103/PhysRevE.64.011114.

Hu, K., Ivanov, P.C., Chen, Z., Hilton, M.F., Stanley, H.E., Shea, S.A., 2004a. Non-random fluctuations and multi-scale dynamics regulation of human activity. Phys. A 337, 307–318.

Hu, K., Ivanov, P.C., Hilton, M.F., Chen, Z., Ayers, R.T., Stanley, H.E., Shea, S.A., 2004b. Endogenous circadian rhythm in an index of cardiac vulnerability independent of changes in behavior. Proc. Natl. Acad. Sci. U S A 101, 18223–18227. https://doi.org/10.1073/pnas.0408243101.

Ivanov, P.C., 2007. Scale-invariant aspects of cardiac dynamics — observing sleep stages and circadian phases. IEEE Eng. Med. Biol. 26 (6), 33–37. https://doi.org/10.1109/MEMB.2007.907093.

Ivanov, P.C., 2003. Long-range dependence in heartbeat dynamics. In: Rangarajan, G., Ding, M. (Eds.), Processes with Long-Range Correlations: Theory and Applications, Lecture Notes in Physics. Springer, Berlin, Heidelberg, pp. 339–372. https://doi.org/10.1007/3-540-44832-2_19.

Ivanov, P.C., Amaral, L.A.N., Goldberger, A.L., Havlin, S., Rosenblum, M.G., Stanley, H.E., Struzik, Z., 2001. From $1/f$ noise to multifractal cascades in heartbeat dynamics. Chaos 11, 641–652.

Ivanov, P.C., Amaral, L.A.N., Goldberger, A.L., Havlin, S., Rosenblum, M.G., Struzik, Z.R., Stanley, H.E., 1999a. Multifractality in human heartbeat dynamics. Nature 399, 461–465. https://doi.org/10.1038/20924.

Ivanov, P.C., Amaral, L.A.N., Goldberger, A.L., Stanley, H.E., 1998a. Stochastic feedback and the regulation of biological rhythms. EPL Europhys. Lett. 43, 363. https://doi.org/10.1209/epl/i1998-00366-3.

Ivanov, P.C., Ashkenazy, Y., Kantelhardt, J.W., Stanley, H.E., 2003. Quantifying heartbeat dynamics by magnitude and sign correlations. AIP Conf. Proc. 665, 383–391. https://doi.org/10.1063/1.1584912.

Ivanov, P.C., Bartsch, R.P., 2014. In: D'Agostino, G., Scala, A. (Eds.), Network Physiology: Mapping Interactions between Networks of Physiologic Networks. Springer International Publishing, pp. 203–222.

Ivanov, P.C., Bernaola-Galvan, P., Amaral, L.A.N., Stanley, H.E., 2002a. Fractal features in the nonstationarity of physiological time series. In: Emergent Nature. World Scientific, pp. 55–63. https://doi.org/10.1142/9789812777720_0005.

Ivanov, P.C., Bunde, A., Amaral, L.A.N., Havlin, S., Fritsch-Yelle, J., Baevsky, R.M., Stanley, H.E., Goldberger, A.L., 1999b. Sleep-wake differences in scaling behavior of the human heartbeat: analysis of terrestrial and long-term space flight data. Eur. Lett. 48, 594–600. https://doi.org/10.1209/epl/i1999-00525-0.

Ivanov, P.C., Goldberger, A.L., Havlin, S., Peng, C.-K., Rosenblum, M.G., Stanley, H.E., 1998b. Wavelets in medicine and physiology. In: van den Berg, J.C. (Ed.), Wavelets in Physics. Cambridge University Press, Cambridge.

Ivanov, P.C., Goldberger, A.L., Stanley, H.E., 2002b. Fractal and multifractal approaches in physiology. In: Bunde, A., Kropp, J., Schellnhuber, H.J. (Eds.), The Science of Disasters: Climate Disruptions, Heart Attacks, and Market Crashes. Springer, Berlin, Heidelberg, pp. 218–257. https://doi.org/10.1007/978-3-642-56257-0_7.

Ivanov, P.C., Hu, K., Hilton, M.F., Shea, S.A., Stanley, H.E., 2007. Endogenous circadian rhythm in human motor activity uncoupled from circadian influences on cardiac dynamics. Proc. Natl. Acad. Sci. U S A 104, 20702–20707. https://doi.org/10.1073/pnas.0709957104.

Ivanov, P.C., Liu, K.K.L., Bartsch, R.P., 2016. Focus on the emerging new fields of network physiology and network medicine. New J. Phys. 18, 100201. https://doi.org/10.1088/1367-2630/18/10/100201.

Ivanov, P.C., Liu, K.K.L., Lin, A., Bartsch, R.P., 2017. Network physiology: from neural plasticity to organ network interactions. In: Mantica, G., Stoop, R., Stramaglia, S. (Eds.), Emergent Complexity from Nonlinearity, in Physics, Engineering and the Life Sciences, Springer Proceedings in Physics. Springer International Publishing, Cham, pp. 145–165. https://doi.org/10.1007/978-3-319-47810-4_12.

Ivanov, P.C., Lo, C.C., 2002. Stochastic approaches to modelling of physiological rhythms. In: Modelling Biomedical Signals. Presented at the Modelling Biomedical Signals, World Scientific, Bari, Italy, pp. 28–50. https://doi.org/10.1142/9789812778055_0003.

Ivanov, P.C., Ma, Q.D.Y., Bartsch, R.P., Hausdorff, J.M., Amaral, L.A.N., Schulte-Frohlinde, V., Stanley, H.E., Yoneyama, M., 2009. Levels of complexity in scale-invariant neural signals. Phys. Rev. E 79, 041920. https://doi.org/10.1103/PhysRevE.79.041920.

Ivanov, P.C., Rosenblum, M.G., Peng, C.-K., Mietus, J., Havlin, S., Stanley, H.E., Goldberger, A.L., 1998c. Scaling and universality in heart rate variability distributions. Phys. A 249, 587–593. https://doi.org/10.1016/S0378-4371(97)00522-0.

Ivanov, P.C., Rosenblum, M.G., Peng, C.-K., Mietus, J., Havlin, S., Stanley, H.E., Goldberger, A.L., 1996. Scaling behaviour of heartbeat intervals obtained by wavelet-based time-series analysis. Nature 383, 323–327.

Iyengar, N., Peng, C.-K., Morin, R., Goldberger, A.L., Lipsitz, L.A., 1996. Age-related alterations in the fractal scaling of cardiac interbeat interval dynamics. Am. J. Physiol. Regul. Integr. Comp. Physiol. 271, 1078–1084.

Kantelhardt, J.W., Ashkenazy, Y., Ivanov, P.C., Bunde, A., Havlin, S., Penzel, T., Peter, J.-H., Stanley, H.E., 2002. Characterization of sleep stages by correlations in the magnitude and sign of heartbeat increments. Phys. Rev. E 65, 051908. https://doi.org/10.1103/PhysRevE.65.051908.

Kantelhardt, J.W., Havlin, S., Ivanov, P.C., 2003a. Modeling transient correlations in heartbeat dynamics during sleep. Eur. Lett. 62, 147–153.

Kantelhardt, J.W., Penzel, T., Rostig, S., Becker, H.F., Havlin, S., Bunde, A., 2003b. Breathing during REM and non-REM sleep: correlated versus uncorrelated behaviour. Phys. A 319, 447–457. https://doi.org/10.1016/S0378-4371(02)01502-9.

Kaplan, D.T., Furman, M.I., Pincus, S.M., Ryan, S.M., Lipsitz, L.A., Goldberger, A.L., 1991. Aging and the complexity of cardiovascular dynamics. Biophys. J. 59, 945–949.

Karasik, R., Sapir, N., Ashkenazy, Y., Ivanov, P.C., Dvir, I., Lavie, P., Havlin, S., 2002. Correlation differences in heartbeat fluctuations during rest and exercise. Phys. Rev. E 66, 062902. https://doi.org/10.1103/PhysRevE.66.062902.

Kitney, R.I., Rompelman, O., 1980. The Study of Heart-Rate Variability. Oxford University Press, USA.

Kiyono, K., Struzik, Z.R., Aoyagi, N., Sakata, S., Hayano, J., Yamamoto, Y., 2004. Critical scale invariance in a healthy human heart rate. Phys. Rev. Lett. 93, 178103. https://doi.org/10.1103/PhysRevLett.93.178103.

Kiyono, K., Struzik, Z.R., Aoyagi, N., Togo, F., Yamamoto, Y., 2005. Phase transition in a healthy human heart rate. Phys. Rev. Lett. 95, 058101.

Kobayashi, M., Musha, T., 1982. $1/f$ fluctuation of heartbeat period. IEEE Trans. Biomed. Eng. 29, 456–457. https://doi.org/10.1109/TBME.1982.324972.

Leeuwen, P.V., Geue, D., Thiel, M., Cysarz, D., Lange, S., Romano, M.C., Wessel, N., Kurths, J., Grönemeyer, D.H., 2009. Influence of paced maternal breathing on fetal-maternal heart rate coordination. Proc. Natl. Acad. Sci. U S A 106, 13661–13666. https://doi.org/10.1073/pnas.0901049106.

Lefebvre, J.H., Goodings, D.A., Kamath, M.V., Fallen, E.L., 1993. Predictability of normal heart rhythms and deterministic chaos. Chaos 3, 267–276. https://doi.org/10.1063/1.165990.

Lin, A., Liu, K.K.L., Bartsch, R.P., Ivanov, P.C., 2020. Dynamic network interactions among distinct brain rhythms as a hallmark of physiologic state and function. Commun. Biol. 3, 1–11. https://doi.org/10.1038/s42003-020-0878-4.

Lin, A., Liu, K.K.L., Bartsch, R.P., Ivanov, P.C., 2016. Delay-correlation landscape reveals characteristic time delays of brain rhythms and heart interactions. Philos. Trans. R. Soc. Math. Phys. Eng. Sci. 374 https://doi.org/10.1098/rsta.2015.0182, 20150182.

Lipsitz, L.A., Goldberger, A.L., 1992. Loss of "complexity" and aging. Potential applications of fractals and chaos theory to senescence. J. Am. Med. Assoc. 267, 1806–1809. https://doi.org/10.1001/jama.267.13.1806.

Lipsitz, L.A., Mietus, J., Moody, G.B., Goldberger, A.L., 1990. Spectral characteristics of heart rate variability before and during postural tilt. Relations to aging and risk of syncope. Circulation 81, 1803–1810. https://doi.org/10.1161/01.cir.81.6.1803.

Liu, K.K.L., Bartsch, R.P., Lin, A., Mantegna, R.N., Ivanov, P.C., 2015a. Plasticity of brain wave network interactions and evolution across physiologic states. Front. Neural Circ. 9, 62. https://doi.org/10.3389/fncir.2015.00062.

Liu, K.K.L., Bartsch, R.P., Ma, Q.D.Y., Ivanov, P.C., 2015b. Major component analysis of dynamic networks of physiologic organ interactions. J. Phys. Conf. Ser. 640, 012013. https://doi.org/10.1088/1742-6596/640/1/012013.

Lo, C.C., Amaral, L.A.N., Havlin, S., Ivanov, P.C., Penzel, T., Peter, J.-H., Stanley, H.E., 2002. Dynamics of sleep-wake transitions during sleep. Eur. Lett. 57, 625–631. https://doi.org/10.1209/epl/i2002-00508-7.

Lo, C.C., Chou, T., Penzel, T., Scammell, T.E., Strecker, R.E., Stanley, H.E., Ivanov, P.C., 2004. Common scale-invariant patterns of sleep-wake transitions across mammalian species. Proc. Natl. Acad. Sci. U S A 101, 17545–17548. https://doi.org/10.1073/pnas.0408242101.

Ma, Q.D., Bartsch, R.P., Bernaola-Galván, P., Yoneyama, M., Ivanov, P.C., 2010. Effect of extreme data loss on long-range correlated and anticorrelated signals quantified by detrended fluctuation analysis. Phys. Rev. E 81, 031101.

Malik, M., Camm, A.J., 1995. Heart Rate Variability. Futura, Armonk NY.

Martinis, M., Knezevic, A., Krstacic, G., Vargovic, E., 2004. Changes in the Hurst exponent of heartbeat intervals during physical activity. Phys. Rev. E 70, 012903.

Moelgaard, H., Soerensen, K.E., Bjerregaard, P., 1991. Circadian variation and influence of risk factors on heart rate variability in healthy subjects. Am. J. Cardiol. 68, 777–784. https://doi.org/10.1016/0002-9149(91)90653-3.

Mohsenin, V., 2001. Sleep-related breathing disorders and risk of stroke. Stroke 32, 1271–1278.

Moorman, J.R., Lake, D.E., Ivanov, P.C., 2016. Early detection of sepsis—a role for network physiology? Crit. Care Med. 44, e312. https://doi.org/10.1097/CCM.0000000000001548.

Mrowka, R., Patzak, A., Rosenblum, M., 2000. Quantitative analysis of cardiorespiratory synchronization in infants. Int. J. Bifurc. Chaos 10, 2479–2488.

Muller, J.E., Stone, P.H., Turi, Z.G., Rutherford, J.D., Czeisler, C.A., Parker, C., Poole, W.K., Passamani, E., Roberts, R., Robertson, T., 1985. Circadian variation in the frequency of onset of acute myocardial infarction. N. Engl. J. Med. 313, 1315–1322.

Otzenberger, H., Gronfier, C., Simon, C., Charloux, A., Ehrhart, J., Piquard, F., Brandenberger, G., 1998. Dynamic heart rate variability: a tool for exploring sympathovagal balance continuously during sleep in men. Am. J. Physiol. 275, H946–H950.

Peng, C.-K., Buldyrev, S.V., Havlin, S., Simons, M., Stanley, H.E., Goldberger, A.L., 1994. Mosaic organization of DNA nucleotides. Phys. Rev. E 49, 1685–1689. https://doi.org/10.1103/PhysRevE.49.1685.

Peng, C.-K., Havlin, S., Stanley, H.E., Goldberger, A.L., 1995. Quantification of scaling exponents and crossover phenomena in nonstationary heartbeat time series. Chaos 5, 82–87. https://doi.org/10.1063/1.166141.

Peng, C.-K., Mietus, J., Hausdorff, J.M., Havlin, S., Stanley, H.E., Goldberger, A.L., 1993. Long-range anticorrelations and non-Gaussian behavior of the heartbeat. Phys. Rev. Lett. 70, 1343–1346. https://doi.org/10.1103/PhysRevLett.70.1343.

Peters Robert, W., Brent, M., Brooks Maria, M., Echt Debra, S., Barker Allan, H., Robert, C., Liebson Philip, R., Leon, G.H., 1994. Circadian pattern of arrhythmic death in patients receiving encainide, flecainide or moricizine in the cardiac arrhythmia. J. Am. Coll. Cardiol. 23, 283–289. https://doi.org/10.1016/0735-1097(94)90408-1.

Pikovsky, A.S., Rosenblum, M.G., Kurths, J., 2001. Synchronization: A Universal Concept in Nonlinear Sciences. Cambridge University Press, Cambridge.

Poon, C.-S., Merrill, C.K., 1997. Decrease of cardiac chaos in congestive heart failure. Nature 389, 492–495. https://doi.org/10.1038/39043.
Prokhorov, M.D., Ponomarenko, V.I., Gridnev, V.I., Bodrov, M.B., Bespyatov, A.B., 2003. Synchronization between main rhythmic processes in the human cardiovascular system. Phys. Rev. E Stat. Nonlinear Soft Matter Phys. 68, 041913.
Rosenblum, M.G., Pikovsky, A.S., Kurths, J., 1996. Phase synchronization of chaotic oscillators. Phys. Rev. Lett. 76, 1804–1807.
Rostig, S., Kantelhardt, J.W., Penzel, T., Cassel, W., Peter, J.H., Vogelmeier, C., Becker, H.F., Jerrentrup, A., 2005. Nonrandom variability of respiration during sleep in healthy humans. Sleep 28, 411–417.
Schäfer, C., Rosenblum, M.G., Kurths, J., Abel, H.H., 1998. Heartbeat synchronized with ventilation. Nature 392, 239–240. https://doi.org/10.1038/32567.
Schmitt, D.T., Ivanov, P.C., 2007. Fractal scale-invariant and nonlinear properties of cardiac dynamics remain stable with advanced age: a new mechanistic picture of cardiac control in healthy elderly. Am. J. Physiol. Regul. Integr. Comp. Physiol. 293 (5), R1923–R1937. https://doi.org/10.1152/ajpregu.00372.2007.
Schmitt, D.T., Stein, P.K., Ivanov, P.C., 2009. Stratification pattern of static and scale-invariant dynamic measures of heartbeat fluctuations across sleep stages in young and elderly. IEEE Trans. Biomed. Eng. 56, 1564–1573. https://doi.org/10.1109/TBME.2009.2014819.
Schumann, A.Y., Bartsch, R.P., Penzel, T., Ivanov, P.C., Kantelhardt, J.W., 2010. Aging effects on cardiac and respiratory dynamics in healthy subjects across sleep stages. Sleep 33, 943–955.
Shea, S.A., Hilton, M.F., Muller, J.E., 2007. Clinical Hypertension and Vascular Disease: Blood Pressure Monitoring in Cardiovascular Medicine and Therapeutics. Totowa NJ Humana, pp. 253–291.
Somers, V.K., Dyken, M.E., Mark, A.L., Abboud, F.M., 1993. Sympathetic-nerve activity during sleep in normal subjects. N. Engl. J. Med. 328, 303–307.
Song, H.-S., Lehrer, P.M., 2003. The effects of specific respiratory rates on heart rate and heart rate variability. Appl. Psychophysiol. Biofeedback 28, 13–23.
Stanley, H.E., 1971. Introduction to Phase Transitions and Critical Phenomena. Oxford University Press, London.
Stanley, H.E., Amaral, L.A.N., Goldberger, A.L., Havlin, S., Ivanov, P.C., Peng, C.-K., 1999. Statistical physics and physiology: monofractal and multifractal approaches. Phys. A 270, 309–324.
Stanley, H.E., Amaral, L.A.N., Gopikrishnan, P., Ivanov, P.C., Keitt, T.H., Plerou, V., 2000. Scale invariance and universality: organizing principles in complex systems. Phys. A 281, 60–68. https://doi.org/10.1016/S0378-4371(00)00195-3.
Stickgold, R., 2005. Sleep-dependent memory consolidation. Nature 437, 1272–1278.
Stickgold, R., Hobson, J.A., Fosse, R., Fosse, M., 2001. Sleep, learning, and dreams: off-line memory reprocessing. Science 294, 1052–1057. https://doi.org/10.1126/science.1063530.
Suki, B., Alencar, A.M., Frey, U., Ivanov, P.C., Buldyrev, S.V., Majumdar, A., Stanley, H.E., Dawson, C.A., Krenz, G.S., Mishima, M., 2003. Fluctuations, noise and scaling in the cardiopulmonary system. Fluctuation Noise Lett. 3, R1–R25.
Theiler, J., Eubank, S., Longtin, A., Galdrikian, B., Garmer, D.J., 1992. Testing for nonlinearity in time series: the method of surrogate data. Phys. D 58, 77–94. https://doi.org/10.1016/0167-2789(92)90102-S.
Tsuji, H., Venditti, F.J., Manders, E.S., Evans, J.C., Larson, M.G., Feldman, C.L., Levy, D., 1994. Reduced heart rate variability and mortality risk in an elderly cohort. The Framingham Heart Study. Circulation 90, 878–883.

Umetani, K., Singer, D.H., McCraty, R., Atkinson, M., 1998. Twenty-four hour time domain heart rate variability and heart rate: relations to age and gender over nine decades. J. Am. Coll. Cardiol. 31, 593–601.
Vicsek, T., 1992. Fractal Growth Phenomena, second ed. World Scientific, Singapore.
Xu, L., Chen, Z., Hu, K., Stanley, H.E., Ivanov, P.C., 2006. Spurious detection of phase synchronization in coupled nonlinear oscillators. Phys. Rev. E 73, 065201.
Xu, L., Ivanov, P.C., Hu, K., Chen, Z., Carbone, A., Stanley, H.E., 2005. Quantifying signals with power-law correlations: a comparative study of detrended fluctuation analysis and detrended moving average techniques. Phys. Rev. E 71, 051101.
Xu, Y., Ma, Q.D., Schmitt, D.T., Bernaola-Galván, P., Ivanov, P.C., 2011. Effects of coarse-graining on the scaling behavior of long-range correlated and anti-correlated signals. Phys. A Stat. Mech. Appl. 390, 4057–4072.

CHAPTER 5

Deep brain stimulation for understanding the sleep-wake phenomena

Francisco J. Urbano[1,2] and Edgar Garcia-Rill[3]
[1]Universidad de Buenos Aires, Facultad de Ciencias Exactas y Naturales, Departamento de Fisiología, Biología Molecular y Celular "Dr. Héctor Maldonado", Ciudad de Buenos Aires, Argentina; [2]CONICET- Instituto de Fisiología, Biología Molecular y Neurociencias (IFIBYNE), Ciudad Universitaria, Ciudad Autónoma de Buenos Aires, Argentina; [3]Emeritus, Center for Translational Neuroscience, Department of Neurobiology and Developmental Sciences, College of Medicine, University of Arkansas for Medical Sciences. Little Rock, AR, United States

Deep brain stimulation for the treatment of sleep disorder in several thalamocortical dysrhythmia pathologies

Deep brain stimulation (DBS) has been used in the treatment of a number of neurological and neurophysiological pathologies. For example, tremor and Parkinson's disease (PD), and related diseases (Benabid et al., 2000, 2001; Lozano, 2001; Mogilner et al., 2001), collectively known as *thalamocortical dysrhythmia syndrome* (Llinás et al., 1999, 2001, 2005; Jeanmonod et al., 2001).

Increased arousal and increased rapid eye movement (REM) sleep drive are widely common symptoms in a number of psychiatric and neurological disorders, including PD (Garcia-Rill et al., 2015a,b, 2019a,b). The standard treatment for PD has been dopamine replacement with L-DOPA, however, symptoms including gait deficits and sleep disorders are unresponsive to L-DOPA. The later symptoms have been linked to aberrant activity in the pedunculopontine nucleus (PPN), a structure known to be modulated by dopamine and other monoamines (Garcia-Rill, 1991; Urbano et al., 2015). The PPN is known to control the manifestation of cortical gamma band activity during waking and REM sleep (Garcia-Rill, 1991; Garcia-Rill et al., 2015a,b, 2019a,b). Fig. 5.1 shows how PPN neurons modulate ascending and descending targets, representing a strategic neuronal center to modulate sensory and locomotor interactions through DBS stimulating electrodes.

Methodological Approaches for Sleep and Vigilance Research
ISBN 978-0-323-85235-7
https://doi.org/10.1016/B978-0-323-85235-7.00011-9

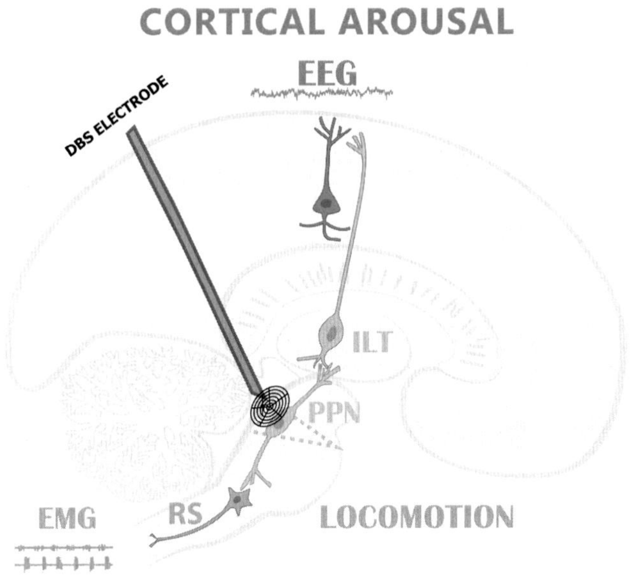

Figure 5.1 Pedunculopontine nucleus (PPN) is located in a strategic brainstem region that allows deep brain stimulation (DBS) to modulate both ascending and descending targets. Such stimulation would optimally activate PPN outputs if it were applied in the 40–60 Hz range. The intrinsic physiological properties of PPN neurons that allow them to fire at gamma band frequencies make this an ideal target for using DBS to boost gamma activity, especially via the intralaminar thalamus (ILT) to drive cortical electroencephalography gamma activity and arousal. The properties of PPN neurons may also underlie the long-known effects of PPN stimulation to induce locomotion primarily using 40–60 Hz stimulation, presumably driven by reticulospinal systems to elicit alternating limb electromyogram signals.

Lower number of PPN neurons have been observed in patients suffering from idiopathic PD and in the parkinsonian syndrome of progressive supranuclear palsy (Hirsch et al., 1987).

Extensive DBS use has been performed to treat the postural, locomotor, and sleep deficits induced by PD. Electrode properties used for DBS have been extensively improved (Gimsa et al., 2005), although large diameter electrodes may produce undesired effects (Mazzone et al., 2011; Aviles-Olmos et al., 2011; Garcia-Rill et al., 2019a). Congruent with PPN involvement in sleep dysfunction, PPN-DBS increases sleep efficiency, REM and stage 2 sleep, and decreases awakenings in PD patients (Romigi et al., 2008). Significant improvement in sleep patterns of patients with PD was observed after DBS of the PPN nucleus (Stefani et al., 2007; Alessandro et al., 2010). PPN stimulation at gamma band frequency (40–60 Hz) has been observed as the optimal range to improve PD (Mazzone et al., 2008, 2009; Moro et al., 2010). Future experiments are still needed to clarify whether PPN-DBS would not only ameliorate sleep alterations in several *thalamo-cortical dysrhythmia syndrome based pathologies*, while modulating higher functions such as arousal, perception, and attention (Garcia-Rill et al., 2019a).

Neuronal mechanisms underlying deep bran stimulation

Classic works in this field have described the activity reduction properties of high-frequency electrical deep brain stimulation (DBS) activation on basal ganglia nuclei (Benazzouz et al., 2000). The initial leading hypothesis was related to a nonspecific suppression of neuronal activity or "depolarization block"; similar to what has been observed during subthalamic nucleus stimulation (Beurrier et al., 2001). Since high-frequency DBS is clinically effective in disease states historically associated with a reduction of neuronal activity (i.e., dystonia and hemiballismus; Vitek, 1998), further suppression of activity may not be consistent with a viable therapeutic mechanism. Other studies have suggested DBS is able to induce a synaptic failure due to transmitter depletion (Urbano et al., 2002).

More recent works have shown a rebound increment in action potential firing after a reduction during DBS stimulation (Xiao et al., 2020). In a PD model of rats treated with 6-hydroxydopamine, subthalamic nucleus stimulation at frequencies below 100 Hz, showed a rebound increase in action potential firing after 10 second-long DBS stimulation. Above 210 Hz and up to 300 Hz frequencies, the increase in neural spike firing decreased (Xiao et al., 2020). Extracellular dopamine concentrations followed firing rate changes (Xiao et al., 2020), suggesting that DBS stimulation also affected intracellular [Ca^{2+}] (Trevathan et al., 2020), and therefore synaptic dynamics at target networks.

Indeed, network interaction between PPN and thalamocortical-basal ganglia nuclei has been suggested to underlie the effects of DBS stimulation in patients. PPN-DBS stimulation has been shown to affect a large majority of subthalamic nucleus neurons (Galati et al., 2008). PPN-DBS stimulation has been shown to modulate basal ganglia-brainstem-spinal cord descending system activity (Pierantozzi et al., 2008), a descending motor pathway from PPN (Skinner et al., 1990).

It has been suggested that synaptic/network interactions would spread fast membrane potential oscillations/pacemaker activity of PPN neurons

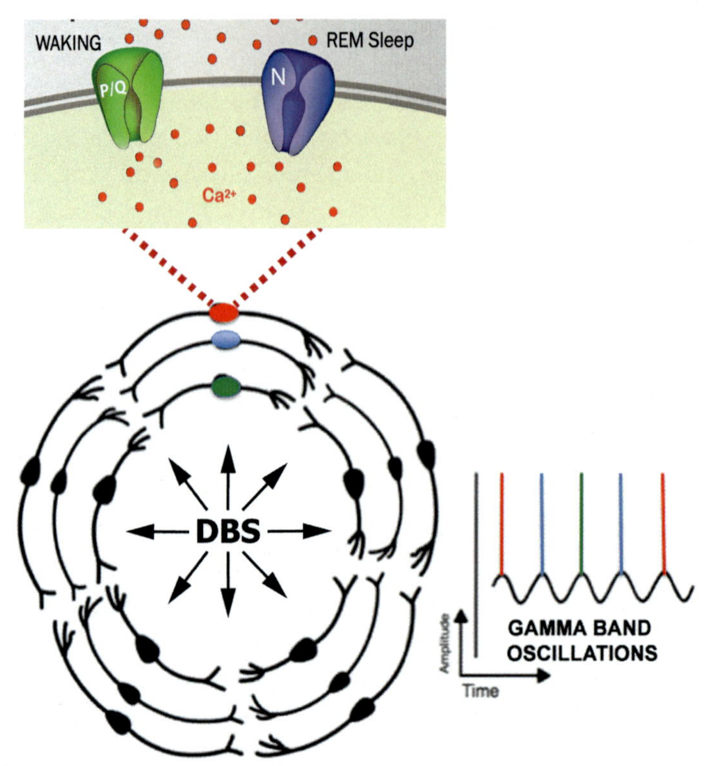

Figure 5.2 In the pedunculopontine nucleus, gamma oscillations have been suggested to emerge from the dynamic interaction between intrinsic N- and P/Q-type calcium channels, as well as synaptic properties. Synchronous activation of several neurons (different colors) during DBS would be essential to maintaining action potential discharge at gamma oscillations peaks (represented as different color action potentials). That is, while not all neurons may actually fire action potentials at gamma frequencies, the population as a whole will maintain gamma frequencies promoted by the peaks of the oscillations facilitating action potential firing.

(Fig. 5.2) during and after DBS stimulation (Garcia-Rill et al., 2019a,b). Every neuronal type of the PPN has been shown to exhibit gamma band activity, i.e., the PPN was proposed to be a *gamma-making machine* (Kezunovic et al., 2011; Garcia-Rill et al., 2015a,b).

The Garcia-Rill lab's findings established that *all* PPN neurons fired maximally at beta/gamma frequencies, and that all PPN neurons manifested beta/gamma frequency intrinsic membrane oscillations mediated by high threshold, voltage-dependent N- and P/Q-type calcium channels (Fig. 5.2) (Kezunovic et al., 2011; Garcia-Rill, 2019; Garcia-Rill et al., 2015a,b; 2019a,b). Beyond a few misleading reports from other groups (see

Garcia-Rill and Urbano, 2019), voltage-gated calcium channels have been described to be distributed along the dendrites of PPN neurons (Hyde et al., 2013). In half of PPN neurons, gamma oscillations were mediated by both N- and P/Q-type calcium channels (Fig. 5.2). However, fewer PPN neurons expressed gamma-band oscillatory activity mediated by only P/Q- or N-type channels (D'Onofrio et al., 2015; Luster et al., 2015).

Pharmacological intervention of histone deacetylase enzymes for the treatment of sleep disorders during deep brain stimulation

DBS stimulation protocols usually last for years. Hence, there is a critical need for understanding of the underlying disease mechanisms that explain its beneficial outcomes for the treatment of sleep disorders. Recent pharmacological strategies have been focused predominantly on inhibitors of histone deacetylase (HDAC) enzymes to ameliorate PD symptoms (Lebbe et al., 2016; Mazzocchi et al., 2020). Indeed, transcriptional dysregulation plays an important role in the progression and development of numerous brain disorders (Haberland et al., 2009; Gupta et al., 2020). HDAC inhibitors have been shown to be neuroprotective in animal models of PD (Monti et al., 2010). Interestingly, the HDAC inhibitor trichostatin A attenuated all the negative effects induced by sleep deprivation (Duan et al., 2016).

Long-term PPN DBS, but not STN DBS, at beta/gamma frequencies improved both nighttime sleep and daytime sleepiness (Peppe et al., 2012; Garcia-Rill et al., 2015a,b). Importantly, single cells in the PPN can oscillate at gamma band frequencies when "recruited" using intracellular current ramps (D'Onofrio et al., 2015; Garcia-Rill, 2019; Garcia-Rill et al., 2015a,b, 2019a,b; Garcia-Rill and Urbano, 2019; Kezunovic et al., 2011). Our group has determined that high-threshold, voltage-dependent N- and P/Q-type calcium channels in PPN neurons are recruited to mediate gamma band oscillation under the appropriate balance of HDAC class I/IIa (Garcia-Rill et al., 2019a), both in vivo (Bisagno et al., 2020) and in vitro (Urbano et al., 2018, 2020). Importantly, HDAC levels have been described as key elements for maintaining proper neuronal morphology through its modulation of both P/Q-type calcium channels and f-actin protein levels in the PPN (Fig. 5.3).

Proteomic and electrophysiological studies by our group have revealed that structural proteins such as F-actin and P/Q-type calcium channels are

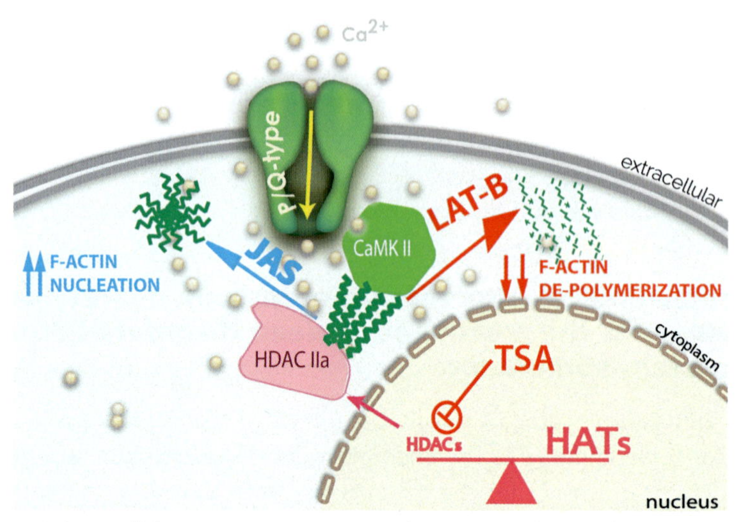

Figure 5.3 Intracellular proteomic interaction between nuclear and cytoplasmic proteins is required for the maintenance of gamma oscillations in the pedunculopontine nucleus. P/Q-type calcium channels (green structure) have been described to be modulated by CaMKII (green hexagon), which is modulated by cytoplasmic HDAC IIa (pink structure). There is a functional link between HDAC IIa/CaMKII and F-actin to shift the balance, either toward greater nucleation by using jasplakinolide (JAS, blue), or toward depolymerization using latrunculin-B (LAT-B, red), which would significantly reduce F-actin-CaMKII interactions necessary for gamma oscillation manifestation. *HAT,* histone acetyltransferase; *TSA,* trichostatin A, and inhibitor of HDAC type I/II.

modulated by histone deacetylation inhibition (Fig. 5.3), which can modulate gamma oscillations in the PPN (Byrum et al., 2019; Urbano et al., 2020). Specific studies are still needed to clearly elucidate PPN changes of the time course of acetylation levels. However, the modulation of HDAC levels may be critical to long-term stability and prolonged efficacy of DBS.

Acknowledgments

This work was supported by grants from FONCYT-Agencia Nacional de Promoción Científica y Tecnológica; Préstamo BID 1728 OC.AR. PICT-2016-1728, PICT 2018-01744, PICT 2019-284, Argentina-Germany collaboration grant, Argentina National Scientific and Technical Research Council (CONICET) - Deutsche Forschungsgemeinschaft (DFG) - MINCYT [Grant number: 2016-23120160100012CO01] and Proyecto Unidad Ejecutora-Idea Proyecto P-UE # 22920170100062CO (to Dr. Urbano). Prof. Garcia-Rill was awarded NIH grant P30 GM110702 from the IDeA program at NIGMS, allowing the Center for Translational Neuroscience, allowing the center to generate over $130 million in grant support for its members over the last 15 years.

References

Alessandro, S., Ceravolo, R., Brusa, L., Pierantozzi, M., Costa, A., Galati, S., et al., 2010. Non-motor functions in parkinsonian patients implanted in the pedunculopontine nucleus: focus on sleep and cognitive domains. J. Neurol. Sci. 289 (1−2), 44−48. https://doi.org/10.1016/j.jns.2009.08.017.

Aviles-Olmos, I., Foltynie, T., Panicker, J., Cowie, D., Limousin, P., Hariz, M., et al., 2011. Urinary incontinence following deep brain stimulation of the pedunculopontine nucleus. Acta Neurochir. 153 (12), 2357−2360. https://doi.org/10.1007/s00701-011-1155-6.

Benabid, A.L., Koudsié, A., Pollak, P., Kahane, P., Chabardes, S., Hirsch, E., et al., 2000. Future prospects of brain stimulation. Neurol. Res. 22 (3), 237−246. https://doi.org/10.1080/01616412.2000.11740666.

Benabid, A.L., Koudsié, A., Benazzouz, A., Vercueil, L., Fraix, V., Chabardes, S., et al., 2001. Deep-brain stimulation of the corpus luysi (subthalamic nucleus) and other targets in Parkinson's disease. Extension to new indications, such as dystonia and epilepsy. J. Neurol. 248 (Suppl. 3), III37−III47. https://doi.org/10.1007/pl00007825.

Benazzouz, A., Gao, D.M., Piallat, B., Bouali-Benazzouz, R., Benabid, A.L., 2000. Effect of high-frequency stimulation of the subthalamic nucleus on the neuronal activities of the substantia nigra pars reticulata and ventrolateral nucleus of the thalamus in the rat. Neuroscience 99 (2), 289−295. https://doi.org/10.1016/s0306-4522(00)00199-8.

Beurrier, C., Bioulac, B., Audin, J., Hammond, C., 2001. High-frequency stimulation produces a transient blockade of voltage-gated currents in subthalamic neurons. J. Neurophysiol. 85, 1351−1356. https://doi.org/10.1152/jn.2001.85.4.1351.

Bisagno, V., Bernardi, M.A., Sanz Blasco, S., Urbano, F.J., Garcia-Rill, E., 2020. Differential effects of HDAC inhibitors on PPN oscillatory activity in vivo. Neuropharmacology 165, 107922. https://doi.org/10.1016/j.neuropharm.2019.107922.

Byrum, S.D., Washam, C.L., Tackett, A.J., Garcia-Rill, E., Bisagno, V., Urbano, F.J., 2019. Proteomic measures of gamma oscillations. Heliyon 5 (8), e02265. https://doi.org/10.1016/j.heliyon.2019.e02265.

D'Onofrio, S., Kezunovic, N., Hyde, J.R., Luster, B., Messias, E., Urbano, F.J., Garcia-Rill, E., 2015. Modulation of gamma oscillations in the pedunculopontine nucleus by neuronal calcium sensor protein-1: relevance to schizophrenia and bipolar disorder. J. Neurophysiol. 113 (3), 709−719. https://doi.org/10.1152/jn.00828.2014.

Duan, R., Liu, X., Wang, T., Wu, L., Gao, X., Zhang, Z., 2016. Histone acetylation regulation in sleep deprivation-induced spatial memory impairment. Neurochem. Res. 41 (9), 2223−2232. https://doi.org/10.1007/s11064-016-1937-6.

Galati, S., Scarnati, E., Mazzone, P., Stanzione, P., Stefani, A., 2008. Deep brain stimulation promotes excitation and inhibition in subthalamic nucleus in Parkinson's disease. Neuroreport 19 (6), 661−666. https://doi.org/10.1097/WNR.0b013e3282fb78af.

Garcia-Rill, E., 1991. The pedunculopontine nucleus. Prog. Neurobiol. 36, 363−389. https://doi.org/10.1016/0301-0082(91)90016-T.

Garcia-Rill, E., 2019. Neuroepigenetics of arousal: gamma oscillations in the pedunculopontine nucleus. J. Neurosci. Res. 97 (12), 1515−1520. https://doi.org/10.1002/jnr.24417.

Garcia-Rill, E., Urbano, F.J., 2019. Concerns regarding Baksa B, Kovacs A, Bayasgalan T, Szentesi P, Koseghy A, Szucs P, Balazs P. Characterization of functional subgroups among genetically identified cholinergic neurons in the pedunculopontine nucleus. Cell. Mol. Life Sci. 76 (23), 4581−4582. https://doi.org/10.1007/s00018-019-03307-x.

Garcia-Rill, E., Hyde, J., Kezunovic, N., Urbano, F.J., Petersen, E., 2015a. The physiology of the pedunculopontine nucleus- implications for deep brain stimulation. J. Neural. Transm. 122, 225−235. https://doi.org/10.1007/s00702-014-1243-x.

Garcia-Rill, E., Luster, B., D'Onofrio, S., Mahaffey, S., Bisagno, V., Urbano, F.J., 2015b. Pedunculopontine arousal system physiology — deep brain stimulation (DBS). Sleep Sci. 8, 153—161. https://doi.org/10.1016/j.slsci.2015.09.001.
Garcia-Rill, E., D'Onofrio, S., Mahaffey, S.C., Bisagno, V., Urbano, F.J., 2019a. Bottom-up gamma and bipolar disorder, clinical and neuroepigenetic implications. Bipolar Disord. 21 (2), 108—116. https://doi.org/10.1111/bdi.12735.
Garcia-Rill, E., Mahaffey, S., Hyde, J.R., Urbano, F.J., 2019b. Bottom-up gamma maintenance in various disorders. Neurobiol. Dis. 128, 31—39. https://doi.org/10.1016/j.nbd.2018.01.010.
Gimsa, J., Habel, B., Schreiber, U., van Rienen, U., Strauss, U., Gimsa, U., 2005. Choosing electrodes for deep brain stimulation experiments–electrochemical considerations. J. Neurosci. Methods 142 (2), 251—265. https://doi.org/10.1016/j.jneumeth.2004.09.001.
Gupta, R., Ambasta, R.K., Kumar, P., 2020. Pharmacological intervention of histone deacetylase enzymes in the neurodegenerative disorders. Life Sci. 243, 117278. https://doi.org/10.1016/j.lfs.2020.117278.
Haberland, M., Montgomery, R.L., Olson, E.N., 2009. The many roles of histone deacetylases in development and physiology: implications for disease and therapy. Nat. Rev. Genet. 10, 32—42. https://doi.org/10.1038/nrg2485.
Hirsch, E.C., Graybiel, A.M., Duyckaerts, C., Javoy-Agid, F., 1987. Neuronal loss in the pedunculopontine tegmental nucleus in Parkinson disease and in progressive supranuclear palsy. Proc. Natl. Acad. Sci. U. S. A. 84 (16), 5976—5980. https://doi.org/10.1073/pnas.84.16.5976.
Hyde, J., Kezunovic, N., Urbano, F.J., Garcia-Rill, E., 2013. Spatiotemporal properties of high-speed calcium oscillations in the pedunculopontine nucleus. J. Appl. Physiol. 115 (9), 1402—1414. https://doi.org/10.1152/japplphysiol.00762.2013 (1985).
Jeanmonod, D., Magnin, M., Morel, A., Siegmund, M., Cancro, A., Lanz, M., et al., 2001. Thalamocortical dysrhythmia. Part II. Clinical and surgical aspects. Thalamus Relat. Syst. 1, 245—254. https://doi.org/10.1016/S1472-9288(01)00026-7.
Kezunovic, N., Urbano, F.J., Simon, C., Hyde, J., Smith, K., Garcia-Rill, E., 2011. Mechanism behind gamma band activity in the pedunculopontine nucleus (PPN). Eur. J. Neurosci. 34, 404—415. https://doi.org/10.1111/j.1460-9568.2011.07766.x.
Labbé, C., Lorenzo-Betancor, O., Ross, O.A., 2016. Epigenetic regulation in Parkinson's disease. Acta Neuropathol. 132 (4), 515—530. https://doi.org/10.1007/s00401-016-1590-9.
Llinás, R., Ribary, U., Jeanmonod, D., Kronberg, E., Mitra, P.P., 1999. Thalamocortical dysrhythmia: a neurological and neuropsychiatric syndrome characterized by magnetoencephalography. Proc. Natl. Acad. Sci. U. S. A. 96 (26), 15222—15227. https://doi.org/10.1073/pnas.96.26.15222.
Llinás, R., Ribary, U., Jeanmonod, D., Cancro, R., Kronberg, E., Schulman, J., et al., 2001. Thalamocortical dysrhythmia. Part I. Functional and imaging aspects. Thalamus Relat. Syst. 1 (3), 237—244. https://doi.org/10.1016/S1472-9288(01)00023-1.
Llinás, R., Urbano, F.J., Leznik, E., Ramírez, R.R., van Marle, H.J., 2005. Rhythmic and dysrhythmic thalamocortical dynamics: GABA systems and the edge effect. Trends Neurosci. 28 (6), 325—333. https://doi.org/10.1016/j.tins.2005.04.006, 2005 Jun.
Lozano, A.M., 2001. Deep-brain stimulation for Parkinson's disease. Park. Relat. Disord. 7, 199—203. https://doi.org/10.1016/s1353-8020(00)00057-2.
Luster, B., D'Onofrio, S., Urbano, F., Garcia-Rill, E., 2015. High-threshold Ca^{2+} channels behind gamma band activity in the pedunculopontine nucleus (PPN). Phys. Rep. 3 (6), e12431. https://doi.org/10.14814/phy2.12431.
Mazzocchi, M., Collins, L.M., Sullivan, A.M., O'Keeffe, G.W., 2020. The class II histone deacetylases as therapeutic targets for Parkinson's disease. Neuronal Signal 4 (2). https://doi.org/10.1042/NS20200001. NS20200001.

Mazzone, P., Sposato, S., Insola, A., DilazzaroV, S.E., 2008. Stereotactic surgery of nucleus tegmenti pedunculopontine. Br. J. Neurosurg. 22 (S1), S33–S40. https://doi.org/10.1080/02688690802448327.
Mazzone, P., Insola, A., Sposato, S., Scarnati, E., 2009. The deep brain stimulation of the pedunculopontine tegmental nucleus. Neuromodulation 191–204. https://doi.org/10.1111/j.1525-1403.2009.00214.x.
Mazzone, P., Sposato, S., Insola, A., Scarnati, E., 2011. The deep brain stimulation of the pedunculopontine tegmental nucleus: towards a new stereotactic neurosurgery. J. Neural Transm. (Vienna) 118 (10), 1431–1451. https://doi.org/10.1007/s00702-011-0593-x.
Mogilner, A.Y., Benabid, A.-L., Rezai, A.R., 2001. Brain stimulation: current applications and future prospects. Thalamus Relat. Syst. 1, 255–267. https://doi.org/10.1016/S1472-9288(01)00024-3.
Monti, B., Gatta, V., Piretti, F., Raffaelli, S.S., Virgili, M., Contestabile, A., 2010. Valproic acid is neuroprotective in the rotenone rat model of Parkinson's disease: involvement of alpha-synuclein. Neurotox. Res. 17 (2), 130–141. https://doi.org/10.1007/s12640-009-9090-5.
Moro, E., Hamani, C., Poon, Y.Y., Al-Khairallah, T., Dostrovsky, J.O., Hutchison, W.D., Lozano, A.M., 2010. Unilateral pedunculopontine stimulation improves falls in Parkinson's disease. Brain 133, 215–224. https://doi.org/10.1093/brain/awp261.
Peppe, A., Pierantozzi, M., Baiamonte, V., Moschella, V., Caltagirone, C., Stanzione, P., et al., 2012. Deep brain stimulation of pedunculopontine tegmental nucleus: role in sleep modulation in advanced Parkinson disease patients: one-year follow-up. Sleep 35 (12), 1637–1642. https://doi.org/10.5665/sleep.2234.
Pierantozzi, M., Palmieri, M.G., Galati, S., Stanzione, P., Peppe, A., Tropepi, D., et al., 2008. Pedunculopontine nucleus deep brain stimulation changes spinal cord excitability in Parkinson's disease patients. J. Neural. Transm. 115 (5), 731–735. https://doi.org/10.1007/s00702-007-0001-8.
Romigi, A., Placidi, F., Peppe, A., Pierantozzi, M., Izzi, F., Brusa, L., et al., 2008. Pedunculopontine nucleus stimulation influences REM sleep in Parkinson's disease. Eur. J. Neurol. 15, e64–e65. https://doi.org/10.1111/j.1468-1331.2008.02167.x.
Skinner, R.D., Kinjo, N., Henderson, V., Garcia-Rill, E., 1990. Locomotor projections from the pedunculopontine nucleus to the spinal cord. Neuroreport 1 (3–4), 183–186. https://doi.org/10.1097/00001756-199011000-00001.
Stefani, A., Lozano, A.M., Peppe, A., Stanzione, P., Galati, S., Tropepi, D., et al., 2007. Bilateral deep brain stimulation of the pedunculopontine and subthalamic nuclei in severe Parkinson's disease. Brain 130 (Pt 6), 1596–1607. https://doi.org/10.1093/brain/awl346.
Trevathan, J.K., Asp, A.J., Nicolai, E.N., Trevathan, J., Kremer, N.A., Kozai, T.D.Y., et al., 2020. Calcium imaging in freely-moving mice during electrical stimulation of deep brain structures. J. Neural. Eng. https://doi.org/10.1088/1741-2552/abb7a4 (in press).
Urbano, F.J., Leznik, E., Llinás, R.R., 2002. Cortical activation patterns evoked by afferent axons stimuli at different frequencies: an in vitro voltage-sensitive dye imaging study. Thalamus Relat. Syst. 1, 371–378. https://doi.org/10.1016/S1472-9288(02)00009-2.
Urbano, F.J., Bisagno, V., González, B., Rivero-Echeto, M.C., Muñiz, J.A., Luster, B., et al., 2015. Pedunculopontine arousal system physiology-Effects of psychostimulant abuse. Sleep Sci. 8 (3), 162–168. https://doi.org/10.1016/j.slsci.2015.09.004.
Urbano, F.J., Bisagno, V., Mahaffey, S., Lee, S.H., Garcia-Rill, E., 2018. Class II histone deacetylases require P/Q-type Ca^{2+} channels and CaMKII to maintain gamma oscillations in the pedunculopontine nucleus. Sci. Rep. 8 (1), 13156. https://doi.org/10.1038/s41598-018-31584-2.

Urbano, F.J., Bisagno, V., Garcia-Rill, E., 2020. Gamma oscillations in the pedunculopontine nucleus are regulated by F-actin: neuroepigenetic implications. Am. J. Physiol. Cell Physiol. 318 (2), C282–C288. https://doi.org/10.1152/ajpcell.00374.2019.

Vitek, J., 1998. Surgery for dystonia. Neurosurg. Clin. 9, 345–366.

Xiao, G., Song, Y., Zhang, Y., Xing, Y., Xu, S., Wang, M., et al., 2020. Dopamine and striatal neuron firing respond to frequency-dependent DBS detected by microelectrode arrays in the rat model of Parkinson's disease. Biosensors 10 (10), E136. https://doi.org/10.3390/bios10100136.

CHAPTER 6

Electroencephalography power spectra and electroencephalography functional connectivity in sleep

Chiara Massullo[1], Giuseppe A. Carbone[1], Eric Murillo-Rodríguez[2,3], Sérgio Machado[2,4,5], Henning Budde[2,6], Tetsuya Yamamoto[2,7], Claudio Imperatori[1]

[1]Cognitive and Clinical Psychology Laboratory, Department of Human Science, European University of Rome, Rome, Italy; [2]Intercontinental Neuroscience Research Group, Rome, Italy; [3]Laboratorio de Neurociencias Moleculares e Integrativas Escuela de Medicina, División Ciencias de la Salud, Universidad Anáhuac Mayab Mérida, Mérida, Yucatán, México; [4]Department of Sports Methods and Techniques, Federal University of Santa Maria, Santa Maria, Brazil; [5]Laboratory of Physical Activity Neuroscience, Neurodiversity Institute, Queimados-RJ, Brazil; [6]Faculty of Human Sciences, Medical School Hamburg, Hamburg, Germany; [7]Graduate School of Technology, Industrial and Social Sciences Tokushima University, Tokushima, Tokushima, Japan

Quantitative electroencephalography

Firstly introduced by Hans Berger (1929), the electroencephalography (abbreviated as EEG) is a neurophysiological technique consisting in the recording of electrical activity generated from the brain (Porjesz and Begleiter, 2003). With a high-temporal resolution (expressed in terms of milliseconds), the EEG allows to record brain electrical activity emerging from pyramidal cortical neurons which are perpendicularly oriented to the surface of the brain (Pizzagalli, 2007). By considering that each of these brain neurons generates small electrical activity, when a cluster of neurons becomes active, EEG records the sum of all these tiny electrical charges by means of small electrodes (which can be placed also by using an electrode-cap) placed on the surface of the scalp. These electrodes amplify electrical activity which is recorded in form of brain waves (Porjesz and Begleiter, 2003).

For its economical and noninvasive features, EEG is considered one of the most neuroscientific and clinical brain imaging technique to explore the electrophysiological correlates of cognitive processes and pathological conditions (Cohen, 2017). Usually, to explore cognition or disease-related electrophysiological dynamics, the EEG can be performed in several conditions such as eyes closed/open resting state (RS), during tasks or after the

Methodological Approaches for Sleep and Vigilance Research
ISBN 978-0-323-85235-7
https://doi.org/10.1016/B978-0-323-85235-7.00005-3

presentation of selected stimuli. Once recorded, the EEG signal can be analyzed from both a qualitative and a quantitative point of view. While qualitative EEG concerns the clinical visual inspection and evaluation of EEG trace, quantitative EEG (QEEG) refers to the process by which, starting from the original EEG trace, it can be extracted numerical values gained by mathematical computations by using digital technology (Pizzagalli, 2007; Novo-Olivas, 2014). Furthermore, while qualitative EEG mainly depends on subjectivity due to the high influence of the technician expertise, QEEG provides more objective measures useful both in clinical and research settings (Thakor and Tong, 2004) and allows to measure electrical configurations of the scalp reflecting the underlying brain activity.

In QEEG studies, there are two main approaches to quantify and analyze brain electrical activity known as linear and nonlinear ones (Pizzagalli, 2007). Nonlinear approaches are usually applied by considering EEG pattern as irregular and transient, and are performed by computing higher order statistics. Conversely, linear approaches assume that brain electrical signals are stationary processes and can be used when EEG features (e.g., frequencies or amplitudes) characterize the physiological status (Thakor and Tong, 2004). Across these approaches, QEEG methods allow to investigate brain electrical activity by considering several parameters such as amplitudes and waveforms frequencies, phase, and coherence (Pizzagalli, 2007). By computing these parameters, the most used linear methods are power spectra and coherence analyses which are used to investigate both at rest and task-related EEG activity (Pizzagalli, 2007).

Electroencephalography power spectra

The power spectral analysis (PSA) is considered one of the principal methods to quantify brain electrical activity (Dressler et al., 2004). The assumption behind the PSA is that any oscillatory activity can be characterized by the sum of different sinusoidal waves with distinct features of frequency and amplitude (Pizzagalli, 2007). Indeed, this analysis computes the amplitude and the scalp distribution of the different frequency bands which compose the EEG waveform (Duffy et al., 2012). In other words, it mainly consists in determining the amount of different frequency bands composing EEG signal. Particularly, in PSA, the EEG signal is transformed from time domain into frequency domain by means of Fast Fourier Transform (FFT; Grass and Gibbs, 1938), and it is further deconstructed into waveforms with different frequency components.

Starting from this index, several power spectrum parameters have been introduced including absolute power, relative power in each band and symmetry ones (Dressler et al., 2004; Pizzagalli, 2007). The absolute power is a simple index to interpret, reflecting how much a specific frequency band is represented in EEG signal. Conversely, the relative power is gained from the computation of the amount of brain electrical activity in selected frequency bands divided by the total power (Pizzagalli, 2007). Lastly, symmetry indices can be obtained by computing the power ratio between symmetrical scalp seeds for each frequency band; the advantage of this parameter is that it can provide relevant data about hemispheric activation (for a review, see Reznik and Allen, 2018).

Electroencephalography functional connectivity

Generally, brain connectivity can be distinguished in structural, functional, and effective connectivity. While structural connectivity is a measure of physical (or anatomical) connections among brain components (e.g., neurons or brain structures; Sporns, 2010; Hahn et al., 2019), effective connectivity is a time-dependent measure causally describing the effects among neural network elements (Friston, 1994). Even if functional connectivity also is highly time-dependent, differently from effective one, it is defined in terms of statistical covariance (or correlation) and, therefore, it does not allow to infer causal relationships (Schoenberg, 2020). Indeed, it is a measure derived from the temporal association between distant neurophysiological events (Friston, 1994, 2007). In other words, functional connectivity is a measure of statistical dependencies between the activity recorded at different neural sites (or more in general, between neural system components) (Eickhoff and Müller, 2015; Sporns, 2010). These bivariate statistical dependencies can be computed in time domain but also in frequency-spatial and temporal-frequency-spatial domain (e.g., coherence, phase synchronization).

Functional connectivity data can be obtained from several neural recording techniques including EEG. This measure is influenced by both external (e.g., a task or a specific stimulation) and/or internal states (Sporns, 2010). In bivariate connectivity field (i.e., functional connectivity), two measures, such as the *coherence* and the *phase synchronization* ones, are usually considered (Schoenberg, 2020). While coherence is a frequency-dependent measure of linear similarity between electrical signals, the phase synchronization is a measure of nonlinear similarity (Pascual-Marqui et al., 2011). Particularly, EEG coherence is a measure of dynamic functional interactions between EEG signals recorded at different scalp locations (Nunez et al., 1997; Schoenberg, 2020). In other words, this measure indicates whereas two or more different regions are active at the same time and in the same frequency band.

Coherence values can extend from 0 to 1 (i.e., no synchronization and the maximum level of synchronization) (Pizzagalli, 2007). This measure is based on the assumption that when different brain regions are together activated during a cognitive process, they show an increase of neural synchronization (i.e., coherence) within certain EEG frequency bands (Weiss and Mueller, 2003). Accordingly, EEG coherence represents an important measure estimating cortical functional interactions in different frequency bands (Srinivasan et al., 2007). From a statistical point of view, coherence is gained by a covariance statistical application to the processed EEG data. Particularly, it is obtained by a linear cross-correlation between the time series from two separate electrodes in a specific frequency band. In other words, coherence is considered a measure of linear covariance between the cross-spectra of two signals. Due to its simplicity, the coherence is the most widely used connectivity index (Faes et al., 2012).

Compared to coherence, the phase synchronization is a bivariate measure which does not depend on the spectra amplitude (Lachaux et al., 1999). According to the experimental procedure (i.e., including hypotheses and signal processing procedure), it can be considered several measures of phase synchronization.

One of the most used index in neurophysiological studies (e.g., Canuet et al., 2012; Olbrich et al., 2014; Ramyead et al., 2015) is the *lagged phase synchronization* (LPS). Phase synchronization is a measure of "*the similarity of two time series by means of the phases of the analyzed signal*" (Olbrich et al., 2014; page 3) and, similarly to coherence, its values range between 0 and 1 (i.e., the absence and the maximum synchronization, respectively). Particularly, the LPS indicates the synchrony of two signals after removing the instantaneous zero-lag component, which is known to be constituted by several artifacts (Hata et al., 2016). Notwithstanding the removal of zero-lag contribution could not entirely eliminate the problem of volume conduction (Palva et al., 2018), it has been estimated that LPS is altered in minimal way by low-spatial resolution, and it is considered to solely reflect data of physiological nature (Hata et al., 2016; Pascual-Marqui et al., 2011).

Source localization methods

Despite the advantages of QEEG, it may suffer from the problem of volume conduction or common sources (Stam et al., 2007). Nevertheless specific algorithms for the identification of signal sources can address this issue (Grech et al., 2008; Jatoi et al., 2014). To localize brain electrical activity recorded from the scalp, several solutions have been proposed which can be classified into two main approaches: the equivalent dipole and the linear distributed one (Pizzagalli, 2007). The first approach considers that brain electromagnetic signals are generated by limited focal and distinct sources, which can be represented as dipoles (e.g., Scherg and Ebersole, 1994). Due to its dependency on the a priori selected number of dipoles, the main limitation of this approach is that this aprioristic selection cannot be performed in many experimental cases (Phillips et al., 2002).

Conversely, the second class of approaches (i.e., linear distributed ones) admits taking into account all possible source sites at the same time. These approaches allow to overcome the limit of a priori assumptions about the numbers of the sources (which, conversely, are required by the first class of approaches) by using localization algorithms implemented with anatomical and functional limits which are defined by considering that, compared to others, some brain structures show a higher likelihood of generating EEG signals (Pizzagalli, 2007).

Among this class of approaches, one of the most used is the exact low-resolution brain electromagnetic tomography (eLORETA) software (Pascual-Marqui et al., 2011) which is a widely validated tool to localize brain electrical activity from the electrical signal recorded on the surface of the scalp (Figs. 6.1 and 6.2). Particularly, it allows to estimate the current distribution throughout the scalp in a three-dimensional (3D) way providing a minimum norm inverse solution which is linear, weighted, and

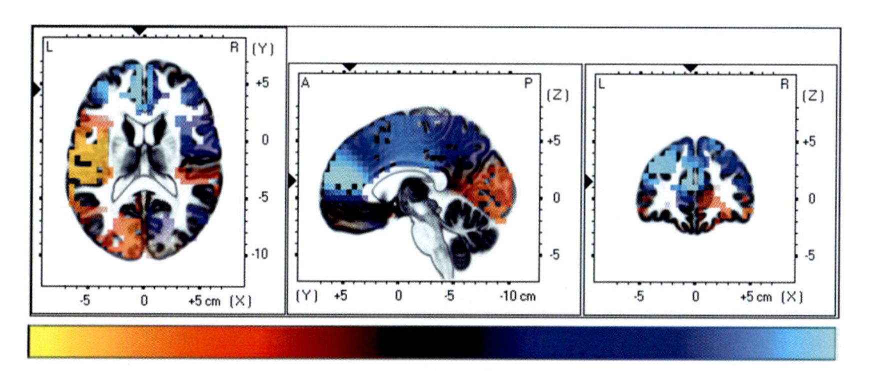

Figure 6.1 Example of power spectra image provided by the eLORETA. At the bottom of the figure is represented the color bar in which red-to-yellow spectrum colors (on the right) represent an increase of EEG power in the specific frequency band and the blue spectrum (on the left) indicate a power reduction.

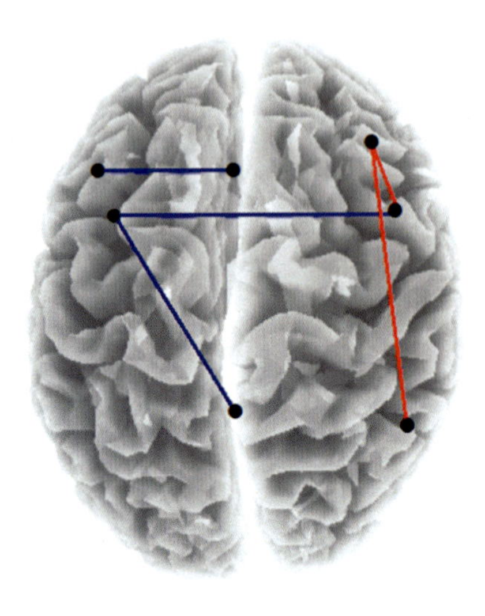

Figure 6.2 Example of coherence image provided by the eLORETA. Blue *lines* indicate decreased coherence while red *lines* indicate increased coherence; black *dots* indicate selected regions of interest.

discrete (Pascual-Marqui et al., 2011; Canuet et al., 2012). The head model used by eLORETA software for the inverse solution considers an electric potential lead field obtained by computations performed according to the boundary element method and by taking into account a realistic head model (Fuchs et al., 2002) determined according to the Montreal Neurological Institute's (MNI) template (Mazziotta et al., 2001). The 3D solution provided by the software includes 6239 voxels, with a spatial resolution of 5 × 5 × 5 mm (Canuet et al., 2012), which are selected according to the probabilistic Talairach atlas as referring exclusively to cortical gray matter (Lancaster et al., 2000). Indeed, the software takes into account exclusively the voxels that, in a nonambiguous way, are identified as cortical gray matter and are localized in the brain area (Imperatori et al., 2020). In this way, the images provided by the eLORETA software exactly represent the electrical activity at all voxels in neuroanatomic MNI space as the precise magnitude of the current density estimated (Takahashi et al., 2013). More deeply, this software assumes that EEG signals are mainly generated by cortical gray matter and that proximal neurons are activated at the same time and show only gradually changing orientations (Pizzagalli, 2007). Moreover, it assumes that connected neurons are synchronously and simultaneously activated (Kreiter and Singer, 1992; Murphy et al., 1992;

Pascual-Marqui et al., 1994). Furthermore, compared to previous versions, the eLORETA is characterized by a correct localization even when structured noise is present (Pascual-Marqui et al., 2011; Jatoi et al., 2014). eLORETA also has the benefit to provide an appropriate localization agreement with other neuroimaging tools (e.g., Kirino, 2017; Horacek et al., 2007; De Ridder et al., 2011).

In conclusion, due to its widely reported advantages, QEEG could be considered an useful method because it allows to investigate EEG activity during sleep in both normal and pathological conditions (e.g., Togo et al., 2006). Accordingly, in sleep research area, EEG has the advantage to be a useful tool in distinguishing different conscious states (e.g., awakening from sleep) and sleep-related states such as rapid-eye-movement (REM) and non-REM (NREM) ones.

Quantitative electroencephalography in normal sleep

In sleep research, QEEG is considered a useful instrument to understand neurophysiological mechanisms underlying several processes that cannot be accessed without an external instrument. For this reason, literature about QEEG correlates of sleep process is extended. In general, literature reports several stable, frequency-specific and locally-occurring QEEG differences among different phases of whole sleep phenomenon [i.e., both between NREM-REM-NREM phases and wake-sleep-wake ones, as well as during sleep; (e.g., Finelli et al., 2001; Marzano et al., 2010)].

Taken together, QEEG literature about sleep has shown data suggesting that sleep is a complex process which could be defined as *local* and *use-dependent* (for a review, see Ferrara and De Gennaro, 2011). The local nature characterizes the whole sleep phenomenon (i.e., from sleep onset to awake). Indeed, QEEG power spectral studies (e.g., Marzano et al., 2013) reported that sleep onset mainly occurs in frontal areas (i.e., the appearance of slow wave EEG activity during sleep onset mainly occurs in anterior regions). This pattern of activation remains during all the homeostatic process which principally characterizes the first part of the night. Accordingly, sleep onset seems to occur not simultaneously in the whole brain (i.e., different brain regions can enter the sleep state at different times) (De Gennaro et al., 2001). Indeed, different brain regions can simultaneously exhibit different sleep intensities indicating that sleep is not a spatially uniform and homogeneous process. This is a common pattern observed in normal sleepers but it can also occur (with different gradients) in abnormal sleep such as some parasomnias or paradoxical insomnia (Marzano et al., 2008; Mahowald and Schenck, 1991). For example, a study on normal sleepers (De Gennaro et al., 2001), reported anterior-posterior neurophysiological differences in the period preceding sleep onset. Indeed, during this phase (i.e., sleep onset), EEG activity is mainly characterized by a prevalence of slow frequencies (i.e., <7 Hz) in anterior regions alongside a prevalence of relatively higher frequencies (i.e., >8 Hz) in occipital regions (no anterior-posterior difference has been detected in >13 Hz frequencies). Furthermore, after sleep-onset, authors detected an increase of slow frequencies power in centrofrontal brain regions as well as an increase of sigma activity in centroparietal areas.

Furthermore, sleep process could be defined also as *use-dependent* (Ferrara and De Gennaro, 2011). This aspect emerges particularly from studies performed in normal sleepers during awakening or prolonged awakening. Similarly to wake-sleep transition, during the inverse phenomenon (i.e., awakening), EEG activity seems to be asynchronous and involving different cortical areas. For example, anterior brain regions seem to be the first to become active during awakening. In the same way, in frontal regions, it has been observed an increase of low-EEG frequencies (the ones associated with sleep state) during protracted waking periods. Accordingly, these data suggest that local nature of sleep is strictly linked to a use-dependency due to the fact that frontal regions are the most active during wake state and for this reason they could have a higher need to sleep (Ferrara and De Gennaro, 2011). Indeed, Finelli and coworkers (2000) observed an increase of EEG power spectra in theta activity during a period of prolonged wake (i.e., 40 h of wake). Interestingly, compared to baseline, authors also observed an increase of slow wave activity (i.e., 0.75–4.5 Hz) during NREM sleep in the recovery night after the prolonged waking. By topographically analyzing these results, it has emerged that both results are more pronounced in frontal brain regions. According to the use-dependent perspective, these results support the hypothesis that the general slow activity detected during prolonged wake and sleep could work as a part of the homeostatic sleep process (Finelli et al., 2000). Similarly, in a full-scalp QEEG study, Marzano and coworkers (2010) reported topographic and frequency-specific changes in both REM and NREM sleep states after 40 h of wake. Particularly, compared to the baseline night, they reported an increase of slow wave frequencies in frontocentral areas in both REM and NREM sleep after the period of sleep deprivation (i.e., increased 0.5–7 Hz in REM and an increase of 0.5–11 Hz in NREM sleep). Authors also reported a decrease of 13–15 Hz power in NREM sleep alongside a decrease of 8–11 Hz power during REM in the night after the sleep deprivation compared to the baseline night. Taken together, these data suggest the existence of a general homeostatic response, which works independently on sleep stage.

Lastly, studies performed on wake state reports that EEG fluctuations observed in theta and alpha frequency bands could be linked to fluctuations in alertness state (Makeig and Jung, 1995; Åkerstedt and Gillberg, 1990), leading to consider them as possible electrophysiological correlates of waking intensity. Furthermore, in healthy people, it has been demonstrated that resting wake EEG power varies with increasing time awake (Finelli et al., 2000; Cajochen et al., 1995). Furthermore, in healthy people, drowsiness has been linked to a decrease of occipital alpha activity alongside an increase of theta activity, particularly in central brain areas (Broughton and Hasan, 1995; Hori, 1985).

In conclusion, QEEG power spectra literature about sleep in healthy people is extended and coherently report information helping to understand neurophysiological mechanisms underlying this complex process. Compared to EEG power spectra, literature about EEG functional connectivity correlates of sleep in healthy people remains less explored, but these studies showed intriguing data about EEG coherence variations in sleep across age. For example, a longitudinal EEG study (Kurth et al., 2013) reported an increase of EEG coherence during sleep across the early childhood window of development (i.e., 2–5 years). This increase of coherence has been observed in a frequency and region-specific way across the whole night: while at interhemispheric level connectivity increased in slow frequency bands and sleep spindles, at intrahemispheric level the connectivity decreased. Modifications of sleep EEG coherence have been reported also in the late developmental phase going from 9 to 23 years old. In this regard, Tarokh and coworkers (2010) documented an increase of sleep EEG intrahemispheric coherence across frequency bands and in all sleep stages occurring with increasing age in this window of growth. Taken together, these data suggest that EEG coherence could reflect the processes underlying brain maturation during sleep. Notwithstanding these data, the literature available about EEG coherence in normal sleep remains limited. Therefore, future studies are needed to understand sleep related brain functional connectivity from an electrophysiological perspective.

Quantitative electroencephalography in abnormal sleep

Wake-sleep cycle QEEG features have been investigated also in pathological sleep conditions, especially primary insomnia, obstructive sleep apnea syndrome (OSAS), NREM sleep parasomnias, and neurodegenerative diseases. These studies are useful in understanding the neurophysiological mechanisms underlying the specific symptomatology of abnormal sleep presentations.

Primary insomnia

A large body of QEEG (mainly power spectral studies) literature targeted primary insomnia, a widespread sleep disturbance which the main distinctive feature is the nonrestorative perception of sleep accompanied by significant daytime impairments (Morin and Benca, 2012). Insomnia-related power spectral studies have been performed in several conditions including wake state and the whole sleep cycle.

QEEG power spectra literature about primary insomnia is extensive and several studies have been performed investigating different phases of wake-sleep cycle (i.e., during wake, sleep onset, and alongside REM and NREM sleep).

Literature about NREM sleep in these patients principally reported arousal-related brain electrical activity (e.g., Hall et al., 2007; Spiegelhalder et al., 2012). For example, a cross-sectional study on a sample of 30 patients affected by chronic primary insomnia (Hall et al., 2007) reported that perceived psychological stress was highly correlated with a decrease of EEG delta power as well as with an increase of beta power (i.e., signs of physiological arousal) during NREM sleep. Furthermore, in this study, it has been observed that these QEEG data were significantly associated with both sleep maintenance (measured as wakefulness after sleep onset) and time spent in delta sleep. Similarly, Spiegelhalder et al. (2012) documented that, during NREM sleep, individuals with primary insomnia had an increase of high frequency (i.e., beta and sigma) in EEG spectral power compared with good sleepers. These results suggest the simultaneous presence of neurophysiological patterns involved in wake fostering and sleep maintaining. Indeed, while beta activity is considered an index of arousal, the presence of sigma (i.e., sleep spindle) is a kind of brain activity which has been reported to be associated with sleep maintenance mechanisms (e.g., Dang-Vu et al., 2010). For these reasons, it has been suggested that the co-occurring presence of these two neurophysiological opposite mechanisms could contribute to explain the feeling of nonrestorative sleep which is typical of primary insomnia.

Intriguingly, Neu et al. (2015) investigated slow wave sleep (SWS) EEG power spectral distribution in quite similar clinical conditions such as primary insomnia and chronic fatigue syndrome (CFS), which are known to be often characterized by sleep nonrestoratively (Wilkinson and Shapiro, 2012). Power spectral results documented that, compared to CFS patients, those with primary insomnia showed an enhanced frontal power ratio of higher frequencies within SWS which resulted to be mainly associated with affective symptoms. Furthermore, compared to controls, both groups of patients exhibited a diminished central ultra-slow power (i.e., 0.3–0.79 Hz) ratio throughout SWS, which has been linked to both poor sleep quality and fatigue. According to these data, the authors suggest that the ratio of ultra-slow power during SWS could reflect a neurophysiological mechanism linked to restorative functions of sleep (Neu et al., 2015).

In subjects with primary insomnia, several QEEG studies have also been performed during wakefulness. These reports generally detected neurophysiological signs of hyperarousal (i.e., increased power in high-frequency bands) (Wołyńczyk-Gmaj and Szelenberger, 2011; Corsi-Cabrera et al., 2016; Lamarche and Ogilvie, 1997). For example, Wołyńczyk-Gmaj and Szelenberger (2011) showed a decrease of prefrontal theta power and higher beta power in individuals with insomnia. Furthermore, these power spectra features were found to be highly correlated (i.e., negatively for theta and positively for beta power) with hyperarousal levels suggesting that this pattern could reflect the neurophysiological correlate of insomnia altered arousal. Similarly, Corsi-Cabrera et al. (2016) reported hyperarousal QEEG data during morning wakefulness in patients with primary insomnia. Coherently, EEG power spectra studies in primary insomnia showed similar higher frequency power (i.e., mainly beta band) in sleep onset associated with hyperarousal (Merica and Gaillard, 1992; Freedman, 1986).

As widely reported, clinical presentations of insomnia are different. Indeed, this disturbance could express itself with different patterns of sleep difficulties such as difficulty in starting sleep (i.e., sleep-onset insomnia), problems in maintaining sleep (i.e., central or sleep-maintenance insomnia) and insomnia with early morning awakening (American Psychiatric Association, 2013). According to these different clinical presentations, several differences have been reported in power spectral QEEG pattern during sleep onset. Indeed, among patients with sleep-onset insomnia and those with central insomnia, different power spectral patterns during sleep onset have been reported (Cervena et al., 2014). For example, it has been observed that, compared to patients with sleep maintenance insomnia, those with sleep-onset presentation showed a decrease of beta2 (i.e., 18–29.75 Hz) power during sleep onset phase. This result seems to suggest that hyperarousal is not primary involved in the etiology of this type of insomnia and, conversely, the main problem at the basis of this subtype of insomnia could be the switching from wake to sleep (Cervena et al., 2014).

In insomnia research area, the literature about EEG functional connectivity is not so extended. Among available data, some studies showed patterns of increased EEG coherence during sleep in insomniac patients. Particularly, Aydın (2011) reported that, compared to normal sleep, insomnia is associated with an increased EEG coherence across the whole night. Similarly, in another QEEG study (Corsi-Cabrera et al., 2012), it has been reported that compared to controls, patients with primary insomnia showed an increased synchronous activity among frontoparietal regions during wake and the first stage of sleep in the high-frequency range (i.e., beta and gamma bands). These data are in accordance with the difficulty of patients with insomnia in deactivate cortical regions involved in wake-related higher order functions (e.g., attention and executive functioning) and can help to understand the neurophysiological aspects associated with this sleep disturbance.

Sleep-related breathing disorders

In sleep-related breathing disorders literature, especially about OSAS, several studies have been performed using QEEG.

OSAS is a condition characterized by a repetitive breathing interruption, or reduction in flow, during sleep which is caused by the upper airway total or partial obstruction (Guilleminault and Quo, 2001; Azagra-Calero et al., 2012). During the night, OSAS symptoms principally include fragmentation of sleep, recurrent and severe snoring, but also excessive sleepiness, headaches, and lack of energy during daytime (Guilleminault and Quo, 2001; Greenberg et al., 2017). OSAS is also linked to multisystemic consequences including abnormal negative oscillations of intrathoracic pressure and decrease of blood oxygenation (Greenberg et al., 2017).

Associated with apneas and hypopneas episodes, the observed frequent hypoxemia can negatively impact on brain health and functions leading to neurocognitive long-term effects if untreated (Gosselin et al., 2019). Furthermore, as widely reported, sleep apnea also lead to daytime symptoms such as prominent drowsiness and poor cognitive performance (Findley et al., 1986; Bédard et al., 1991). For these reasons, QEEG studies prove to be extremely important for understanding the neurophysiological mechanisms of this sleep disorder.

From a power spectral point of view, literature shows that compared to good sleepers, patients with obstructive sleep apnea are more characterized by a slow waves EEG (i.e., computed by the ratio of slow-to-fast frequencies) during wake state (Mathieu et al., 2007; Morisson et al., 1998). It has been demonstrated that, this specific pattern of slow waves is strictly associated to both sleepiness episodes and dysfunctional attentional processes, which get worsen with increasing age (Mathieu et al., 2007).

Furthermore, in a recent power spectral density EEG study, it has been reported that, compared to individuals with simple snoring, those with OSA showed higher power in both delta and beta frequency bands during NREM sleep (Kang et al., 2020). A significant association between beta power and apnea-hypopnea index was also detected (Kang et al., 2020).

Intriguingly, Toth and coworkers (2009), in a LORETA QEEG eye closed RS study, reported several power spectral alterations in OSAS patients.

Specifically, PSA showed that compared to controls, patients with OSAS had an increase of activity in alpha2 (10.5—12 Hz) frequency band in several regions such as precuneus, paracentral, and posterior cingulate cortex during RS potentially reflecting the correlate of chronic/intermittent hypoxia on cortical regions. These regions are known to be highly involved in emotion regulation, memory (i.e., autobiographical) and internally oriented mental processes (Maddock et al., 2003; Andrews-Hanna et al., 2014). Furthermore, these cortical regions are known to be core hubs of the default mode network (DMN) a brain network highly involved in RS processes including mentalization, self-referential thoughts, and autobiographical retrieval (Andrews-Hanna et al., 2014). According to these results, it has been reported that changes in alpha2 frequency band are strictly associated with these core posterior regions of the DMN (Chen, 2007).

Recently, Fortin and coworkers (2020) investigated QEEG waking power spectra and functional connectivity in middle-aged and old adult patients with OSA. Particularly, they recorded a full-night polysomnography and the following waking EEG recording. Furthermore, all subjects underwent a neuropsychological assessment. While no results have been detected in PSA, functional connectivity (computed as *imaginary coherence*) results showed that compared to control group, patients with OSA exhibit a decrease of connectivity between frontal and parietal regions in both high and slow frequency bands (i.e., delta and beta1). Particularly, while these results were not associated with neuropsychological performance, in patients with OSA these connectivity changes were associated with higher apnea-hypopnea index and poor sleep parameters (i.e., lower total sleep duration and poor sleep efficiency).

Sleep parasomnias

Sleep parasomnias are considered a category of disorders whose main features are undesirable events which can include abnormal behaviors, movements and/or vocalizations, misperceptions occurring during sleep, which can have impairing consequences on sleep, health, and psychosocial domain (Kazaglis and Bornemann, 2016). These events can occur in different sleep stages such as wake-sleep transition, REM, and NREM sleep by showing different neurophysiological mechanisms (Singh et al., 2018). Particularly, sleep parasomnias can occur primarily (i.e., the sleep disorder is the main cause of parasomnias) or secondarily as a manifestation of disturbances of other nature (Mahowald, 2010).

Accordingly, REM sleep behavior disorder (RBD) is a sleep disturbance characterized by parasomnias accompanied by a loss of normal muscle atonia during REM sleep alongside an increase of phasic muscular activity leading to complex nocturnal motor behaviors linked to dream-enactment which, in turn, can result in physical injury of the patients and their partners (Olson et al., 2000; Schenck and Mahowald, 2002). The important role of EEG studies is highlighted by the involvement of brainstem's structures, considered responsible for RBD, in cortical activation (Fantini et al., 2003).

For example, in a QEEG study (Fantini et al., 2003), it has been reported that, compared to controls, individuals with idiopathic RBD exhibit lower beta power in occipital cortical regions during REM sleep. Similarly, during wakefulness, these patients showed a decrease of beta power in the occipital cortical region alongside an increase in theta power in widespread brain regions such as frontal, temporal, and occipital ones. These results suggest that the slowing of EEG rhythm could be linked to sign of mild central nervous system and cognitive dysfunction (Fantini et al., 2003).

Furthermore, several QEEG studies have also been performed about NREM sleep parasomnias (e.g., Castelnovo et al., 2016; Desjardins et al., 2017) such as sleepwalking.

In this regard, EEG studies coherently reported that sleepwalking episodes are preceded by a partial arousal from NREM sleep. According to this, in an eLORETA study (Januszko et al., 2016), the association between sleepwalking motor episodes and local arousal-related activity has been detected. Particularly, the authors observed an increase of beta3 (24.0–30.0 Hz) current density in cingulate motor area (Januszko et al., 2016). Similarly, in a QEEG study, Desjardins and coworkers (2017) investigated the EEG pattern of a sample of adult sleepwalkers by comparing the EEG activity immediately prior to the episode with that occurring in the 2 min preceding the episode for each subject. From a power spectral point of view, results showed that, compared to EEG activity 2 min prior to the episode, that one immediately before the sleepwalking episode was characterized by an increase of both delta and theta power. From a functional connectivity point of view, the same authors reported a decreased connectivity in delta frequency band in posterior brain regions (i.e., parietal and occipital regions) alongside an increase of connectivity in alpha and beta frequency bands in a distributed network involving anteroposterior brain regions (Desjardins et al., 2017). Considering these results together, this EEG functional connectivity pattern suggests that there are complex neurophysiological processes involving parallel signs of arousal and deep sleep preceding sleepwalking.

Furthermore, in a high-density EEG study (Castelnovo et al., 2016), several dissimilarities have been reported in local sleep among patients with sleep arousal disorders (i.e., night terrors and sleepwalking) and controls. Particularly, it has been reported that, compared to controls, subjects affected by sleep arousal disorders showed a reduction of slow wave activity across the sleep/wake cycle in several cortical regions such as limbic, associative and both motor and sensory-motor associative ones.

Neuropsychiatric and neurodegenerative diseases

Some QEEG studies about wake and sleep states have been performed also in neuropsychiatric (e.g., depression and schizophrenia) and neurodegenerative diseases due to sleep disturbances that can occur in these conditions.

For example, among neurodegenerative disorders, some EEG studies about wake and sleep states have been performed in patients affected by Huntington's disease (HD) (e.g., Piano et al., 2017a, 2017b) due to sleep related motor symptoms often observed in these patients (Neutel et al., 2015). Indeed, HD is a genetic neurodegenerative disorder where an anomalous expansion of repeated sequences of CAG occurs in the gene involved in huntingtin protein encoding on chromosome 4 (Walker, 2007). This neurodegeneration leads to serious primary clinical alterations such as dysfunctional movements, cognitive degeneration, and psychiatric symptoms where sleep and circadian problems are also common (Morton, 2013).

HD-related sleep alterations have also been investigated by considering EEG power spectra parameters. For example, Piano et al. (2017b) investigated EEG power spectra modifications during circadian phases such as wake, NREM, and REM sleep. They reported that, compared to controls, individuals affected by HD exhibit power spectra abnormalities differences in both motor and premotor brain areas across all targeted stages. Alterations in sensory-motor network in HD during wake, NREM and REM sleep have also been documented through the use of EEG connectivity (Piano et al., 2017a).

Sleep abnormalities have also been observed in association with other neurological disorders such as Parkinson's disease (PD). For example, a QEEG study performed on PD (without symptoms of dementia) showed that, compared to controls and to PD patients without RBD, PD patients with comorbid RBD exhibited an EEG slowing (i.e., higher theta power) during wake in distributed brain regions such as frontal, temporal, parietal, and occipital regions (Gagnon et al., 2004). Therefore, these data indicate the specific contribution of RBD to the slowing brain electrical activity in PD patients without dementia.

Furthermore, several EEG coherence alterations have been reported in psychiatric disorders such as depression, schizophrenia, and autism spectrum (e.g., Wamsley et al., 2012; Léveillé et al., 2010; Lázár et al., 2010; Wichniak et al., 2013). For example, in patients with schizophrenia, it has been reported a reduction of sleep spindles and sleep spindles coherence during the sleep stage 2 which have been suggested to be linked to alterations in the thalamocortical network (Wamsley et al., 2012). Particularly, in these patients, the reduction of density and number of sleep spindles have been found in association to a reduction of memory consolidation through the night (investigated by means of a motor procedural paradigm of finger tapping motor sequence task). Taken together, these data could contribute to understand cognitive symptoms in schizophrenia (Wamsley et al., 2012). Furthermore, in a preliminary report, it has been observed a decreased coherence in high-frequency bands (i.e., beta and gamma) during the wake in patients with schizophrenia (particularly, first-episode patients) highlighting the abnormalities previously observed in neural synchronization in these patients (Yeragani et al., 2006). Intriguingly, in a further study (Chaparro-Vargas et al., 2016) it has been targeted the quantification of interdependencies among EEG-related and heart rate variability signals across individuals with insomnia, schizophrenia and controls, respectively. Accordingly, results showed interactions among neural oscillations and cardiac signal which were specific for sleep stage and pathological condition, supporting the hypothesis that there are widespread physiological dynamics involved in these conditions.

Furthermore, in autism research area, it has been reported (Lázár et al., 2010) that, compared to controls, individuals with autism spectrum disorder (i.e., patients previously defined as having Asperger syndrome) showed an intrahemispheric coherence reduction in frontocentral regions in several frequency bands (i.e., sigma, alpha, theta, and delta). Similarly, during REM sleep, individuals with autism showed lower coherence in frontal sites of the right hemisphere alongside an increased coherence among left visual cortex and widespread brain regions both close and spatially far from occipital cortex (Léveillé et al., 2010).

In conclusion, while EEG functional connectivity literature is not so extended in sleep disorders, these data can help in understanding neurophysiological mechanisms underlying these disorders and their related symptomatology and can help in addressing further studies.

Conclusions and future directions

QEEG is a useful instrument to understand neurophysiological mechanisms underlying both normal and abnormal sleep. Compared to EEG power spectra studies, up to now EEG connectivity literature is not so extended. For this reason, future studies are needed to further investigate EEG functional connectivity among distributed brain networks to highlight neurophysiological mechanisms underlying sleep and pathological sleep conditions. Indeed, while several functional connectivity studies have been performed in these conditions using functional Magnetic Resonance Imaging (fMRI) (e.g., Sämann et al., 2011; Koike et al., 2011; Li et al., 2016) few EEG connectivity reports are available. Particularly, it could be interesting to investigate how distributed brain networks such as the default mode network work across sleep stages and frequency bands both in normal and pathological sleep conditions by means of QEEG. Indeed, compared to fMRI, which offers spatially relevant data in terms of functional interactions among brain structures, the EEG provides significant information about how neural systems interact in different frequency bands with each other (Neuner et al., 2014; Thatcher et al., 2014; Whitton et al., 2018), revealing valuable evidence under a clinical and neurophysiological point of view. Moreover, it has been documented (Liu et al., 2018) the robustness and the suitability of EEG in detecting large-scale brain networks.

Lastly, investigating QEEG in both normal and abnormal sleep may also have important clinical implications. Indeed, starting from QEEG studies, some techniques focused on self-neuromodulation, such as neurofeedback (NF), might be used to improve sleep quality and reduce abnormal sleep symptoms in sleep disorders. NF is an endogenous operant-conditioning neuromodulation technique that allows individuals to learn how to regulate neurophysiological activity in response to real-time feedback (Weiskopf, 2012). NF is currently used as additional treatment modality for different psychiatric disorders (Schoenberg and David, 2014; Imperatori et al., 2018; Trudeau, 2005; Tolin et al., 2020), and the clinical usefulness of NF techniques for sleep disorders, especially in insomnia (Melo et al., 2019), is recently increased.

References

Åkerstedt, T., Gillberg, M., 1990. Subjective and objective sleepiness in the active individual. Int. J. Neurosci. 52, 29–37.

American Psychiatric Association, 2013. Diagnostic and Statistical Manual of Mental Disorders, 5 edition. American Psychiatric Publishing, Washington/London.

Andrews-Hanna, J.R., Smallwood, J., Spreng, R.N., 2014. The default network and self-generated thought: component processes, dynamic control, and clinical relevance. Ann. N Y Acad. Sci. 1316, 29–52.

AYDın, S., 2011. Computer based synchronization analysis on sleep EEG in insomnia. J. Med. Syst. 35, 517–520.

Azagra-Calero, E., Espinar-Escalona, E., Barrera-Mora, J.M., Llamas-Carreras, J.M., Solano-Reina, E., 2012. Obstructive sleep apnea syndrome (OSAS). Review of the literature. Med. Oral Patol. Oral Cir. Bucal 17, e925–e929.

Bédard, M.-A., Montplaisir, J., Richer, F., Rouleau, I., Malo, J., 1991. Obstructive sleep apnea syndrome: pathogenesis of neuropsychological deficits. J. Clin. Exp. Neuropsychol. 13, 950–964.

Berger, H., 1929. Über das elektrenkephalogramm des menschen. Archiv für Psychiatrie und Nervenkrankheiten 87, 527–570.

Broughton, R., Hasan, J., 1995. Quantitative topographic electroencephalographic mapping during drowsiness and sleep onset. J. Clin. Neurophysiol. 12, 372–386.

Cajochen, C., Brunner, D.P., Krauchi, K., Graw, P., Wirz-Justice, A., 1995. Power density in theta/alpha frequencies of the waking EEG progressively increases during sustained wakefulness. Sleep 18, 890–894.

Canuet, L., Tellado, I., Couceiro, V., Fraile, C., Fernandez-Novoa, L., Ishii, R., Takeda, M., Cacabelos, R., 2012. Resting-state network disruption and APOE genotype in Alzheimer's disease: a lagged functional connectivity study. PLoS One 7, e46289.

Castelnovo, A., Riedner, B.A., Smith, R.F., Tononi, G., Boly, M., Benca, R.M., 2016. Scalp and source power topography in sleepwalking and sleep terrors: a high-density EEG study. Sleep 39, 1815–1825.

Cervena, K., Espa, F., Perogamvros, L., Perrig, S., Merica, H., Ibanez, V., 2014. Spectral analysis of the sleep onset period in primary insomnia. Clin. Neurophysiol. 125, 979–987.

Chaparro-Vargas, R., Schilling, C., Schredl, M., Cvetkovic, D., 2016. Sleep electroencephalography and heart rate variability interdependence amongst healthy subjects and insomnia/schizophrenia patients. Med. Biol. Eng. Comput. 54, 77–91.

Chen, A.C. EEG default mode network in the human brain: spectral field power, coherence topology, and current source imaging. In: 2007 Joint Meeting of the 6th International Symposium on Noninvasive Functional Source Imaging of the Brain and Heart and the International Conference on Functional Biomedical Imaging, 2007. IEEE, 215–218.

Cohen, M.X., 2017. Where does EEG come from and what does it mean? Trends Neurosci. 40, 208–218.

Corsi-Cabrera, M., Figueredo-Rodríguez, P., Del Río-Portilla, Y., Sánchez-Romero, J., Galán, L., Bosch-Bayard, J., 2012. Enhanced frontoparietal synchronized activation during the wake-sleep transition in patients with primary insomnia. Sleep 35, 501–511.

Corsi-Cabrera, M., Rojas-Ramos, O.A., Del Río-Portilla, Y., 2016. Waking EEG signs of non-restoring sleep in primary insomnia patients. Clin. Neurophysiol. 127, 1813–1821.

Dang-Vu, T.T., Mckinney, S.M., Buxton, O.M., Solet, J.M., Ellenbogen, J.M., 2010. Spontaneous brain rhythms predict sleep stability in the face of noise. Curr. Biol. 20, R626–R627.

De Gennaro, L., Ferrara, M., Curcio, G., Cristiani, R., 2001. Antero-posterior EEG changes during the wakefulness–sleep transition. Clin. Neurophysiol. 112, 1901–1911.

De Ridder, D., Vanneste, S., Kovacs, S., Sunaert, S., Dom, G., 2011. Transient alcohol craving suppression by rTMS of dorsal anterior cingulate: an fMRI and LORETA EEG study. Neurosci. Lett. 496, 5–10.

Desjardins, M.-È., Carrier, J., Lina, J.-M., Fortin, M., Gosselin, N., Montplaisir, J., Zadra, A., 2017. EEG functional connectivity prior to sleepwalking: evidence of interplay between sleep and wakefulness. Sleep 40, 1–8.

Dressler, O., Schneider, G., Stockmanns, G., Kochs, E., 2004. Awareness and the EEG power spectrum: analysis of frequencies. Br. J. Anaesth. 93, 806–809.

Duffy, F.H., Iyer, V.G., Surwillo, W.W., 2012. Clinical Electroencephalography and Topographic Brain Mapping: Technology and Practice. Springer-Verlag, New York.

Eickhoff, S.B., Müller, V.I., 2015. Functional connectivity. In: Toga, A.W. (Ed.), Brain Mapping. Academic Press, Waltham, pp. 187–201.

Faes, L., Erla, S., Nollo, G., 2012. Measuring connectivity in linear multivariate processes: definitions, interpretation, and practical analysis. Comput. Math. Methods Med. 2012.

Fantini, M.L., Gagnon, J.F., Petit, D., Rompré, S., Décary, A., Carrier, J., Montplaisir, J., 2003. Slowing of electroencephalogram in rapid eye movement sleep behavior disorder. Ann. Neurol. 53, 774–780.

Ferrara, M., DE Gennaro, L., 2011. Going local: insights from EEG and stereo-EEG studies of the human sleep-wake cycle. Curr. Top. Med. Chem. 11, 2423–2437.

Findley, L.J., Barth, J.T., Powers, D.C., Wilhoit, S.C., Boyd, D.G., Suratt, P.M., 1986. Cognitive impairment in patients with obstructive sleep apnea and associated hypoxemia. Chest 90, 686–690.

Finelli, L.A., Baumann, H., Borbély, A.A., Achermann, P., 2000. Dual electroencephalogram markers of human sleep homeostasis: correlation between theta activity in waking and slow-wave activity in sleep. Neuroscience 101, 523–529.

Finelli, L.A., Borbély, A.A., Achermann, P., 2001. Functional topography of the human nonREM sleep electroencephalogram. Eur. J. Neurosci. 13, 2282–2290.

Fortin, M., Lina, J.-M., Desjardins, M.-È., Gagnon, K., Baril, A.-A., Carrier, J., Gosselin, N., 2020. Waking EEG functional connectivity in middle-aged and older adults with obstructive sleep apnea. Sleep Med. 75, 88–95.

Freedman, R., 1986. EEG power spectra in sleep-onset insomnia. Electroencephalogr. Clin. Neurophysiol. 63, 408–413.

Friston, K.J., 1994. Functional and effective connectivity in neuroimaging: a synthesis. Hum. Brain Mapp. 2, 56–78.

Friston, K.J., 2007. Functional integration. In: Friston, K.J., Ashburner, J.T., Kiebel, S.J., Nichols, T.E., Penny, W.D. (Eds.), Statistical Parametric Mapping: The Analysis of Functional Brain Image. Academic Press, Burlington, Massachusetts, pp. 471–491.

Fuchs, M., Kastner, J., Wagner, M., Hawes, S., Ebersole, J.S., 2002. A standardized boundary element method volume conductor model. Clin. Neurophysiol. 113, 702–712.

Gagnon, J.-F., Fantini, M., Bédard, M.-A., Petit, D., Carrier, J., Rompré, S., Décary, A., Panisset, M., Montplaisir, J., 2004. Association between waking EEG slowing and REM sleep behavior disorder in PD without dementia. Neurology 62, 401–406.

Gosselin, N., Baril, A.-A., Osorio, R.S., Kaminska, M., Carrier, J., 2019. Obstructive sleep apnea and the risk of cognitive decline in older adults. Am. J. Respir. Crit. Care Med. 199, 142–148.

Grass, A.M., Gibbs, F.A., 1938. A Fourier transform of the electroencephalogram. J. Neurophysiol. 1, 521–526.

Grech, R., Cassar, T., Muscat, J., Camilleri, K.P., Fabri, S.G., Zervakis, M., Vanrumste, B., 2008. Review on solving the inverse problem in EEG source analysis. J. Neuroeng. Rehabil. 5 (1), 1–33.

Greenberg, H., Lakticova, V., Scharf, S.M., 2017. Obstructive sleep apnea: clinical features, evaluation, and principles of management. In: Kryger, M.H., Roth, T., Dement, W.C. (Eds.), Principles and Practice of Sleep Medicine, sixth ed. Elsevier, Philadelphia, pp. 1110–1124.

Guilleminault, C., Quo, S.D., 2001. Sleep-disordered breathing. A view at the beginning of the new Millennium. Dent. Clin. 45, 643–656.

Hahn, A., Lanzenberger, R., Kasper, S., 2019. Making sense of connectivity. Int. J. Neuropsychopharmacol. 22, 194–207.

Hall, M., Thayer, J.F., Germain, A., Moul, D., Vasko, R., Puhl, M., Miewald, J., Buysse, D.J., 2007. Psychological stress is associated with heightened physiological arousal during NREM sleep in primary insomnia. Behav. Sleep Med. 5, 178–193.

Hata, M., Kazui, H., Tanaka, T., Ishii, R., Canuet, L., Pascual-Marqui, R.D., Aoki, Y., Ikeda, S., Kanemoto, H., Yoshiyama, K., 2016. Functional connectivity assessed by resting state EEG correlates with cognitive decline of Alzheimer's disease—An eLORETA study. Clin. Neurophysiol. 127, 1269–1278.

Horacek, J., Brunovsky, M., Novak, T., Skrdlantova, L., Klirova, M., Bubenikova-Valesova, V., Krajca, V., Tislerova, B., Kopecek, M., Spaniel, F., 2007. Effect of low-frequency rTMS on electromagnetic tomography (LORETA) and regional brain metabolism (PET) in schizophrenia patients with auditory hallucinations. Neuropsychobiology 55, 132–142.

Hori, T., 1985. Spatiotemporal changes of EEG activity during waking-sleeping transition period. Int. J. Neurosci. 27, 101–114.

Imperatori, C., Mancini, M., Della Marca, G., Valenti, E.M., Farina, B., 2018. Feedback-based treatments for eating disorders and related symptoms: a systematic review of the literature. Nutrients 10.

Imperatori, C., Massullo, C., Carbone, G.A., Panno, A., Giacchini, M., Capriotti, C., Lucarini, E., Ramella Zampa, B., Murillo-Rodríguez, E., Machado, S., 2020. Increased resting state triple network functional connectivity in undergraduate problematic cannabis users: a preliminary EEG coherence study. Brain Sci. 10, 136.

Januszko, P., Niemcewicz, S., Gajda, T., Wołyńczyk-Gmaj, D., Piotrowska, A.J., Gmaj, B., Piotrowski, T., Szelenberger, W., 2016. Sleepwalking episodes are preceded by arousal-related activation in the cingulate motor area: EEG current density imaging. Clin. Neurophysiol. 127, 530–536.

Jatoi, M.A., Kamel, N., Malik, A.S., Faye, I., 2014. EEG based brain source localization comparison of sLORETA and eLORETA. Australas. Phys. Eng. Sci. Med. 37, 713–721.

Kang, J.M., Kim, S.T., Mariani, S., Cho, S.-E., Winkelman, J.W., Park, K.H., Kang, S.-G., 2020. Difference in spectral power density of sleep EEG between patients with simple snoring and those with obstructive sleep apnoea. Sci. Rep. 10, 1–8.

Kazaglis, L., Bornemann, M.A.C., 2016. Classification of parasomnias. Curr. Sleep Med. Rep. 2, 45–52.

Kirino, E., 2017. Three-dimensional stereotactic surface projection in the statistical analysis of single photon emission computed tomography data for distinguishing between Alzheimer's disease and depression. World J. Psychiatr. 7, 121–127.

Koike, T., Kan, S., Misaki, M., Miyauchi, S., 2011. Connectivity pattern changes in default-mode network with deep non-REM and REM sleep. Neurosci. Res. 69 (4), 322–330.

Kreiter, A., Singer, W., 1992. Oscillatory neuronal responses in the visual cortex of the awake macaque monkey. Eur. J. Neurosci. 4, 369–375.

Kurth, S., Achermann, P., Rusterholz, T., Lebourgeois, M.K., 2013. Development of brain EEG connectivity across early childhood: does sleep play a role? Brain Sci. 3, 1445–1460.

Lachaux, J.P., Rodriguez, E., Martinerie, J., Varela, F.J., 1999. Measuring phase synchrony in brain signals. Hum. Brain Mapp. 8, 194–208.

Lamarche, C.H., Ogilvie, R.D., 1997. Electrophysiological changes during the sleep onset period of psychophysiological insomniacs. Psychiatric insomniacs, and normal sleepers. Sleep 20, 726–733.

Lancaster, J.L., Woldorff, M.G., Parsons, L.M., Liotti, M., Freitas, C.S., Rainey, L., Kochunov, P.V., Nickerson, D., Mikiten, S.A., Fox, P.T., 2000. Automated Talairach atlas labels for functional brain mapping. Hum. Brain Mapp. 10, 120–131.

Lázár, A.S., Lázár, Z.I., Bíró, A., Győri, M., Tárnok, Z., Prekop, C., Keszei, A., Stefanik, K., Gádoros, J., Halász, P., 2010. Reduced fronto-cortical brain connectivity during NREM sleep in Asperger syndrome: an EEG spectral and phase coherence study. Clin. Neurophysiol. 121, 1844–1854.

Léveillé, C., Barbeau, E.B., Bolduc, C., Limoges, É., Berthiaume, C., Chevrier, É., Mottron, L., Godbout, R., 2010. Enhanced connectivity between visual cortex and other regions of the brain in autism: a REM sleep EEG coherence study. Autism Res. 3, 280–285.

Li, H.J., Nie, X., Gong, H.H., Zhang, W., Nie, S., Peng, D.C., 2016. Abnormal resting-state functional connectivity within the default mode network subregions in male patients with obstructive sleep apnea. Neuropsychiatr. Dis. Treat. 12, 203.

Liu, Q., Ganzetti, M., Wenderoth, N., Mantini, D., 2018. Detecting large-scale brain networks using EEG: impact of electrode density, head modeling and source localization. Front. Neuroinf. 12 (4).

Maddock, R.J., Garrett, A.S., Buonocore, M.H., 2003. Posterior cingulate cortex activation by emotional words: fMRI evidence from a valence decision task. Hum. Brain Mapp. 18, 30–41.

Mahowald, M.W., 2010. Other parasomnias. In: Kryger, M.H., Roth, T., Dement, W.C. (Eds.), Principles and Practice of Sleep Medicine, fifth ed. Elsevier Saunders, St. Louis, Missouri, pp. 1098–1105.

Mahowald, M.W., Schenck, C.H., 1991. Status dissociatus—a perspective on states of being. Sleep 14, 69–79.

Makeig, S., Jung, T.-P., 1995. Changes in alertness are a principal component of variance in the EEG spectrum. Neuroreport 7, 213–216.

Marzano, C., Ferrara, M., Curcio, G., De Gennaro, L., 2010. The effects of sleep deprivation in humans: topographical electroencephalogram changes in non-rapid eye movement (NREM) sleep versus REM sleep. J. Sleep Res. 19, 260–268.

Marzano, C., Ferrara, M., Sforza, E., De Gennaro, L., 2008. Quantitative electroencephalogram (EEG) in insomnia: a new window on pathophysiological mechanisms. Curr. Pharmaceut. Des. 14, 3446–3455.

Marzano, C., Moroni, F., Gorgoni, M., Nobili, L., Ferrara, M., De Gennaro, L., 2013. How we fall asleep: regional and temporal differences in electroencephalographic synchronization at sleep onset. Sleep Med. 14, 1112–1122.

Mathieu, A., Mazza, S., Petit, D., Décary, A., Massicotte-Marquez, J., Malo, J., Montplaisir, J., 2007. Does age worsen EEG slowing and attention deficits in obstructive sleep apnea syndrome? Clin. Neurophysiol. 118, 1538–1544.

Mazziotta, J., Toga, A., Evans, A., Fox, P., Lancaster, J., Zilles, K., Woods, R., Paus, T., Simpson, G., Pike, B., Holmes, C., Collins, L., Thompson, P., Macdonald, D., Iacoboni, M., Schormann, T., Amunts, K., Palomero-Gallagher, N., Geyer, S., Parsons, L., Narr, K., Kabani, N., Le Goualher, G., Boomsma, D., Cannon, T., Kawashima, R., Mazoyer, B., 2001. A probabilistic atlas and reference system for the human brain: international Consortium for Brain Mapping (ICBM). Philos. Trans. R. Soc. Lond. Ser. B Biol. Sci. 356, 1293–1322.

Melo, D.L.M., Carvalho, L.B.C., Prado, L.B.F., Prado, G.F., 2019. Biofeedback therapies for chronic insomnia: a systematic review. Appl. Psychophysiol. Biofeedback 44, 259–269.

Merica, H., Gaillard, J.-M., 1992. The EEG of the sleep onset period in insomnia: a discriminant analysis. Physiol. Behav. 52, 199–204.

Morin, C.M., Benca, R., 2012. Chronic insomnia. Lancet 379, 1129–1141.

Morisson, F., Lavigne, G., Petit, D., Nielsen, T., Malo, J., Montplaisir, J., 1998. Spectral analysis of wakefulness and REM sleep EEG in patients with sleep apnoea syndrome. Eur. Respir. J. 11, 1135–1140.

Morton, A.J., 2013. Circadian and sleep disorder in Huntington's disease. Exp. Neurol. 243, 34–44.

Murphy, T.H., Blatter, L.A., Wier, W.G., Baraban, J.M., 1992. Spontaneous synchronous synaptic calcium transients in cultured cortical neurons. J. Neurosci. 12, 4834–4845.

Neu, D., Mairesse, O., Verbanck, P., Le Bon, O., 2015. Slow wave sleep in the chronically fatigued: power spectra distribution patterns in chronic fatigue syndrome and primary insomnia. Clin. Neurophysiol. 126, 1926–1933.

Neuner, I., Arrubla, J., Werner, C.J., Hitz, K., Boers, F., Kawohl, W., Shah, N.J., 2014. The default mode network and EEG regional spectral power: a simultaneous fMRI-EEG study. PLoS One 9, e88214.

Neutel, D., Tchikviladzé, M., Charles, P., Leu-Semenescu, S., Roze, E., Durr, A., Arnulf, I., 2015. Nocturnal agitation in Huntington disease is caused by arousal-related abnormal movements rather than by rapid eye movement sleep behavior disorder. Sleep Med. 16, 754–759.

Novo-Olivas, C.A., 2014. Diagnosing and treating closed head injury: exposing and defeating the mild huge monster. In: Cantor, D.S., Evans, J.R. (Eds.), Clinical Neurotherapy: Application of Techniques for Treatment. Academic Press, London, pp. 191–211.

Nunez, P.L., Srinivasan, R., Westdorp, A.F., Wijesinghe, R.S., Tucker, D.M., Silberstein, R.B., Cadusch, P.J., 1997. EEG coherency: I: statistics, reference electrode, volume conduction, Laplacians, cortical imaging, and interpretation at multiple scales. Electroencephalogr. Clin. Neurophysiol. 103, 499–515.

Olbrich, S., Tränkner, A., Chittka, T., Hegerl, U., Schönknecht, P., 2014. Functional connectivity in major depression: increased phase synchronization between frontal cortical EEG-source estimates. Psychiatr. Res. Neuroimaging 222, 91–99.

Olson, E.J., Boeve, B.F., Silber, M.H., 2000. Rapid eye movement sleep behaviour disorder: demographic, clinical and laboratory findings in 93 cases. Brain 123, 331–339.

Palva, J.M., Wang, S.H., Palva, S., Zhigalov, A., Monto, S., Brookes, M.J., Schoffelen, J.-M., Jerbi, K., 2018. Ghost interactions in MEG/EEG source space: a note of caution on inter-areal coupling measures. Neuroimage 173, 632–643.

Pascual-Marqui, R., Lehmann, D., Koukkou, M., Kochi, K., Anderer, P., Saletu, B., Tanaka, H., Hirata, K., John, E.R., Prichep, L., 2011. Assessing interactions in the brain with exact low-resolution electromagnetic tomography. Phil. Trans. Math. Phys. Eng. Sci. 369, 3768–3784.

Pascual-Marqui, R., Michel, C.M., Lehmann, D., 1994. Low-resolution electromagnetic tomography—a new method for localizing electrical activity in the brain. Int. J. Psychophysiol. 18, 49—65.
Phillips, C., Rugg, M.D., Friston, K.J., 2002. Anatomically informed basis functions for EEG source localization: combining functional and anatomical constraints. Neuroimage 16, 678—695.
Piano, C., Imperatori, C., Losurdo, A., Bentivoglio, A.R., Cortelli, P., Della Marca, G., 2017a. Sleep-related modifications of EEG connectivity in the sensory-motor networks in Huntington Disease: an eLORETA study and review of the literature. Clin. Neurophysiol. 128, 1354—1363.
Piano, C., Mazzucchi, E., Bentivoglio, A.R., Losurdo, A., Calandra Buonaura, G., Imperatori, C., Cortelli, P., Della Marca, G., 2017b. Wake and sleep EEG in patients with Huntington disease: an eLORETA study and review of the literature. Clin. EEG Neurosci. 48, 60—71.
Pizzagalli, D.A., 2007. Electroencephalography and high-density electrophysiological source localization. In: Cacioppo, J.T., Tassinary, L.G., Berntson, G.G. (Eds.), Handbook of Psychophysiology, third ed. Cambridge University Press, New York, pp. 56—84.
Porjesz, B., Begleiter, H., 2003. Alcoholism and human electrophysiology. Alcohol Res. Health 27, 153.
Ramyead, A., Kometer, M., Studerus, E., Koranyi, S., Ittig, S., Gschwandtner, U., Fuhr, P., Riecher-Rössler, A., 2015. Aberrant current source-density and lagged phase synchronization of neural oscillations as markers for emerging psychosis. Schizophr. Bull. 41, 919—929.
Reznik, S.J., Allen, J.J., 2018. Frontal asymmetry as a mediator and moderator of emotion: an updated review. Psychophysiology 55, e12965.
Sämann, P.G., Wehrle, R., Hoehn, D., Spoormaker, V.I., Peters, H., Tully, C., Czisch, M., 2011. Development of the brain's default mode network from wakefulness to slow wave sleep. Cereb. Cortex 21 (9), 2082—2093.
Schenck, C.H., Mahowald, M.W., 2002. REM sleep behavior disorder: clinical, developmental, and neuroscience perspectives 16 years after its formal identification in SLEEP. Sleep 25, 120—138.
Scherg, M., Ebersole, J., 1994. Brain source imaging of focal and multifocal epileptiform EEG activity. Neurophysiol. Clin. 24, 51—60.
Schoenberg, P., David, A., 2014. Biofeedback for psychiatric disorders: a systematic review. Appl. Psychophysiol. Biofeedback 39, 109—135.
Schoenberg, P.L., 2020. Linear and nonlinear EEG-based functional networks in anxiety disorders. In: Kim, Y.-K. (Ed.), Anxiety Disorders. Advances in Experimental Medicine and Biology. Springer Nature Singapore, pp. 35—59.
Singh, S., Kaur, H., Singh, S., Khawaja, I., 2018. Parasomnias: a comprehensive review. Cureus 10, e3807.
Spiegelhalder, K., Regen, W., Feige, B., Holz, J., Piosczyk, H., Baglioni, C., Riemann, D., Nissen, C., 2012. Increased EEG sigma and beta power during NREM sleep in primary insomnia. Biol. Psychol. 91, 329—333.
Sporns, O., 2010. Networks of the Brain. MIT press, Cambridge/London.
Srinivasan, R., Winter, W.R., Ding, J., Nunez, P.L., 2007. EEG and MEG coherence: measures of functional connectivity at distinct spatial scales of neocortical dynamics. J. Neurosci. Methods 166, 41—52.
Stam, C.J., Nolte, G., Daffertshofer, A., 2007. Phase lag index: assessment of functional connectivity from multi channel EEG and MEG with diminished bias from common sources. Hum. Brain Mapp. 28 (11), 1178—1193.
Takahashi, H., Rissling, A.J., Pascual-Marqui, R., Kirihara, K., Pela, M., Sprock, J., Braff, D.L., Light, G.A., 2013. Neural substrates of normal and impaired preattentive sensory discrimination in large cohorts of nonpsychiatric subjects and schizophrenia patients as indexed by MMN and P3a change detection responses. Neuroimage 66, 594—603.

Tarokh, L., Carskadon, M.A., Achermann, P., 2010. Developmental changes in brain connectivity assessed using the sleep EEG. Neuroscience 171, 622–634.

Thakor, N.V., Tong, S., 2004. Advances in quantitative electroencephalogram analysis methods. Annu. Rev. Biomed. Eng. 6, 453–495.

Thatcher, R.W., North, D.M., Biver, C.J., 2014. LORETA EEG phase reset of the default mode network. Front. Hum. Neurosci. 8, 529.

Togo, F., Cherniack, N.S., Natelson, B.H., 2006. Electroencephalogram characteristics of autonomic arousals during sleep in healthy men. Clin. Neurophysiol. 117, 2597–2603.

Tolin, D.F., Davies, C.D., Moskow, D.M., Hofmann, S.G., 2020. Biofeedback and neurofeedback for anxiety disorders: a quantitative and qualitative systematic review. Adv. Exp. Med. Biol. 1191, 265–289.

Toth, M., Faludi, B., Wackermann, J., Czopf, J., Kondakor, I., 2009. Characteristic changes in brain electrical activity due to chronic hypoxia in patients with obstructive sleep apnea syndrome (OSAS): a combined EEG study using LORETA and omega complexity. Brain Topogr. 22, 185–190.

Trudeau, D.L., 2005. EEG biofeedback for addictive disorders—the state of the art in 2004. J. Adult Dev. 12, 139–146.

Walker, F.O., 2007. Huntington's disease. Semin. Neurol. 27 (2), 143–150.

Wamsley, E.J., Tucker, M.A., Shinn, A.K., Ono, K.E., Mckinley, S.K., Ely, A.V., Goff, D.C., Stickgold, R., Manoach, D.S., 2012. Reduced sleep spindles and spindle coherence in schizophrenia: mechanisms of impaired memory consolidation? Biol. Psychiatr. 71, 154–161.

Weiskopf, N., 2012. Real-time fMRI and its application to neurofeedback. Neuroimage 62, 682–692.

Weiss, S., Mueller, H.M., 2003. The contribution of EEG coherence to the investigation of language. Brain Lang. 85, 325–343.

Whitton, A.E., Deccy, S., Ironside, M.L., Kumar, P., Beltzer, M., Pizzagalli, D.A., 2018. Electroencephalography source functional connectivity reveals abnormal high-frequency communication among large-scale functional networks in depression. Biol. Psychiatr. 3, 50–58.

Wichniak, A., Wierzbicka, A., Jernajczyk, W., 2013. Sleep as a biomarker for depression. Int. Rev. Psychiatr. 25, 632–645.

Wilkinson, K., Shapiro, C., 2012. Nonrestorative sleep: symptom or unique diagnostic entity? Sleep Med. 13, 561–569.

Wołyńczyk-Gmaj, D., Szelenberger, W., 2011. Waking EEG in primary insomnia. Acta Neurobiol. Exp. 71, 387–392.

Yeragani, V.K., Cashmere, D., Miewald, J., Tancer, M., Keshavan, M.S., 2006. Decreased coherence in higher frequency ranges (beta and gamma) between central and frontal EEG in patients with schizophrenia: a preliminary report. Psychiatr. Res. 141, 53–60.

CHAPTER 7

Optogenetics in sleep and integrative systems research

Brook L.W. Sweeten, Laurie L. Wellman, Larry D. Sanford
Sleep Research Laboratory, Center for Integrative Neuroscience and Inflammatory Diseases, Department of Pathology and Anatomy, Eastern Virginia Medical School, Norfolk, VA, United States

Background

Optogenetics is one of the most notable of the extensive advancements in neuroscience techniques that have been developed over the last 2 decades. It combines genetic and optical techniques to control cellular events, which allows for the interrogation of neural circuitry with increased cell type specificity, spatial and temporal selectivity compared to previous techniques (Deisseroth, 2011, 2015). Optogenetics has been widely implemented across multiple fields, is readily combined with sleep recording techniques and has provided a means to assess the neural circuitry important for regulating sleep and arousal states (Tyree and de Lecea, 2017). It is also a powerful technique for examining interactions between sleep circuitry and that of other systems that impact or are affected by alterations in sleep.

A complete discussion of optogenetics and the technical issues pertaining to light delivery into the brain is well beyond the scope of this chapter. However, there are numerous reviews that provide in depth coverage of core principles and advances in optogenetics (Johansen et al., 2012; Dugue et al., 2012; Lee et al., 2020) as well as a variety of sources that provide technical information on light delivery (Sileo et al., 2018; Mohanty and Lakshminarayananan, 2015) and control (Wentz et al., 2011; Mohanty and Lakshminarayananan, 2015) using optic fibers and various attempts to bypass the limitations posed by implanted optic fibers and external light sources (Wu et al., 2019; Yu et al., 2019). There are also very nice descriptions of the confluence of events that led researchers to attempt, and succeed, in bringing what would become optogenetics into neuroscience, as well as the broader influences it has had (Boyden, 2011; Deisseroth, 2015). Indeed, the field is rapidly evolving with newly engineered opsins and new approaches to presenting and controlling light that may dramatically impact its use in the study of neural circuits. It has also been extended

Methodological Approaches for Sleep and Vigilance Research
ISBN 978-0-323-85235-7
https://doi.org/10.1016/B978-0-323-85235-7.00003-X

to include applications designed to induce protein—protein interactions, manipulate enzyme activity, control subcellular localization, regulate gene transcription (Deisseroth, 2011; de Mena et al., 2018; Hulsemann et al., 2020), and work in different tissues (Chemi et al., 2019; Crocini et al., 2017; Koopman et al., 2017).

Our laboratory utilizes optogenetics in combination with telemetric and cabled sleep recording to investigate the role(s) of specific regions, cell types, and circuits in the regulation of sleep and arousal. We also use these methods in combination with behavioral stress paradigms to assess the involvement of limbic circuitry in regulating stress and fear memory induced alterations in sleep and arousal with behavioral, stress and neuroimmune readouts. Optogenetics provides a way to understand neural regulation of sleep and its interaction with multiple other systems with temporal and spatial precision that was previously impossible. Our goal in this chapter is to provide a discussion of some of the practical factors that need to be considered in using optogenetics in combination with sleep recording, and procedures we have found valuable in using these methods to assess interactions between sleep regulation and behavior. Many of these procedures and considerations should remain pertinent for studies of sleep and its interactions with other systems regardless of future advances in optogenetic methodologies.

Basics of optogenetics

Optogenetics is the combination of genetic and optical methods to induce or inhibit well-defined events in tissue or in freely behaving animals. Extensive advances have been made increasing efficacy and the stimulation parameters that can be controlled (Deisseroth, 2015). Optogenetics involves three core features: the microbial opsin, methods for targeting the cell populations, and methods for light delivery.

Opsins are membrane bound proteins that are activated with light, which results in cell activation, inhibition, or modulation. Channelrhodopsins (ChRs) are light-gated ion channels first discovered in *Chlamydomonas reinhardtii*, a unicellular green alga. Specifically, ChR2 is a light-gated nonspecific cation channel, which when illuminated with blue light results in the channel opening and allowing an efflux of cations into the cell causing depolarization (Boyden et al., 2005; Ishizuka et al., 2006). Halorhodopsin (NpHR) is an inhibitory opsin from the archaeon *Natronomonas pharaonic*. When activated by 590 nm wavelength of light (yellow, green, or red light), chloride ions are pumped into the cell resulting in hyperpolarization (Schobert and Lanyi, 1982; Lanyi and Oesterhelt, 1982). Another silencing tool, archaerhodopsin-3 (Arch) from *Halorubrum sodomense* works as a proton pump and leads to hyperpolarization of the target cell (Chow et al., 2010). In addition to these core opsins, variants, which have been engineered to possess specific properties, provide a wide range of tools to understand the specific function of targeted neural populations.

Opsins are generally placed under the control of a promoter, which will drive its expression in cells. These could be universal promoters such as elongation factor 1α (ELF-1α) or synapsin which will provide strong opsin expression in almost any cell type in which the construct is present. They could also be cell-type specific promoters such as α-calcium/calmodulin-dependent kinase II (αCamKII) which would target the opsin expression to specific cells. Viral transduction using an adeno associated virus or lentivirus is typically used to introduce the promoter-opsin construct into cells in vivo studies. The development of Cre recombinase—dependent opsin-expressing viruses and growing number of mouse lines that selectively express Cre recombinase in defined cell types provide tremendous tools for targeting specific cells (Deisseroth, 2011). These factors (opsin, promotor, virus, and mouse line/subject species) are central considerations for design of any experiment.

Once the optogenetic construct is incorporated into the desired cell types, light must be delivered in a targeted manner to control the opsin expressing cells. Our first work with optogenetics used blue and green lasers and individually purchased components to route light to the animal. We have now gravitated to light-emitting diode (LED) light sources and matched components from single source suppliers. There are now many commercially available light delivery systems, and one of these may be the simplest way to begin. These light systems will be paired to a fiber optic implant aimed at the region of interest, allowing for light delivery to be targeted to the cells expressing the opsins. These systems are typically paired with software that allows for adjustment of intensity, frequency, and duration of light delivery, as different cell types may require different activation frequencies. When one knows the characteristics of the target cell population, it can be possible to mimic certain aspects of "normal" physiological activity.

Optogenetics versus chemical or electrical stimulation techniques

Chemical and electrical stimulation approaches have long been used to interrogate the role brain regions and cell populations have in regulating sleep. While these approaches remain important, optogenetics provides increased temporal and spatial selectivity. For example, the effects of drugs can be slow, and it generally is not possible to alter effect duration. Electrical stimulation may activate both local cell bodies and fibers of passage (Dugue et al., 2012), thereby complicating interpretation of results. With optogenetics, a specific cell type can be targeted using a promotor and light delivered to the area to solely activate that specific cell subpopulation at time scales compatible with neural activity (Deisseroth, 2011).

Potential disadvantages of optogenetic

In their most basic form, in vivo optogenetic experiments require tethering to optic fibers for light delivery which requires the animals to be habituated to not just the tethering but the handling required to attach the external fibers to the implanted optic probes. When conducted in combination with cabled sleep recording, even more habituation may be needed. It is also desirable to have both electrical cable and optic fibers rotate uniformly as an animal moves in its cage. Though commutators that have both light and electrical channels have been produced, it is possible to achieve acceptable results with separate light rotary joints and electrical commutators by ensuring they rotate along the same axis. We have had success in mounting light rotary joints (Doric Lenses, model LEDFRJ-B/A_FC) above electrical commutators [P1 Technologies (formerly Plastics One) models SL12C and SL6C] and routing the optic fibers through the hole in the commutator. This works reasonably well in rats; however, when conducted in combination with cabled sleep recording in small animals such as mice, the fibers, though lightweight, would have the potential to further limit the free movement of the animals. One should also consider the potential effect of patch cords with metal jackets intended to prevent damage by chewing. This is a concern because higher weight tethering can significantly alter sleep and activity (Tang et al., 2004). For that reason, most of our optogenetic work in mice has been conducted using telemetry with simple lightweight patch cords. We have used metal jacketed patch cords in work with rats, but have generally found running lightweight patch cords along the recording cable sufficient to prevent damage. Regardless, the patch cords will eventually be damaged by freely moving animals, and it is critical to ensure that each experiment is begun with fully functioning equipment.

Optogenetics requires surgical techniques to deliver the viral vector to the region of interest as well as implanting the optic cannulas for light delivery. These steps require careful preparation and skill with stereotaxic surgery to implement, and it can take significant time for expression of the opsins within the neural population (~2–4 weeks), which can be an issue for time or age sensitive studies. The viral vectors used to convey the opsins into cells may also have undesirable effects on neural activity (Jackman et al., 2014). Other limitations include potential damage from overheating, light delivery to large brain regions can be difficult, lack of tissue specificity and low expression can be a problem when using minimal tissue specific promoters in combination with viral transduction (Johansen et al., 2012). The surgical implant as well as the incorporated opsins may have effects that can alter output measures. This may be a particular concern for studies examining neuroimmune responses and indicates the need for surgical controls and care in interpreting results.

Applications of optogenetics in sleep research

Optogenetics allows for detailed assessment of the neural circuits that regulate sleep. One of the initial studies using optogenetics to directly investigate sleep circuitry targeted the orexin-hypocretin neurons in the lateral hypothalamus. Deficiency in these neurons results in narcolepsy (Nishino et al., 2000). This study directly established a causal link between the hypocretin expression neurons and sleep-wake transitions as light stimulation of these neurons at 5 Hz resulted in a decrease in sleep to wake latencies from nonrapid eye movement (NREM) sleep to rapid eye movement (REM) sleep in vivo (Adamantidis et al., 2007). Because optogenetic methods allow for millisecond timescale precision, it can be used to assess sleep to wake transitions. Specific phasic stimulation at 25 Hz of tyrosine hydroxylase positive neurons in the ventral tegmental area induced a very quick transition from NREM to wakefulness in mice expressing ChR2 even after sleep deprivation (Eban-Rothschild et al., 2016, 2018). Furthermore, activation of these neurons can even induce wakefulness following anesthesia (Taylor et al., 2016).

Our work has been primarily aimed at understanding how sleep may be influenced by other systems, particularly those regulating fear and stress.

Previous work in our lab determined that inescapable footshock stress could reduce REM sleep, and this was regulated by the basolateral nucleus of the amygdala (BLA) as inhibition of the BLA with muscimol, a $GABA_A$ receptor agonist, blocked footshock induced reductions in REM (Wellman et al., 2014). Using an inhibitory opsin targeting the glutamatergic cells of the BLA, we inhibited these neurons around the time of each footshock presentation and found that this blocked stress-induced reductions in REM (Machida et al., 2017). Thus, optogenetic methods are applicable to questions regarding direct sleep regulation and to questions regarding how sleep is influenced by other systems. The next sections of this chapter will describe for implementing optogenetics in conjunction with sleep recording.

Design/construction of optic cannulas

It is possible to directly route optic fibers from the light source into the region of interest via implanted cannulas and using housing and caps to hold them in place inside the cannulas. The components needed to utilize this methodology are commercially available from P1 Technologies and are described in various papers (Carter and de Lecea, 2011; Carter et al., 2010; Zhang et al., 2010). This method has an advantage in that it does not allow light loss that can happen when fibers are joined via ferrules. Another advantage of using implanted cannulas to guide the optic fiber is that both opsin and optic fiber could be routed to the same spot without the potential for error inherent in performing two stereotaxic drops into the same site. This would not be a consideration when the location of the opsin and the optic cannulas are different (e.g., in origination and projection sites). A disadvantage in this method is that it can be relatively more fragile, an important consideration when one is conducting prolonged recordings that require optic fibers to remain attached in freely moving animals, and in the case of many of our studies, when optogenetic simulation is provided during shock presentation in fear conditioning paradigms (Machida et al., 2017). For that reason, we switched early on in our studies to implanted optic cannulas that used the ferrule and sleeve system for connection to the optic fiber coming from the light source.

Implantable optic cannulas with ferrules attached are available from various commercial sources. However, they can be made within the lab much more inexpensively using bulk purchased ferrules and optic fiber and a few specialized tools and materials. Essentially everything that is needed to produce optic cannulas, except the optic fiber and ferrules, can be obtained in kit form (e.g., Fiber Optic Termination Kits from Thorlabs, Inc). These kits can be a useful start as they contain instructions regarding how to cleave, strip, mount, and polish optic fibers. Subsequently, individual tools and materials can be purchased as needed. An optical power and intensity meter will also be needed to determine the intensity of light transmitted locally to brain tissue. Beyond that, one needs to consider the characteristics of the fibers most appropriate for the application, and the connectors needed to convey light from the source to the region of interest.

Optic fibers come in a variety of diameters and numerical apertures, a measure of the light gathering ability of a fiber that has consequence for light spread. They need to be matched to appropriate sized ferrules which may be plastic, ceramic, or stainless steel. We prefer stainless steel ferrules for our studies because the bond of dental acrylic to ceramic ferrules is more likely to fail when animals are kept for extended periods, resulting in incomplete experiments. Plastic can also be more prone to breakage. Stainless steel ferrules are sturdy, and it is possible to notch or score the portion of the ferrule that is covered with dental acrylic in the implant and greatly reduce the potential for lost optic cannulas. Using a simple jig that holds multiple ferrules in line, it is possible to notch 50 or more at the same time thereby greatly increasing efficiency. This is crucial as we often perform studies with 10–20 implanted animals with bilateral optic cannulas.

Detailed below is a description of the steps we follow in preparing optic cannulas.

(1) Determine the necessary length of optical fiber based on region of interest. This can be determined using a brain stereotaxic atlas. For example, targeting the BLA of mice will require that the optic fiber extend approximately 6 mm beyond the ferrule to reach the desired depth during implantation.

(2) Strip the end of the optic fiber using a fiber stripping tool approximately 0.5–1 mm. Insert the stripped in of the optic fiber into the open end of the ferrule so that a small portion of the stripped cable sticks out of the end of the ferrule. Cleave the optic fiber at the desired length.

(3) Secure the optic fiber within the ferrule using epoxy. Keep the optic fiber in the center of the ferrule. Let dry. Note: we often use fast setting dental acrylic for this step to speed up the assembly process.

(4) Using epoxy, cover the bottom of the optic fiber to where the tip sticking out is no longer visible. Place optic cannulas into an oven for 3–5 h or until the epoxy has dried completely.
(5) Polish the ends of the optic cannulas using polishing paper until smooth. This is performed by moving the ferrules in a figure eight pattern until the desired result is achieved.
(6) Test the optic cannulas using the light source to make sure the light is transmitted through the optic fiber and can be produced in the desired intensity.

Surgical procedures

Key factors for successful use of optogenetics in sleep research are the targeted injection of the viral vector for opsin expression, implantation of the optic cannulas for light delivery, and implantation of electrodes and/or devices necessary for sleep recording. Therefore, a successful surgery leading to fully recovered healthy animals and optimal viral expression are necessary. A major factor in stereotaxic surgery is identifying the appropriate coordinates for the region of interest. Rodents are generally robust and allow for successful surgeries with relatively small numbers of failures.

Proper surgical technique, post-surgical care, and adequate recovery time are crucial for obtaining the best results for optogenetic studies in freely behaving animals. Following is a detailed procedure for viral prep and surgical procedures that we use in our lab.

We use isoflurane as the surgical anesthetic, and we also secure the optic cannulas in place with dental acrylic which results in an exposed head cap. We often pair our optic cannulas with skull screws to enable electroencephalography (EEG) or implantable telemetry devices for sleep recording. The procedures described in this section should be easily used or adapted by researchers with experience in animal surgery. Surgery requires two people, the surgeon and the assistant who manages anesthesia and assists the surgeon as required. A specific procedure describing the delivery of viral vector and implantation of optic cannulas is described for mice below.

Preparation of viral vector

A large variety of optogenetic viral vectors are now available and the list keeps growing. Our lab typically obtains ours from the University of North Carolina Vector Core facility, but several core facilities and commercial entities act as providers.

The procedure below describes the method of preparing 100 uL volume of 1.5 × 10^12 virus molecules/mL for microinjection into the BLA.

(1) Dissolve 50 mg of sorbitol in sterile saline

(2) Calculate the dilution

  a. For example: Viral stock is 4 × 10^12 mol/mL stock

  b. 4/1.5 = 2.67

  c. 100 uL/2.67 = 37.45 uL

(3) Take viral stock from storage (should remained stored at −80°C until needed), thaw by spinning at 1000 g for 90 s

(4) Pipette the calculated value of viral stock into a clean tube, and add 5% sorbitol in sterile saline to make a final volume of 100 uL

(5) If performing surgery on a smaller number of animals, this can be halved to only use 50 uL

(6) Viral vector can be used for up to 24 h if stored at 4°C. A new batch should be prepared for each surgical day

(7) Spin down tube before placing into the microinjection tubing.

Preoperative procedures

Preoperatively, each mouse is prepared for surgery by providing it with analgesic. For each mouse, we provide ad libitum access to ibuprofen (30 mg/kg; 4.75 mL children's ibuprofen in 500 mL water) in its water bottle for 24–48 h prior to surgery and continuing for at least 72 h post-surgery. Antibiotics (gentamicin 5–8 mg/kg and potassium penicillin 100,000 IU/kg) and anti-inflammatory (dexamethasone (0.4 mg; 0.2 mL total dosage)) are administered subcutaneously preoperatively and may be continued if necessary postoperatively.

For the surgical microinjection of the viral vector and implantation of the optic cannulas, the mouse is initially anesthetized with isoflurane (5% induction, 2%–3% maintenance; the requirement for maintenance of anesthesia may vary depending on how the animal responds) in an induction chamber. In our case, this is a plastic cylindrical chamber (diameter 12 cm and height 11 cm). Following attainment of surgical anesthesia, the head site is shaved. Eye lubricant (Paralube sterile ocular ointment) is then applied to the eyes.

Operative procedures

After it reaches a surgical plane of anesthesia, the animal is placed into the stereotaxic device to prevent movement of the head during surgery for stereotaxic accuracy. Following placement into the ear and incisor bars, the head site is cleaned using gauze soaked in Betadine (10% povidone-iodine) followed by cleaning with gauze soaked in 70% ethanol. Afterward, an incision approximately 1 cm in length is made using microdissection scissors. A scalpel blade is used to scrape away the membrane of the connective tissue covering the skull. A cautery may be used to stop excessive bleeding if necessary. A drop of phosphoric acid is placed on the skull and a cotton swab is used to "scrub" the skull (avoiding the skin and muscle) to ensure all connective tissue has been removed. The skull is then rinsed three times with saline.

Once the skull is clear of tissue and blood, the microinjection cannula is targeted to the region of interest using stereotaxic coordinates from Bregma. For our projects, we often target the BLA bilaterally (coordinates; −1.5 mm AP, ± 2.9 mm ML, −4.7 mm DV). Once the proper location on the skull has been determined, a small hole is gently drilled to enable insertion of the microinjection cannula. At this time, holes for anchor screws are also drilled and the screws placed. Prior to lowering the microinjection needle, the injection pumps are checked to ensure functioning and that fluid is exiting the cannula. Subsequently, the microinjection cannula is lowered into the brain to the proper coordinates, and the injection is begun and volume is monitored based on the flow rates of the injector pump. For our setup, 1 mm of movement of solution through the tubing equates to 0.1 uL of fluid and we typically run the pumps until 0.5 uL of viral vector has been injected. We then wait at least 5 min before removing the microinjection cannula to allow for optimal absorption of vector into injection site. Bilateral injections can be done at the same time with the use of a two armed stereotaxic device or repeated on the opposite side using a one armed stereotaxic device. Next, the optic cannulas are stereotaxically implanted into the bilateral BLA through the same holes used to inject the viral vector.

Once the optic cannulas(s) are in place, the hole(s) are gently filled with gel foam using needle nose forceps and a dental pick. Next, the optic cannulas are secured with using 3 M dual cure dental glue (activated by UV light (Patterson Dental, Model TCL490 Plus, visible curing unit)). Dental acrylic is placed over the skull and approximately halfway up the optic cannulas ferrules to secure to the head. There needs to be enough space left to attach the fiber optic connector to the head to delivery light.

Simple variations of this procedure can be used to deliver viral vector or implant optic cannulas in different locations of the brain. Implantation of intraperitoneal telemetry transmitters can be done in combination with optic cannulas implantation to measure EEG and other activity. Extensive detail regarding surgical telemetry implants and recordings are provided in (Sanford et al., 2011; Tang and Sanford, 2002) and are not repeated here.

Postoperative procedures

Once the animal is taken off anesthesia, it is observed until its respiration rate returns to normal levels and movement is witnessed. The time it takes the animal to wake up from anesthesia is recorded. Then it is returned to its home cage and kept warm by an under cage (water circulating) heating pad. The animal is monitored every 15 min for at least 2 h post-operatively, a record is made of the animal's coloring, returning reflexes, and pain level (posture, movement, vocalization, etc.). We have found that providing the animals with highly palatable food and supplemental high calorie gel (Nutri-Cal) can improve recovery.

Postoperative recovery period

The animals are closely monitored for at least eight days post-operatively where weights, eating and drinking behavior, and attitude are recorded and assessments are made regarding whether additional care is needed. An adequate postsurgery recovery period also is essential for sufficient viral expression of the opsin in the area of interest as well as recovery from implantation of the optic cannulas and telemetry transmitters. When including telemetry implants, we allow a minimum of 28 days for the mice to recover from surgery. This may appear to be a long recovery period, but we have found that the longer recovery periods provide more consistent sleep data across animals. Fully recovered and stabilized animals are important to insure that stressful manipulations or behavioral tests that we use in our stress-related work is not interacting with the recovery from surgery. Other types of experiments and recording methods may require less time before experiments can be conducted. Surgical recovery may also vary with species. Rats with subcutaneously transmitter showed stabilized NREM and total sleep amounts within two to three days (compared to sleep at 15 days postsurgery) whereas REM amounts may require seven or more days (Tang et al., 2007).

Sample protocol for optogenetics in sleep research

The precise temporal control provided by optogenetics readily lends itself to manipulations within sleep states and/or during specific sleep-related activity, and to manipulations of cell populations important for regulating sleep and arousal. Critically for our work, these properties also allow temporally constrained manipulations during behavior and learning tasks and thereby reduce the potential for confounding effects on subsequent sleep that could occur with longer lasting manipulations such as drug microinjections. The following is an example of a paradigm using optogenetics in our research on fear conditioning and sleep.

Subjects

The protocol below was developed in adult male C57BL/6J mice.

Optogenetic light-emitting diode light delivery system

There are many options in terms of light delivery systems. Here we employ LEDs to generate the appropriate wavelength of light to target the particular opsin of interest. For the procedure described below, inhibitory green light (532 nm) was produced via LEDs (Model: LEDFRJ-B/G_FC; Doric Lenses, Inc., Quebec Canada). Light output was measured by an optical power meter and adjusted to approximately 10 mW at the optic fiber tip. From the LED system, bilateral optic fibers are attached to the animal's head via optic cannula connectors for light delivery directly at the region of interest.

Experimental procedures

(1) Mice are surgically prepared for viral expression of the opsin (in this case, NpHR) and implanted with optic cannulas for light delivery and a telemetry transmitter for sleep recording and allowed to recover as described above. When the animals are not on study, the transmitters are inactivated to preserve battery life. The recording room is kept on a 12:12 light:dark cycle, and ambient temperature is maintained at $24 \pm 1.5°C$. Cages are changed two days prior to recording onset and timed so they do not disturb subsequent phases of the study.

(2) A minimum of 20 h of uninterrupted baseline sleep is recorded for each mouse. For recording, the mice are housed in individual cages with food and water available ad libitum. Individual home cages are placed on a DSI telemetry receiver (RPC-1), and the transmitter is activated with a magnetic switch.

(3) After baseline recording is obtained, the mice are connected bilaterally to optic fibers and placed in a shock chamber (Coulbourn Instruments) for shock training (ST). The ST procedure lasts 30 min with a 5 min pre-ST period; 20 shock presentations (0.5 mA, 0.5 s duration, 1-min interval) and a 5 min postshock period. Inhibitory light stimulation to activate NpHR starts 5 s prior to each shock onset and continues for 5 s after shock onset for peri-shock light stimulation. This training occurs in a different room than the colony/recording room. Immediately following the 5 min postshock period, the animals are disconnected from the optic fibers and returned to their home cages for subsequent sleep recording for at least 20 h.

(4) One week after ST, at the same circadian time as the ST procedure, the animals are placed back into the shock chambers for 30 min without footshock or light presentation to test long-term memory of the stressful experience. Immediately following context re-exposure, the animals are returned to their home cages for sleep recording.

(5) Three weeks after ST (two weeks after context re-exposure), at the same circadian time as the initial ST procedure, the animals are once again placed back into the shock chambers for 30 min without footshock or light presentation to further assess long-term memory of the stressful experience. Immediately following the context re-exposure, animals are returned to their home cages for sleep recording.

Data collected

(1) Freezing behavior: Freezing is a measurement of fear memory in rodents and is classified as an absence of body movement except for respiration. Freezing is scored in 5-s intervals over the course of the initial 5 min pre-ST period and the course of each of the re-exposure session. From these data, the percentage of time spent freezing is calculated (FT%: freezing time/observed time × 100) for each animal for each observation period.

(2) Telemetry data: Telemetric data collected include EEG, core body temperature, and activity measurements. These signals can be used to provide relevant information on their own. For example, the EEG can be subjected to a Fourier analysis, and the temperature measure can be used to determine stress-induced hyperthermia. With appropriate training, the EEG signal can be used to differentiate sleep states, number of episodes of sleep states, and the average duration of sleep episode.

Confirmation of opsin expression at injection site

Optogenetic vectors often contain a fluorescent tag to allow confirmation of opsin expression in the tissue. Therefore, through the use of relatively simple histological techniques, demonstration of opsin expression can be obtained.

(1) Following behavioral experiments, mice are deeply anesthetized with isoflurane (5%) and transcardially perfused with 40 mL of cold phosphate buffered saline (PBS), pH7.4 followed by 40 mL of 4% paraformaldehyde (PFA) in PBS.

(2) Carefully remove the brains from the animals and postfix them in 4% PFA for 24 h and then transfer for 30% sucrose solution for 24—48 h to cryoprotect.

(3) Once cryoprotected, section out the region of interest (based on the surgical section, this will cover the BLA region) and snap freeze the issue in isopentane placed on liquid nitrogen. Coronal sections are made on a cryostat.

(4) For detection of virus expression in the region of interest, serial sections (40 um thick) are made through the amygdala region and mounted on slides.

(5) Using a fluorescent microscope with the necessary filter to detect the fluorescent marker (we used eYFP) examine the region of interest for opsin expression. Any animals not showing fluorescent expression in the area of interest should be excluded from analysis.

(6) This method can also be used to confirm optic fiber placement accuracy.

Optogenetic methods can be further combined with immunohistochemistry to confirm that the opsin was expressed in the cell types targeted. In our studies, we targeted the excitatory pyramidal neurons of the BLA using a CaMKII promotor.

(1) Animals would be perfused, and tissue fixed and sectioned as described above.

(2) Coronal sections (25 um thick) are collected for free-floating immunofluorescence methods.

(3) Sections would be labeled with anti-CaMKII antibodies (rabbit monoclonal anti-CaMKII 1:250), followed by fluorescent secondary antibodies (goat anti-rabbit Alexa Fluor 568 1:200). It is important to use a fluorescent antibody with a different fluorophore than the one identifying the opsin.

(4) All blocking steps are done in PBS containing 3% normal goal serum (or serum from the animal for the secondary antibody), 2% nonfat dry milk, and 0.1% triton X-100.
(5) Sections need to be incubated in the primary antibody for 24 h at 4°C and incubated in the secondary antibody for 1 h at room temperature in the dark.
(6) Sections would then be mounted and coverslipped with an antifade reagent.
(7) It is important to include both positive and negative controls during the immunofluorescence process.
(8) Immunoreactive cells could then be visualized using a fluorescent microscope and analyzed further with ImageJ software to confirm co-localization.

Experimental variants

The experiment described above was specifically focused on determining how inhibiting activity of BLA during shock presentation in the acquisition phase of fear memory would impact subsequent sleep and behavior. However, these basic procedures are readily adaptable to assess fear memory consolidation and extinction, and the roles that sleep may play in them across strains and species. For example, we have conducted studies in which BLA was optogenetically activated or inhibited during NREM and REM during the period when fear memory would be consolidated. They are also applicable across strains and with some modifications across other brain regions, e.g., we have conducted similar studies in C57BL/6-Tg(Grik4-cre) G32-4Stl/J (Grik4- Cre) mice to enable us to selectively manipulate BLA inputs to glutamatergic neurons in the ventral hippocampus (vHPC); Grik4-Cre mice have high expression in CA3 and limited or no expression in CA1 (Nakazawa et al., 2002). In these studies, the mice were injected into BLA with AAV5-EF1a-DIO-hChR2(H134R)-EYFP (DIO-hChR2) for optogenetic excitation and an optic probe for light stimulation in the vHPC. This vector codes for double floxed and inverted open reading frame hChR2 fused with EYFP and Cre activity in BLA neurons will flip the opsin gene into its correct orientation and allow its translation. Expression of the opsin gene then allowed specific excitation of BLA neuron terminals in the vHPC. Control mice were infected AAV5-EF1a-DIO-EYFP (DIO-EYFP, no opsin) to check for injection/recombinant protein side effects and to control for the effects of light. Significant findings were that REM specific stimulation of BLA inputs into glutamatergic neurons in vHPC during the first 3 h after ST reduced subsequent freezing on CTX re-exposure at 24 h post-ST similar to that after stimulation of BLA. In contrast to direct BLA stimulation, stimulation in Cre mice did not significantly reduce theta during REM. The mice still showed significant reductions in REM and increased SIH after CTX. A complementary strategy using AAV5-EF1a-DIO-eNpHR-EYFP (DIO- eNpHR) could be employed for optogenetic inhibition of BLA terminals in vHPC.

Similar circuit specificity can be achieved in rats. For example, to specifically target projections of the central nucleus of the amygdala (CNA) to the locus coeruleus (LC), we infected CNA with AAV-EF1a-DIO-hChR2(H134R)-EYFP and/or AAV-EF1a-DIO-eNpHR3.0-EYFP and LC with AAV-EF1a-mCherry-IRES-WGA-Cre that mediates bicistronic expression of mCherry and WGA-Cre (Gradinaru et al., 2010). Vectors in CNA coded for double floxed and inverted open reading frame opsins, hChR2 or eNpHR3.0. When a neuron in the LC/peri-LC zone synapses with neurons in CNA and expresses WGA-Cre, it is transneurally transferred to CNA neurons, where Cre activity flips the opsin gene(s) into its correct orientation thereby allowing its translation. Subsequent expression of the opsin(s) gene enables excitation of ChR2 expressing neurons/projections by blue light or inhibition of the NpHR expressing neurons/projections from CNA by yellow light. We have used these vectors to target CNA to LC and other pontine REM regulatory regions (Wellman et al., 2017; Williams et al., 2017).

Other labs have used the strengths of optogenetics to target sleep neurocircuits to assess their roles in regulating sleep, and then to assess the role of sleep in cognitive functioning. Direct optogenetic targeting of hypocretin neurons in the hypothalamus increased the probability of transitions to wakefulness from either NREM or REM sleep with the effect dependent on stimulation rate and influenced by sleep pressure (Adamantidis et al., 2007; Carter et al., 2009). Subsequent work used optogenetic manipulations of these neurons to selectively fragment sleep without affecting overall sleep amount or intensity during the learning phase of the novel object recognition (NOR) task (Rolls et al., 2011). Optogenetically induced sleep fragmentation at 60 s intervals, but not 120 s intervals significantly decreased mouse NOR performance the next day. Together, these studies demonstrate the power of incorporating the utility of optogenetics with functional circuit and system level considerations in experimental design.

Conclusions

Optogenetics can provide cell type specificity and temporal and spatial precision in determining neural circuits involved in regulating sleep and waking, and in how those circuits interact with other functional systems. The available and expanding variety of optogenetic tools and options for integrating with sleep recording and behavioral methodologies make it very powerful methodology for basic sleep and integrative research. Drawbacks can include the required and sometimes complicated surgery, subsequent lengthy recovery times, and the potential impact of optic fibers on the already substantial effects of tethering on sleep. Viral expression systems may also produce undesirable effects. However, with proper consideration of these factors in experimental design and interpretation, optogenetics can be a tremendously valuable tool for research on neurocircuits that regulate sleep and its interactions with other systems.

Acknowledgment

This work was supported by NIH research grant MH64827. We thank Dr. Karl Deisseroth for sharing the optogenetic constructs with our lab.

References

Adamantidis, A.R., Zhang, F., Aravanis, A.M., Deisseroth, K., De Lecea, L., 2007. Neural substrates of awakening probed with optogenetic control of hypocretin neurons. Nature 450, 420–424.

Boyden, E.S., 2011. A history of optogenetics: the development of tools for controlling brain circuits with light. F1000 Biol. Rep. 3, 11.

Boyden, E.S., Zhang, F., Bamberg, E., Nagel, G., Deisseroth, K., 2005. Millisecond-timescale, genetically targeted optical control of neural activity. Nat. Neurosci. 8, 1263–1268.

Carter, M.E., Adamantidis, A., Ohtsu, H., Deisseroth, K., De Lecea, L., 2009. Sleep homeostasis modulates hypocretin-mediated sleep-to-wake transitions. J. Neurosci. Off. J. Soc. Neurosci. 29, 10939–10949.

Carter, M.E., de Lecea, L., 2011. Optogenetic investigation of neural circuits in vivo. Trends Mol. Med. 17, 197–206.

Carter, M.E., Yizhar, O., Chikahisa, S., Nguyen, H., Adamantidis, A., Nishino, S., Deisseroth, K., De Lecea, L., 2010. Tuning arousal with optogenetic modulation of locus coeruleus neurons. Nat. Neurosci. 13, 1526–1533.

Chemi, G., Brindisi, M., Brogi, S., Relitti, N., Butini, S., Gemma, S., Campiani, G., 2019. A light in the dark: state of the art and perspectives in optogenetics and opto-pharmacology for restoring vision. Future Med. Chem. 11, 463–487.

Chow, B.Y., Han, X., Dobry, A.S., Qian, X., Chuong, A.S., Li, M., Henninger, M.A., Belfort, G.M., Lin, Y., Monahan, P.E., Boyden, E.S., 2010. High-performance genetically targetable optical neural silencing by light-driven proton pumps. Nature 463, 98–102.

Crocini, C., Ferrantini, C., Pavone, F.S., Sacconi, L., 2017. Optogenetics gets to the heart: a guiding light beyond defibrillation. Prog. Biophys. Mol. Biol. 130, 132–139.
de Mena, L., Rizk, P., Rincon-Limas, D.E., 2018. Bringing light to transcription: the optogenetics repertoire. Front. Genet. 9, 518.
Deisseroth, K., 2011. Optogenetics. Nat. Methods 8, 26–29.
Deisseroth, K., 2015. Optogenetics: 10 years of microbial opsins in neuroscience. Nat. Neurosci. 18, 1213–1225.
Dugue, G.P., Akemann, W., Knopfel, T., 2012. A comprehensive concept of optogenetics. Prog. Brain Res. 196, 1–28.
Eban-Rothschild, A., Appelbaum, L., De Lecea, L., 2018. Neuronal mechanisms for sleep/wake regulation and modulatory drive. Neuropsychopharmacol. Off. Pub. Am. Coll. Neuropsychopharmacol. 43, 937–952.
Eban-Rothschild, A., Rothschild, G., Giardino, W.J., Jones, J.R., De Lecea, L., 2016. VTA dopaminergic neurons regulate ethologically relevant sleep-wake behaviors. Nat. Neurosci. 19, 1356–1366.
Gradinaru, V., Zhang, F., Ramakrishnan, C., Mattis, J., Prakash, R., Diester, I., Goshen, I., Thompson, K.R., Deisseroth, K., 2010. Molecular and cellular approaches for diversifying and extending optogenetics. Cell 141, 154–165.
Hulsemann, M., Verkhusha, P.V., Guo, P., Miskolci, V., Cox, D., Hodgson, L., 2020. Optogenetics: Rho GTPases activated by light in living macrophages. Methods Mol. Biol. 2108, 281–293.
Ishizuka, T., Kakuda, M., Araki, R., Yawo, H., 2006. Kinetic evaluation of photosensitivity in genetically engineered neurons expressing green algae light-gated channels. Neurosci. Res. 54, 85–94.
Jackman, S.L., Beneduce, B.M., Drew, I.R., Regehr, W.G., 2014. Achieving high-frequency optical control of synaptic transmission. J. Neurosci. Off. J. Soc. Neurosci. 34, 7704–7714.
Johansen, J.P., Wolff, S.B., Luthi, A., Ledoux, J.E., 2012. Controlling the elements: an optogenetic approach to understanding the neural circuits of fear. Biol. Psychiatr. 71, 1053–1060.
Koopman, C.D., Zimmermann, W.H., Knopfel, T., De Boer, T.P., 2017. Cardiac optogenetics: using light to monitor cardiac physiology. Basic Res. Cardiol. 112, 56.
Lanyi, J.K., Oesterhelt, D., 1982. Identification of the retinal-binding protein in halorhodopsin. J. Biol. Chem. 257, 2674–2677.
Lee, C., Lavoie, A., Liu, J., Chen, S.X., Liu, B.H., 2020. Light up the brain: the application of optogenetics in cell-type specific dissection of mouse brain circuits. Front. Neural Circ. 14, 18.
Machida, M., Wellman, L.L., Fitzpatrick Bs, M.E., Hallum Bs, O., Sutton Bs, A.M., Lonart, G., Sanford, L.D., 2017. Brief optogenetic inhibition of the basolateral amygdala in mice alters effects of stressful experiences on rapid eye movement sleep. Sleep 40.
Mohanty, S.K., Lakshminarayananan, V., 2015. Optical techniques in optogenetics. J. Mod. Opt. 62, 949–970.
Nakazawa, K., Quirk, M.C., Chitwood, R.A., Watanabe, M., Yeckel, M.F., Sun, L.D., Kato, A., Carr, C.A., Johnston, D., Wilson, M.A., Tonegawa, S., 2002. Requirement for hippocampal CA3 NMDA receptors in associative memory recall. Science 297, 211–218.
Nishino, S., Ripley, B., Overeem, S., Lammers, G.J., Mignot, E., 2000. Hypocretin (orexin) deficiency in human narcolepsy. Lancet 355, 39–40.
Rolls, A., Colas, D., Adamantidis, A., Carter, M., Lanre-Amos, T., Heller, H.C., De Lecea, L., 2011. Optogenetic disruption of sleep continuity impairs memory consolidation. Proc. Natl. Acad. Sci. U S A 108, 13305–13310.

Sanford, L.D., Yang, L., Wellman, L.L., 2011. Telemetry in mice: applications in studies of stress and anxiety disorders. In: GOULD, T.D. (Ed.), Mood and Anxiety Related Phenotypes in Mice: Characterization Using Behavioral Tests, Volume II. Humana Press, Totowa, NJ.

Schobert, B., Lanyi, J.K., 1982. Halorhodopsin is a light-driven chloride pump. J. Biol. Chem. 257, 10306–10313.

Sileo, L., Bitzenhofer, S.H., Spagnolo, B., Popplau, J.A., Holzhammer, T., Pisanello, M., Pisano, F., Bellistri, E., Maglie, E., De Vittorio, M., Ruther, P., Hanganu-Opatz, I.L., Pisanello, F., 2018. Tapered fibers combined with a multi-electrode array for optogenetics in mouse medial prefrontal cortex. Front. Neurosci. 12, 771.

Tang, X., Orchard, S.M., Liu, X., Sanford, L.D., 2004. Effect of varying recording cable weight and flexibility on activity and sleep in mice. Sleep 27, 803–810.

Tang, X., Sanford, L.D., 2002. Telemetric recording of sleep and home cage activity in mice. Sleep 25, 691–699.

Tang, X., Yang, L., Sanford, L.D., 2007. Sleep and EEG spectra in rats recorded via telemetry during surgical recovery. Sleep 30, 1057–1061.

Taylor, N.E., Van Dort, C.J., Kenny, J.D., Pei, J., Guidera, J.A., Vlasov, K.Y., Lee, J.T., Boyden, E.S., Brown, E.N., Solt, K., 2016. Optogenetic activation of dopamine neurons in the ventral tegmental area induces reanimation from general anesthesia. Proc. Natl. Acad. Sci. U S A 113, 12826–12831.

Tyree, S.M., de Lecea, L., 2017. Optogenetic investigation of arousal circuits. Int. J. Mol. Sci. 18.

Wellman, L.L., Fitzpatrick, M.E., Machida, M., Sanford, L.D., 2014. The basolateral amygdala determines the effects of fear memory on sleep in an animal model of PTSD. Exp. Brain Res. 232, 1555–1565.

Wellman, L.L., Sutton, A.M., Kim, M.H., Koech, O.K., Fitzpatrick, M.E., Sutton, A.M., Williams, B.L., Machida, M., Yoon, H., Lonart, G., Sanford, L.D., 2017. Optogenetic stimulation and inhibition of the central nucleus of the amygdala (CNA) alters firing in locus coeruleus (LC) neurons. Sleep 40 (Suppl. 1). A51–A51.

Wentz, C.T., Bernstein, J.G., Monahan, P., Guerra, A., Rodriguez, A., Boyden, E.S., 2011. A wirelessly powered and controlled device for optical neural control of freely-behaving animals. J. Neural. Eng. 8, 046021.

Williams, B.L., Sutton, A.M., Fitzpatrick, M.E., Machida, M., Wellman, L.L., Sanford, L.D., 2017. Amygdalar regulation of pontine REM regulatory regions: effects on sleep. Sleep 40, A51.

Wu, X., Zhu, X., Chong, P., Liu, J., Andre, L.N., Ong, K.S., Brinson Jr., K., Mahdi, A.I., Li, J., Fenno, L.E., Wang, H., Hong, G., 2019. Sono-optogenetics facilitated by a circulation-delivered rechargeable light source for minimally invasive optogenetics. Proc. Natl. Acad. Sci. U S A 116 (52), 26332–26342. https://doi.org/10.1073/pnas.1914387116.

Yu, N., Huang, L., Zhou, Y., Xue, T., Chen, Z., Han, G., 2019. Near-infrared-light activatable nanoparticles for deep-tissue-penetrating wireless optogenetics. Adv. Healthc. Mater. 8, e1801132.

Zhang, F., Gradinaru, V., Adamantidis, A.R., Durand, R., Airan, R.D., De Lecea, L., Deisseroth, K., 2010. Optogenetic interrogation of neural circuits: technology for probing mammalian brain structures. Nat. Protoc. 5, 439–456.

CHAPTER 8

Immunohistochemical analysis and sleep studies: some recommendations to improve analysis data

Fabio García-García[1], Luis Beltrán Parrazal[2], Armando Jesús Martínez[3]
[1]Biomedicine Department, Health Science Institute, Veracruzana University, Xalapa, Veracruz, Mexico; [2]Brain Research Center, Veracruzana University, Xalapa, Veracruz, Mexico; [3]Neuroethology Institute, Veracruzana University, Xalapa, Veracruz, Mexico

Introduction

For several decades, an experimental strategy commonly used to characterize the brain circuits and nuclei in the sleep-wake regulation has been immunohistochemistry. The immunohistochemical characterization of the brain circuits and the electrophysiological studies has made it possible to understand the wake-sleep cycle's neuroanatomy and neurochemistry. Studies carried out in the 90s using a neuronal activity marker, the c-Fos gene, allowed corroborating electrophysiological studies and identifying how specific brain nuclei activated or not in response to external stimuli, and in particular, those regions that participated in the sleep-wake cycle. Currently, the c-Fos marker continues widely used in neuroscience studies.

One of the difficulties that the first immunohistochemical studies faced was regarding how to count the positive cells for the specific marker. The first strategy consisted of the manual counting of positive cells. The counting was carried out with the direct observation of the histological sections under the microscope or photographs of interest areas. In the first studies, the counting area expressed by μm^2 and the volume did not include the analysis. For many years, these aspects were not considered by researchers. With the advancement of technological tools and the implementation of stereology and specialized software, the problems involved in counting cells in nervous tissue began to be improved. However, one aspect that receives little attention is related to the statistical analysis of cell counts. This chapter describes technical recommendations to improve the

Methodological Approaches for Sleep and Vigilance Research
ISBN 978-0-323-85235-7
https://doi.org/10.1016/B978-0-323-85235-7.00014-4

labeling and resolution of the results obtained using the immunohistochemical technique. Besides, recommendations are given to refine the statistical analysis and avoid common mistakes in most cell count studies, pseudoreplicas.

Cell markers and sleep

The anatomical brain circuit that participates in the sleep-wake cycle regulation was described using a wide variety of antibodies and animal models that have made it possible to identify the neurons and the neurotransmitters that participate. At the end of the 1980s, the discovery of the early expression gen c-fos as a marker of neuronal activity became a tool that strongly impacted the investigation of the sleep-wake cycle (Table 8.1). Currently, labeling with c-fos and co-expression with other cell markers is still widely used.

Table 8.1 Most representative studies related with sleep and c-fos marker published between 1992 and 2000.

Author	Title of the publication
Merchant-Nancy et al. (1992)	c-fos proto-oncogene changes in relation to rapid eye movement (REM) sleep duration.
Yamuy et al. (1993)	c-fos expression in the pons and medulla of the cat during carbachol-induced active sleep.
Shiromani et al. (1995)	Time course of Fos-like immunoreactivity associated with cholinergically induced REM sleep
Merchant-Nancy et al. (1995)	Brain distribution of c-fos expression as a result of prolonged REM sleep period duration.
Shiromani et al. (1996)	Pontine cholinergic neurons show Fos-like immunoreactivity associated with cholinergically induced REM sleep.
Yamuy et al. (1998)	c-fos expression in mesopontine noradrenergic and cholinergic neurons of the cat during carbachol-induced active sleep: a double-labeling study.
Basheer et al. (1999)	Adenosine and behavioral state control: adenosine increases c-Fos protein and AP1 binding in basal forebrain of rats.
Maloney et al. (2000)	c-Fos expression in GABAergic, serotonergic, and other neurons of the pontomedullary reticular formation and raphe after paradoxical sleep deprivation and recovery.
Shiromani et al. (2000)	Sleep and wakefulness in c-fos and fos B gene knockout mice.

In current studies, the statiscal analysis of the counting cells represents a great challenge. The main difficulty is obtaining an accurate estimate of positive cells. In part, this problem was solved with the implementation of stereology (see review Gundersen et al., 1988). However, this kind of resources was not available in all laboratories.

Immunohistochemistry and sleep studies

The balance between the duration of sleep and wakefulness is a fundamental feature of the sleep-wake cycle. During wakefulness, sleep-inducing factors (SIFs) accumulate and cause an increase in sleep pressure or our need for sleep. Decades of research have identified neurotransmitters, biochemical processes, and genes involved in regulating sleep homeostasis. Identifying the cells responsible for the synthesis of SIF, the anatomical localization to SIF receptors, and the networks between neuronal nuclei that integrate the circuits that allow the emergence of the sleep-wake cycle continues to be a challenge for the field of neurosciences.

Immunohistochemical analysis of the brain has been allowed to identify the neuronal nuclei involved in the physiology and pathophysiology process of the sleep-wake cycle. For example, combined immunohistochemistry and in situ hybridization against preprohypocretin (protein and mRNA, respectively) demonstrated that preprohypocretin neurons are distributed in a restricted area of the tuberal region of the hypothalamus in rats (Peyron et al., 1998). Also, with double immunostaining analysis strategy against c-Fos protein and neurotransmitters or their synthetic enzymes, it has been used to evaluate cholinergic, serotonergic, noradrenergic, and GABAergic neurons in the pontomesencephalic tegmentum of animals under conditions of deprivation and recovery of REM sleep. In the same study, it was reported that GABAergic neurons are active during REM sleep and could thus be responsible for inhibiting neighboring monoaminergic neurons that are essential in the generation of REM sleep (Maloney et al., 1999).

Immunohistochemical analysis has also been used to analyze neuronal nuclei's participation and the type of neurotransmitter that neurons use to maintain or facilitate the waking state. For example, it has been shown that injection of muscimol injection in the Dorsal Paragigantocellular Reticular Nucleus (DPGi) induces the activation of the noradrenergic neurons of the locus coeruleus (LC) and abolishing during 3 and 5 h the slow-wave sleep and REM sleep, respectively; concluding that DPGi GABAergic neurons would be responsible for the inhibition of noradrenergic and adrenergic neurons during REM sleep, with a minor contribution of the GABAergic neurons of the ventrolateral periaqueductal gray (Clément et al., 2014).

Immunohistochemistry mapping c-Fos has also been used to explore sleep architecture changes caused by activities performed during wakefulness and the neuronal nuclei activated during the sleep phase that proceeds. For example, behavioral manipulations during wakefulness, such as forced wakefulness induced by gentle handling, forced wakefulness associated with a stressful condition such as immobilization, or forced wakefulness associated with excessive intake of appetizing foods, result in a variation of the immunoreactivity of Fos in the brain structures and also in different patterns of the EEG power density, associated with the behavior carried out during forced wakefulness. The results of these studies have shown that the sleep-wake cycle of the rats after all the experimental manipulations was different not only concerning the control group but also between them (García-García et al., 1998).

The immunocytochemical analysis has also been used in the sleep-wake cycle study to understand the impact of induced chronic sleep fragmentation (ICSF) on the synaptic circuits and architectures. For example, a recent quantitative immunohistochemistry analysis reveals an increase of positive cells to the amyloid-beta peptide in the cortex and hippocampus after ICSF in mice. Furthermore, the authors show with immunohistochemical evidence that ICFS increases the number of cells positive to autophagy regulatory factors (Beclin1 and UVRAG), suggesting an endosome-autophagosome-lysosome dysfunction (Xie et al., 2019). These results suggest a molecular mechanism of how the sleep-wake cycle's perturbance may be related to the early stages of neurodegenerative diseases.

Although techniques such as lesion analysis, functional magnetic resonance imaging, 2-deoxyglucose studies, and induction of gene expression have helped determine the brain areas related to the sleep-wake cycle, these methods are technically limited. There is currently no method that allows for the identification and electrophysiological characterization of individual neurons associated with a particular function in the sleep-wake cycle. Local brain activity measurements during the sleep-wake cycle behaving in mammals can be made with electrodes and fluorescent calcium indicators. However, such approaches provide information regarding only a tiny fraction of the brain's existing neurons.

The use of immunohistochemistry against c-Fos and other immediate-early genes (IEGs) linked to recent neuronal activity is a more spatially comprehensive technique to measure population activity of neurons related to an emergent behavior (Guzowski et al., 2001). While it lacks the time resolution of electrophysiological recordings or calcium imaging, it does have the potential of providing a complete view of recent whole-brain activity. Using these whole-brain IEG-based maps of neuronal activity related to behavior, we can design experiments to study a potential structure-function circuit probed by high-resolution recordings and optogenetic and chemogenetic methods (Atasoy and Sternson, 2018; Rickgauer et al., 2014).

Achieving rigorous and reproducible immunohistochemistry results against c-Fos and other IEGs is a challenge between laboratories in the neurosciences field and, particularly, in the sleep-wake study. Unfortunately, a lack of standardized methods to quantify c-Fos positive cells is impeding research reproducibility across laboratories.

Quantitative analysis of cell numbers, cell density, and volume of brain regions has been an essential part of neuroscience research. However, quantification obtained from a few three-dimensional (3D) brain sections is highly variable and inconsistent due to the uneven distribution of cells and the distortion and shrinkage resulting from tissue processing. For example, counting positive cells in which one thin representative section of tissue of a region of interest can lead to different results depending on which tissue level sections are chosen. Stereology was developed to systematically and reproducibly perform quantitative assessments of both subtle and large differences between control and experimental conditions. Several critical principles for design-based stereology ensure the accuracy of the data. These include:

1. Systematic and random sampling (SRS),
2. Calculation of total cell numbers instead of densities,
3. Counting of cells, not cell profiles,
4. Use of thick tissue sections to visualize cells throughout 3D probes in a known tissue volume, and
5. Sensitive and specific staining to identify the cells of interest.

An accurate quantitative brain mapping of IEGs analysis requires optimal experimental design, high quality, well-planned tissue collection, and good tissue processing practice. Here, we provide some guides to analyze the immunocytochemistry of IEGs:

1. Use animals of the same age.
2. Be sure to substitute cleansing the blood with physiological saline at room temperature to prevent high background histological staining,

followed by cleansing the blood with saline, perfuse with paraformaldehyde prepared phosphate buffer to minimize polymerization that can affect the quality of tissue fixation.

3. Cutting brain tissues into sections with defined thickness. Modern design-based stereology uses 3D stereological probes and optical dissector method, which requires minimal shrinkage and the use of thick rather than thin sections. However, if the sections are too thick, antibody penetration becomes problematic; therefore, 40–50 μm brain sections are optimal.
4. Design-based stereology requires that the experimenter access the entire region of interest, which is achieved by SRS of sections from the exhaustive section series encompassing the entire region.
5. The experimenter to be precise and consistent. To aid random sampling of tissues, use well plates to store sequential serial sections.
6. Select the brain sections that contain the regions to be quantified.

Design-based stereology requires SRS to be achieved at two levels: intersection and introsection (Zhao and van Praag, 2020). At the intersection level, all sections containing the neuronal nuclei to analyze must have an equal probability to be selected. To achieve this, select randomly from the well plates the sections to analyze. The intrasection level of SRS is achieved by selecting microscopic fields in a systematic random manner that guarantees that all parts of the neuronal nuclei have the same chance of contributing to the sampling. In addition to the suggestions mentioned above, it is crucial to select antibodies that have been previously validated and preferably have a strong signal when revealed. We suggest some acceptable practices for the selection and use of antibodies.

A common mistake using antibodies is not verified their specificity before their experimental use, especially when antibodies are purchased from a large vendor. The researchers assume that the vendor has verified the reagent's performance and that their reputation is a sufficient assurance. This lack of vigilance has resulted in the widespread use of cross-reactive antibodies and inaccurate data acquisition. To avoid these troubles, the antibody must be validated following the next recommendations: (1). Always include relevant positive and negative controls for validation of each batch. (2). Always first repeat the product datasheet results to make sure the antibody has not lost its integrity. (3). Validate in the application in which the antibody is used. (4). Validate in the tissue type or cell type in which the antibody is used. (5). Use the validated antibodies at their optimal dilutions/concentrations. (6). Compare results obtained with different antibodies from different sources (Voskuil et al., 2020).

In summary, we believe that the use of technics as transgenic mice and rats c–FOS–GFP (Cifani et al., 2012; Kim et al., 2015), 3D stereology based on optical fractionator (Ferrucci et al., 2018), clarity-based tissue clearing techniques (Du et al., 2018; Kanda et al., 2016), and the quantification of neuronal activity using iDISCO and light-sheet fluorescence microscopy (Corsetti et al., 2019; Perens et al., 2020) in combination with the use of chemogenetic using DREADDS (designer receptors exclusively activated by designer drugs) (Masseck et al., 2011; Takata et al., 2018), and optogenetic stimulation and silencing of the specific neuronal population; will allow understanding the function of sleep and will be to helpful to find to cure sleep disorders and sleep-associated neuropsychiatric disorders (Adamantidis et al., 2010; Collins et al., 2020). However, it is important to understand the limitation and caveats of the use of antibodies, molecular manipulation, detection, and optogenetics to resolve critical questions in the sleep-wake cycle study.

Statistical considerations when analyzing immunoreactive cells

In studies that use rodents, it is common to perform ordinal measurements (counts). Notably, in neuroscience studies, the cell count is common (Abbadie et al., 1994; Maloney et al., 1999; Pflüger et al., 2020). The specific designs used in the neuroscience field are characterized to contrast different factors and test a set of statistical hypotheses (Underwood et al., 1997). In this sense, a condition that strengthens a publication's impact depends largely on a correct statistical analysis (Habeck and Brickman, 2018; Makin and De Xivry, 2019; Lazic, 2010; Wilcox and Rousselet, 2018). A common mistake is to assume that the study data meet the statistical assumptions and make a type I or II error in the inferences obtained (Scariano and Davenport, 1987; Nimon, 2012; Dean et al., 2017). Hence, statistical analysis should be a reasoned process based on the research questions, hypotheses, and experimental design. The researchers also have to consider the reproducibility of the results.

There are assumptions and statistical considerations in research that delimit the experimental design, for example, if the statistical model fulfills the respective assumptions. We identify three potential causes that can lead to the publication of results that lack an adequate statistical analysis:

1. Lack of knowledge of statistical tools from the authors of the publication.
2. The journal's reviewers did not request the authors if the data analyzed to fulfill the statistical assumptions.
3. The associate editor omits during review process request database.

In this regard, in the already published article, a common mistake is the lack of information regarding the assumptions of the statistical analyses applied in the data (Abbadie et al., 1994; Maloney et al., 1999, 2002; Ni et al., 2020; Pflüger et al., 2020; Romero et al., 2012). In these studies, the statistical tests used are parametric models such as ANOVA, Linear regression, ANCOVA, MANOVA, repeated measures ANOVA, and Student t- and other tests. The use of these statistical tests requires that the data meet with the assumptions of the normal distribution, homogeneity of variances, and the lack of independence between data (pseudoreplicates) (Kenny and Judd, 1986; Bathke et al., 2018; Caldwell and Cheuvront, 2019; Nimon, 2012).

To avoid incurring in the situation aforementioned is essential to identify if the dependent variable (y) is continuous or ordinal; this helps to adjust the type of statistical analysis. Furthermore, it is necessary to verify whether or not there is a normal distribution with tests such as Kolmogorov–Smirnov, Shapiro–Wilk (Razali and Wah, 2011), and Jarque–Bera (Jarque and Bera, 1980). Also, it is recommended to graph the distribution of observed and expected frequencies (Gaussian curve), test the hypothesis contrast, in which Ho: refers to the equality between observed and expected frequencies ($P > .05$), which indicates that there is a normal distribution in the dependent variable. It also has to check the homogeneity of variances Ho: $\sigma^2 = \sigma^2 = \ \ldots \ \sigma^2$ accepted with its respective value of $P > .05$, with tests such as Levens's and Bartlett's test (Dean et al., 2017). Nevertheless, if the hypothesis Ha: is accepted ($P < .05$) for the two cases, then it is suggested to apply an analysis such as transforming the data into a logarithm or square root (Dean et al., 2017; Zar, 2010; Sokal and Rohlf, 1981), although it is also appropriate to use the Box–Cox transformation (Box and Cox, 1964).

This statical analysis solves the problems related to the lack of normality and homogeneity of variances in the dependent variable (Hall et al., 2017). However, if the experimental design does not involve more than one factor, nonparametric tests can be used with the dependent variable (Conover and Iman, 1981; Conover, 1998; Siegel and Castellan, 1988; Sheskin, 2020). Although neuroscience studies, it is common to have experimental designs with various factors and levels according to the questions and research aim. Therefore, it is recommendable to combine the rank on the response variable (nonparametric test) and use it as a dependent variable in the parametric model, according to (Conover and Iman, 1981). These analyses are usually useful in solving the limitations when these two assumptions are not fulfilled. Another option is nonlinear models, in which it is also necessary to describe the type of frequency distribution of the dependent variable. Examples of nonlinear distributions are Poisson, log-normal, exponential, binomial, curvilinear fit, and generally refer to frequency conditions on ordinal scales (Crawley, 1993; Bolker et al., 2009).

Pseudoreplication

In nature, it is common to identify individuals who have a hierarchical arrangement. Simple concerning conceptualizing the hierarchical design is observing the shape of a tree. The main trunk is distinguished (central axis), from which first-order branches originate, and these, in turn, grow second, third, and fourth-order branches up to the leaves, flowers, and reproductive structures. In this sense, in the branching of a tree, each branch is not independent. This hierarchical organization is similar when using animal models to do experiments. For example, in rats, immunoreactive neurons can be counted in the ventral or dorsal area in one of the cerebellum lobes, which are not independent. The counting cells in these areas are subordinated to a hierarchical unit representing the individual, the cerebellum. Therefore, the hierarchical designs include pseudoreplication, which is not a new concept, and there is sufficient evidence that has delimited its effect in statistical analyses (Hurlbert, 1984; Millar and Anderson, 2004; Schank and Koehnle, 2009; Lazic et al., 2020).

In this respect, the following example helps define how to apply the statistical model appropriate to each situation based on the research hypothesis and the various associated questions. In this regard, the article by Lazic (2010) points out precisely the implications of applying an erroneous analysis when pseudoreplication is implicit and indicates various ways to analyze the data.

The hierarchical designs (nested model) consider that the assumptions of the normal distribution of the dependent variable (number of immunoreactive neurons) and the homogeneity of variances were fulfilled after the specific analysis. The following example defines one of the frequent scenarios in experimental and observational designs. Two groups of rats, a control group (n = 6) and an experimental group (n = 6). After 7 days of treatment, the rats were deeply anesthetized via CO_2 inhalation, then transcardially perfused with phosphate-buffered saline (PBS), 4% paraformaldehyde in 0.1 M PBS (pH 7.4) and postfixed in the 4% paraformaldehyde fixative overnight. Subsequently, the brains of the rats were obtained and transferred to a 30% sucrose PBS solution until they sank; then quickly frozen in cooled isopentane and stored at −80°C until sectioning.

Each brain was sectioned serially in the coronal plane, using a sliding microtome at an instrument setting of 40 μm throughout the brain's entire extent. Then the tissue was processed to Fos immunohistochemistry. Five glass slides with 12 slices each were analyzed; immunoreactive neurons were counted in each rat's six brain sections (Fig. 8.1). In total, 8640 data were obtained, which are not independent (pseudoreplication) since independence resides in each rat corresponding to the groups (total $n = 12$). In appearance, a two-way ANOVA model could be used.

$$y = G2 + BS6 + G \times BS + error,$$

where y corresponds to the number of neurons immunoreactive to Fos, G is the rats' group, and BS stands for brain sections and the respective interaction. In this model, there is no independence and, consequently, origins the error type II, leading to incorrect inferences. One "solution" to control the effect of pseudoreplication is to obtain the averages at the different levels of the hierarchy, as an average of the 120 observations from the boxes/ovals of the five slides for each brain section per each rat (Fig. 8.1).

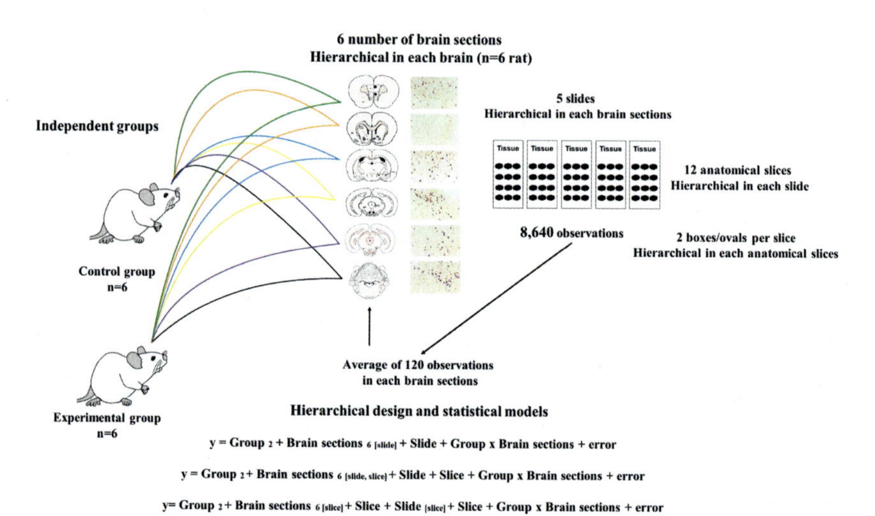

Figure 8.1 Schematic representation of the hierarchical design and statistical model. To details review the text. IL infralimbic cortex; NAc nucleus accumbens core; NAc Shell; AC anterior commissure; LH lateral hypothalamus; CeA central nucleus of the amygdala; VTA ventral tegmental area; PPT pedunculo pontine tegmental nucleus; LC locus coeruleus nucleus.

The average over the observations reduces the pseudoreplication. However, the two-way ANOVA with these averages would still have pseudoreplication since the database would have 72 data (Fig. 8.1). In this regard, the use of hierarchical models that include random effects is an excellent approach to include pseudoreplicate. A recommendation is to use a model with one to two nests and even a set of triple nesting or more (Fig. 8.1). These models are described in detail in several books (Sokal and Rohlf, 1981; Zar, 2010; Dean et al., 2017) and implemented in articles (Paredes-Ramos et al., 2011; García et al., 2015; Hernández-Briones et al., 2017; Rodríguez et al., 2018). Furthermore, almost all software includes hierarchical models (linear and nonlinear), allowing modeling analyses with experimental designs.

Conclusions

Immunoreactive cell counting continues to be one of the challenges facing researchers in the neuroscience field. The proper use of antibodies and the technical procedures required by each protocol are essential to obtain good results. We currently have potent technological tools and software; however, the key element lies in how the data is analyzed. In this sense, the publications must include the details and justification for using a specific statistical test. Assuming a data sample's statistical assumptions can lead not only to wrong analysis but also to a misinterpretation of the information.

Acknowledgments

This work was supported by CONACYT 254264 to F.G.G. We thank you Mario Acosta-Hernández Ph.D. for the edition of the manuscript.

References

Abbadie, C., Besson, J.M., Calvino, B., 1994. c-Fos expression in the spinal cord and pain-related symptoms induced by chronic arthritis in the rat are prevented by pretreatment with Freund adjuvant. J. Neurosci. 14 (10), 5865–5871.

Adamantidis, A., Carter, M.C., De Lecea, L., 2010. Optogenetic deconstruction of sleep-wake circuitry in the brain. Front. Mol. Neurosci. 2, 31. https://doi.org/10.3389/neuro.02.031.2009.

Atasoy, D., Sternson, S.M., 2018. Chemogenetic tools for causal cellular and neuronal biology. Physiol. Rev. 98 (1), 391–418. https://doi.org/10.1152/physrev.00009.2017.

Basheer, R., Porkka-Heiskanen, T., Stenberg, D., McCarley, R.W., 1999. Adenosine and behavioral state control: adenosine increases c-Fos protein and AP1 binding in basal forebrain of rats. Brain Res. Mol. Brain Res. 73 (1–2), 1–10. https://doi.org/10.1016/s0169-328x(99)00219-3.

Bathke, A.C., Friedrich, S., Pauly, M., Konietschke, F., Staffen, W., Strobl, N., Höller, Y., 2018. Testing mean differences among groups: multivariate and repeated measures analysis with minimal assumptions. Multivariate Behav. Res. 53 (3), 348–359.

Bolker, B.M., Brooks, M.E., Clark, C.J., Geange, S.W., Poulsen, J.R., Stevens, M.H.H., White, J.S.S., 2009. Generalized linear mixed models: a practical guide for ecology and evolution. Trends Ecol. Evol. 24 (3), 127–135.

Box, G.E., Cox, D.R., 1964. An analysis of transformations. J. Roy. Stat. Soc. B 26 (2), 211–243.

Caldwell, A.R., Cheuvront, S.N., 2019. Basic statistical considerations for physiology: the journal temperature toolbox. Temperature 6 (3), 181–210.

Cifani, C., Koya, E., Navarre, B.M., Calu, D.J., Baumann, M.H., Marchant, N.J., Liu, Q.-R., Khuc, T., Pickel, J., Lupica, C.R., Shaham, Y., Hope, B.T., 2012. Medial prefrontal cortex neuronal activation and synaptic alterations after stress-induced reinstatement of palatable food seeking: a study using c-fos-GFP transgenic female rats. J. Neurosci. Off. J. Soc. Neurosci. 32 (25), 8480–8490. https://doi.org/10.1523/JNEUROSCI.5895-11.2012.

Clément, O., Valencia Garcia, S., Libourel, P.-A., Arthaud, S., Fort, P., Luppi, P.-H., 2014. The inhibition of the dorsal paragigantocellular reticular nucleus induces waking and the activation of all adrenergic and noradrenergic neurons: a combined pharmacological and functional neuroanatomical study. PLoS One 9 (5). https://doi.org/10.1371/journal.pone.0096851.

Collins, B., Pierre-Ferrer, S., Muheim, C., Lukacsovich, D., Cai, Y., Spinnler, A., Herrera, C.G., Wen, S., Winterer, J., Belle, M.D.C., Piggins, H.D., Hastings, M., Loudon, A., Yan, J., Földy, C., Adamantidis, A., Brown, S.A., 2020. Circadian VIPergic neurons of the suprachiasmatic nuclei sculpt the sleep-wake cycle. Neuron 108 (3), 486–499.e5. https://doi.org/10.1016/j.neuron.2020.08.001.

Conover, W.J., 1998. Practical Nonparametric Statistics, vol. 350. John Wiley & Sons.

Conover, W.J., Iman, R.L., 1981. Rank transformations as a bridge between parametric and nonparametric statistics. Am. Statistician 35 (3), 124–129.

Corsetti, S., Gunn-Moore, F., Dholakia, K., 2019. Light sheet fluorescence microscopy for neuroscience. J. Neurosci. Methods 319, 16–27. https://doi.org/10.1016/j.jneumeth.2018.07.011.

Crawley, M.J., 1993. GLIM for Ecologists. Blackwell Scientific Publications, Oxford.

Dean, A., Voss, D., Draguljić, D., 2017. Design and Analysis of Experiments, second ed. International Publishing AG: Springer.

Du, H., Hou, P., Zhang, W., Li, Q., 2018. Advances in CLARITY-based tissue clearing and imaging. Exp. Ther. Med. 16 (3), 1567–1576. https://doi.org/10.3892/etm.2018.6374.

Ferrucci, M., Lazzeri, G., Flaibani, M., Biagioni, F., Cantini, F., Madonna, M., Bucci, D., Limanaqi, F., Soldani, P., Fornai, F., 2018. In search for a gold-standard procedure to count motor neurons in the spinal cord. Histol. Histopathol. 33 (10), 1021–1046. https://doi.org/10.14670/HH-11-983.

García, L.I., García-Bañuelos, P., Aranda-Abreu, G.E., Herrera-Meza, G., Coria-Avila, G.A., Manzo, J., 2015. Activation of the cerebellum by olfactory stimulation in sexually naive male rats. Neurologia 30 (5), 264–269.

García-García, F., Beltrán-Parrazal, L., Jiménez-Anguiano, A., Vega-González, A., Drucker-Colín, R., 1998. Manipulations during forced wakefulness have differential impact on sleep architecture, EEG power spectrum, and Fos induction. Brain Res. Bull. 47 (4), 317–324. https://doi.org/10.1016/s0361-9230(98)00071-9.

Gundersen, H.J.G., Bendtsen, T.F., Korbo, L., Marcussen, N., Møller, A., Nielsen, K., et al., 1988. Some new, simple and efficient stereological methods and their use in pathological research and diagnosis. APMIS 96 (1-6), 379–394.
Guzowski, J.F., Setlow, B., Wagner, E.K., Mcgaugh, J.L., 2001. Experience-dependent gene expression in the rat Hippocampus after spatial learning: a comparison of the immediate-early GenesArc, c-fos, and zif268. J. Neurosci. 21 (14), 5089–5098. https://doi.org/10.1523/JNEUROSCI.21-14-05089.2001.
Habeck, C., Brickman, A., 2018. A common statistical misunderstanding in Psychology and Neuroscience: do we need normally distributed independent or dependent variables for linear regression to work? BioRxiv 305946.
Hall, S., Deurveilher, S., Ko, K.R., Burns, J., Semba, K., 2017. Region-specific increases in FosB/ΔFosB immunoreactivity in the rat brain in response to chronic sleep restriction. Behav. Brain Res. 322, 9–17.
Hernández-Briones, Z.S., García-Bañuelos, P., Hernández, M.E., López, M.L., Chacón, A.M., Carrillo, P., et al., 2017. Olfactory stimulation induces cerebellar vermis activation during sexual learning in male rats. Neurobiol. Learn. Mem. 146, 31–36.
Hurlbert, S.H., 1984. Pseudoreplication and the design of ecological field experiments. Ecol. Monogr. 54 (2), 187–211.
Jarque, C.M., Bera, A.K., 1980. Efficient tests for normality, homoscedasticity and serial independence of regression residuals. Econ. Lett. 6 (3), 255–259.
Kanda, T., Tsujino, N., Kuramoto, E., Koyama, Y., Susaki, E.A., Chikahisa, S., Funato, H., 2016. Sleep as a biological problem: an overview of frontiers in sleep research. J. Physiol. Sci. (JPS) 66 (1), 1–13. https://doi.org/10.1007/s12576-015-0414-3.
Kenny, D.A., Judd, C.M., 1986. Consequences of violating the independence assumption in analysis of variance. Psychol. Bull. 99 (3), 422.
Kim, Y., Venkataraju, K.U., Pradhan, K., Mende, C., Taranda, J., Turaga, S.C., Arganda-Carreras, I., Ng, L., Hawrylycz, M.J., Rockland, K.S., Seung, H.S., Osten, P., 2015. Mapping social behavior-induced brain activation at cellular resolution in the mouse. Cell Rep. 10 (2), 292–305. https://doi.org/10.1016/j.celrep.2014.12.014.
Lazic, S.E., 2010. The problem of pseudoreplication in neuroscientific studies: is it affecting your analysis? BMC Neurosci. 11 (1), 5.
Lazic, S.E., Mellor, J.R., Ashby, M.C., Munafo, M.R., 2020. A Bayesian predictive approach for dealing with pseudoreplication. Sci. Rep. 10 (1), 1–10.
Makin, T.R., De Xivry, J.J.O., 2019. Science forum: ten common statistical mistakes to watch out for when writing or reviewing a manuscript. eLife 8, e48175.
Maloney, K.J., Mainville, L., Jones, B.E., 1999. Differential c-Fos expression in cholinergic, monoaminergic, and GABAergic cell groups of the pontomesencephalic tegmentum after paradoxical sleep deprivation and recovery. J. Neurosci. 19 (8), 3057–3072.
Maloney, K.J., Mainville, L., Jones, B.E., 2000. c-Fos expression in GABAergic, serotonergic, and other neurons of the pontomedullary reticular formation and raphe after paradoxical sleep deprivation and recovery. J. Neurosci. 20 (12), 4669–4679. https://doi.org/10.1523/JNEUROSCI.20-12-04669.2000.
Maloney, K.J., Mainville, L., Jones, B.E., 2002. c-Fos expression in dopaminergic and GABAergic neurons of the ventral mesencephalic tegmentum after paradoxical sleep deprivation and recovery. Eur. J. Neurosci. 15 (4), 774–778.
Masseck, O.A., Rubelowski, J.M., Spoida, K., Herlitze, S., 2011. Light- and drug-activated G-protein-coupled receptors to control intracellular signalling. Exp. Physiol. 96 (1), 51–56. https://doi.org/10.1113/expphysiol.2010.055517.
Merchant-Nancy, H., Vázquez, J., Aguilar-Roblero, R., Drucker-Colín, R., 1992. c-fos proto-oncogene changes in relation to REM sleep duration. Brain Res. 579 (2), 342–346. https://doi.org/10.1016/0006-8993(92)90072-h.

Merchant-Nancy, H., Vázquez, J., García, F., Drucker-Colín, R., 1995. Brain distribution of c-fos expression as a result of prolonged rapid eye movement (REM) sleep period duration. Brain Res. 681 (1–2), 15–22. https://doi.org/10.1016/0006-8993(95)00275-u.

Millar, R.B., Anderson, M.J., 2004. Remedies for pseudoreplication. Fish. Res. 70 (2–3), 397–407.

Ni, R.J., Wang, J., Shu, Y.M., Xu, L., Zhou, J.N., 2020. Mapping of c-Fos expression in male tree shrew forebrain. Neurosci. Lett. 714, 134603.

Nimon, K.F., 2012. Statistical assumptions of substantive analyses across the general linear model: a mini-review. Front. Psychol. 3, 322.

Paredes-Ramos, P., Pfaus, J.G., Miquel, M., Manzo, J., Coria-Avila, G.A., 2011. Sexual reward induces Fos in the cerebellum of female rats. Physiol. Behav. 102 (2), 143–148.

Perens, J., Salinas, C.G., Skytte, J.L., Roostalu, U., Dahl, A.B., Dyrby, T.B., Wichern, F., Barkholt, P., Vrang, N., Jelsing, J., Hecksher-Sørensen, J., 2020. An optimized mouse brain atlas for automated mapping and quantification of neuronal activity using iDISCO+ and light sheet fluorescence microscopy. Neuroinform. https://doi.org/10.1007/s12021-020-09490-8.

Peyron, C., Tighe, D.K., Van Den Pol, A.N., DE Lecea, L., Heller, H.C., Sutcliffe, J.G., Kilduff, T.S., 1998. Neurons containing hypocretin (orexin) project to multiple neuronal systems. J. Neurosci. Off. J. Soc. Neurosci. 18 (23), 9996–10015.

Pflüger, P., Pinnell, R.C., Martini, N., Hofmann, U.G., 2020. Chronically implanted microelectrodes cause c-fos expression along their trajectory. Front. Neurosci. 13, 1367.

Razali, N.M., Wah, Y.B., 2011. Power comparisons of Shapiro-Wilk, Kolmogorov-Smirnov, Lilliefors and Anderson-Darling tests. J. Stat. Model. Anal. 2 (1), 21–33.

Rickgauer, J.P., Deisseroth, K., Tank, D.W., 2014. Simultaneous cellular-resolution optical perturbation and imaging of place cell firing fields. Nat. Neurosci. 17 (12), 1816–1824. https://doi.org/10.1038/nn.3866.

Rodríguez, A.T., Vásquez-Celaya, L., Coria-Avila, G.A., Pérez, C.A., Aranda-Abreu, G.E., Carrillo, P., et al., 2018. Changes in multiunit activity pattern in cerebellar cortex associated to olfactory cues during sexual learning in male rats. Neurosci. Lett. 687, 241–247.

Romero, A., Gonzalez-Cuello, A., Laorden, M.L., Campillo, A., Vasconcelos, N., Romero-Alejo, E., Puig, M.M., 2012. Effects of surgery and/or remifentanil administration on the expression of pERK1/2, c-Fos and dynorphin in the dorsal root ganglia in mice. N. Schmied. Arch. Pharmacol. 385 (4), 397–409.

Scariano, S.M., Davenport, J.M., 1987. The effects of violations of independence assumptions in the one-way ANOVA. Am. Statistician 41 (2), 123–129.

Schank, J.C., Koehnle, T.J., 2009. Pseudoreplication is a pseudoproblem. J. Comp. Psychol. 123 (4), 421.

Sheskin, D.J., 2020. Handbook of Parametric and Nonparametric Statistical Procedures. crc Press.

Shiromani, P.J., Basheer, R., Thakkar, J., Wagner, D., Greco, M.A., Charness, M.E., 2000. Sleep and wakefulness in c-fos and fos B gene knockout mice. Brain Res. Mol. Brain Res. 80 (1), 75–87. https://doi.org/10.1016/s0169-328x(00)00123-6.

Shiromani, P.J., Mallik, M., Wiston, S., McCarley, R.W., 1995. Time course of Fos-like immunoreactivity associated with cholinergically induced REM sleep. J. Neurosci. 15 (5 Pt 1), 3500–3508. https://doi.org/10.1523/JNEUROSCI.15-05-03500.1995.

Shiromani, P.J., Wiston, S., McCarley, R.W., 1996. Pontine cholinergic neurons show Fos-like immunoreactivity associated with cholinergically induced REM sleep. Brain Res. Mol. Brain Res. 38 (1), 77–84. https://doi.org/10.1016/0169-328x(95)00325-m.

Siegel, S., Castellan JR., N.J., 1988. Nonparametric Statistics for the Behavioral Sciences, second ed. Mcgraw-Hill Book Company, New York, Ny, England.

Sokal, R.R., Rohlf, F.J., 1981. Biometry: The Principles and Practice of Statistics in Biological Research, 3d ed. (New York).

Takata, Y., Oishi, Y., Zhou, X.-Z., Hasegawa, E., Takahashi, K., Cherasse, Y., Sakurai, T., Lazarus, M., 2018. Sleep and wakefulness are controlled by ventral medial midbrain/pons GABAergic neurons in mice. J. Neurosci. Off. J. Soc. Neurosci. 38 (47), 10080–10092. https://doi.org/10.1523/JNEUROSCI.0598-18.2018.

Underwood, A.J., Underwood, A.L., Underwood, A.J., Wnderwood, A.J., 1997. Experiments in Ecology: Their Logical Design and Interpretation Using Analysis of Variance. Cambridge University Press.

Voskuil, J.L.A., Bandrowski, A., Begley, C.G., Bradbury, A.R.M., Chalmers, A.D., Gomes, A.V., Hardcastle, T., Lund-Johansen, F., Plückthun, A., Roncador, G., Solache, A., Taussig, M.J., Trimmer, J.S., Williams, C., Goodman, S.L., 2020. The Antibody Society's antibody validation webinar series. mAbs 12 (1). https://doi.org/10.1080/19420862.2020.1794421.

Wilcox, R.R., Rousselet, G.A., 2018. A guide to robust statistical methods in neuroscience. Curr. Protoc. Neurosci. 82 (1), 8–42.

Xie, Y., Ba, L., Wang, M., Deng, S., Chen, S., Huang, L., Zhang, M., Wang, W., Ding, F., 2019. Chronic sleep fragmentation shares similar pathogenesis with neurodegenerative diseases: endosome-autophagosome-lysosome pathway dysfunction and microglia-mediated neuroinflammation. CNS Neurosci. Ther. 26 (2), 215–227. https://doi.org/10.1111/cns.13218.

Yamuy, J., Mancillas, J.R., Morales, F.R., Chase, M.H., 1993. C-fos expression in the pons and medulla of the cat during carbachol-induced active sleep. J. Neurosci. 13 (6), 2703–2718. https://doi.org/10.1523/JNEUROSCI.13-06-02703.1993.

Yamuy, J., Sampogna, S., Morales, F.R., Chase, M.H., 1998. c-fos Expression in mesopontine noradrenergic and cholinergic neurons of the cat during carbachol-induced active sleep: a double-labeling study. Sleep Res. Online 1 (1), 28–40.

Zar, J.H., 2010. Biostatistical Analysis, fifth ed. Pearson Prentice Hall, Upper Saddle River, New Jersey.

Zhao, X., van Praag, H., 2020. Steps towards standardized quantification of adult neurogenesis. Nat. Commun. 11 (1), 4275. https://doi.org/10.1038/s41467-020-18046-y.

CHAPTER 9

Wireless vigilance state monitoring

Paul-Antoine Libourel
Neurosciences Research Center of Lyon, Inserm U1028 – CNRS UMR5292 – UCBL, Bron, France

Vigilance states: wake states, non-rapid eye movement (NREM) sleep, and rapid eye movement (REM) sleep can be defined by a combination of behavioral and physiological, neurological traits (Aserinsky and Kleitman, 1953; Blumberg et al., 2020; Borbély, 1982; Jouvet et al., 1959; Murali et al., 2003; Parmeggiani, 2003; Piéron, 1913; Roebuck et al., 2014; Schmidt et al., 1994; Snyder et al., 1964). Those parameters can be geographic (sleep sites), postural, kinematic, cerebral, metabolic, homeostatic, cellular, ... Thus, evaluated across time, the combination of a set of parameters enable to assess the vigilance state and sub-states (e.g., active/quiet wake, locomotion, drinking, phasic REM sleep, drowsiness, ...) of animals, including humans.

A comprehensive understanding of animal cognition, behavior, and physiology will not be possible without our ability to measure natural sleep and wakefulness. However, the way we measure the parameters that define these states, in particular, the restraint imposed by recording tethers, might affect the expression of these states, and limit such recordings to unnatural cage environments. In this chapter, we will first review the classical methods used to measure vigilance states and their limitations. Then, we will show how these states can be measured in untethered animals using telemetry or biologging. And, finally, we will review the emerging scientific opportunities to study vigilance states afforded by this new wireless technology.

Measuring states of vigilance

Electroencephalography

The physiologist Richard Caton was the first in 1875 to measure with a galvanometer, the electrical brain changes related to wake, sleep, and death in animals (Caton, 1875). In 1925, the psychiatrist Hans Berger discovered Alpha waves in humans (Berger, 1929). Since these important scientific discoveries, electroencephalography (EEG)—the recording of electrical

Methodological Approaches for Sleep and Vigilance Research
ISBN 978-0-323-85235-7
https://doi.org/10.1016/B978-0-323-85235-7.00009-0

brain potentials—is widely used in neuroscience to measure vigilance states. NREM sleep (or slow wave sleep) in mammals and birds could be identified, thanks to the recording of the cortical high-amplitude slow waves, and/or sleep spindles, whereas a low-amplitude desynchronization is typical from wakefulness and REM sleep like the hippocampal theta of active wake and REM sleep in rodents. This constitutes the basis EEG marker used to identify, score, and quantify primary vigilance states.

Usually, brain potentials can be assessed by using surface, subcutaneous, epidural (cortical), or intracortical electrodes. These different recording methods measure brain activity at different spatial scales (from cortical regions to the neuronal level) (Buzsáki et al., 2012). To identify vigilance states, EEG potentials are usually recorded from the surface of the scalp in humans or epidurally in animals; subcutaneous electrodes have also been used in a few studies of animals in the wild (Rattenborg et al., 2008; Scriba et al., 2013c). One major difference between these three levels of recording is obviously the degree of invasiveness (Buzsáki et al., 2012; Scriba et al., 2013b). In addition, unlike epidural electrodes, signals detected by cutaneous and subcutaneous electrodes are diminished and spectrally filtered by the tissue between the electrode and brain. The EEG recording method also influences the stability of the signals. Whereas epidural electrodes are usually anchored to the skull, they are less susceptible to movement artifacts than cutaneous and subcutaneous electrodes which are susceptible to artifacts as the electrodes can move on and under the skin. Finally, unlike epidural and subcutaneous electrodes, the progressive degradation of contact and impedance between the electrode and skin resulting from desiccating conductive paste renders cutaneous electrodes only suitable for short-term recordings in humans. The way the electrodes are fixed is therefore a major point to consider when recording vigilance states.

In addition to the electrode placement and fixation, one of the biggest limitations while recording the EEG is the signal conditioning (filtering and amplification) and digitalization. To quantify the vigilance states, one needs to get a good signal to noise ratio at a good temporal resolution. This requires an electronic device capable to amplify at least 500 times the signals and to sample them at a very minimum of 100 samples per second. As a consequence, to be recorded, animals and humans are, most of time, connected, thanks to cables (one per electrode and a reference) to the amplifier. This constitutes a great constraint when recording brain activity, especially for small animals that need to be kept in a small enclosures with a cable connected to their head. In addition, the tethered connection avoids the researcher to put any enrichment accessory like tubes or boxes, where the animal normally likes to hide.

Electrophysiology

Thanks to the discovery of bioelectricity by Luigi Galvani in 1791 (Galvani, 1791), the same technology used to record the EEG can be used to record many other biopotentials related to vigilance states. Indeed, not all of the vigilance states can be distinguished from one another by measuring only brain activity. The electrooculogram (EOG) and electromyogram (EMG; often nuchal muscles) are used to differentiate REM sleep from wakefulness (Aserinsky and Kleitman, 1953; Jouvet et al., 1959). To record ocular activity, electrodes are often inserted in the supraorbital bone or under the eyelids, whereas the EMG is recorded by inserting electrodes into the nuchal muscle in animals or by putting electrodes on the chin in humans. These two biopotentials provide information on the density of eye movements (high during REM sleep and wake) and muscle activity which is atonic during REM sleep. Another common parameter also related to vigilance state is heart rate. By fixing electrodes on the chest, the cardiovascular function can be measured (Murali et al., 2003). The electrocardiogram (ECG) gives pertinent information to estimate the level of vigilance. Heart rate and its variability provide a measure of changes in autonomic nervous system activity during sleep and its substates (Chouchou and Desseilles, 2014; Snyder et al., 1964; Toscani et al., 1996). However, because of its invasiveness in animal, the ECG is rarely recorded in animal studies. By contrast, in humans, the ECG is usually recorded to measure the individual's health status and the consequences that sleep disorders, such as sleep apnea have on cardiac function.

As we saw, many electrophysiological parameters need to be recorded to identify vigilance states. However, the methods to record these parameters suffer from the same limitations as the EEG recording: the electrode fixation, the connection to the amplifier, and the digitalizer which impacts the subject freedom for long-term recording.

Other measures related to vigilance states

Vigilance states can be characterized by cerebral and electrophysiological changes, as well as other parameters. Parameters like location, position, speed, temperature, breathing, and arousal threshold are also of interest. Indeed, sleep states are often characterized by a specific behavioral position in a specific location like the stereotypic position of the giraffe during REM sleep (Tobler and Schwierin, 1996). Myoclonic movements (twitches) that occur during REM sleep could also serve to its identification. In addition, during wakefulness position, movement, and speed identify substates, like locomotion, grooming, flying, diving, jumping, eating, drinking, ... The temperature (brain and body) and the breathing rate are also physiological parameters that are correlated with vigilance states (Parmeggiani, 2003; Roebuck et al., 2014; Snyder et al., 1964; Ungurean et al., 2020). However, except in human polysomnography where some of these parameters are recorded during a night, it is rare to record them to asses vigilance states in research. One explanation is still the difficulty to record them chronically, simultaneously with electrophysiological parameters, as this will increase the number of cables to the recording device. This is the case for temperature, breathing rate, and oxygen saturation. Location, position, and speed are accessible by video, but their precise quantification is difficult if the acquisition is not standardized (see next part). Finally, arousal threshold is another behavioral parameter linked to vigilance state. It is low during wake and high during sleep, and can differ between sleep states, torpor, and hibernation. Unfortunately, this parameter is extremely difficult to measure, in part, because the sensory systems and their sensitivity vary across species. Some species are more sensitive to visual stimuli, others to olfactory, auditory, or tactile stimuli. In addition, all animals do not respond in the same manner to stimulation. If the stimuli cause a large, fearful response, assessing arousal thresholds can lead to sleep deprivation. Alternatively, if the stimulus only evokes a small response, the animal might quickly habituate to the stimuli. Either case, poses a problem for investigations of arousal thresholds.

As we saw, parameters other than EEG, ECG, EOG, EMG are of broad interest; however, for chronic recordings of vigilance state, they are rarely used, particularly, in animal research.

Wireless monitoring of vigilance states

Historically, the large size of the recording device has been the main obstacle to assessing vigilance states in freely behaving animals, unhindered by a recording cable. However, multiple solutions now exist to overcome this obstacle. Video, telemetry, and biologging are more and more widespread and constitute three valuable approaches.

Video

Video cameras have widely been used to monitor animal behavior. Before the emergence of hard drives and digital cameras, the durations of the movies were highly limited by the storage technology (Super8, VHS, ...). Since the discovery of charge coupled device (CCD) and Complementary Metal Oxide Semiconductor (CMOS) sensors (digital sensors), video recorders were widely used for behavior monitoring. The improvement of compression algorithms as the increase of the data storage capacity, largely also contributed to the use of video in long-term animal and human behavioral assessments. Indeed, thanks to the temporal (frame rate) and spatial (number of pixels) information provided, video recorders can provide important measures of the vigilance states. The position, the location of the subjects can be identified over long periods of time. The occurrence of twitches during REM sleep can also be observed and quantified. Generally, researchers and clinicians who are interested in monitoring behavior during the night time, use near infrared light (typical wavelength between 800 and 1000 nm), light which is not seen by most animal species. The video camera used has black and white sensors (without Bayer filter) that are sensitive to near infrared wavelengths. This type of setup has been used for assessing animal behavior, animal tracking, and measuring activity (number of pixels changing) (Aguiar et al., 2007; Crispim Junior et al., 2012; Delcourt et al., 2009; Land, 1992; Mathis et al., 2018; Miller and Gerlai, 2012; Patel et al., 2014; Pereira et al., 2020; Salem et al., 2015; Samson et al., 2015; Tort et al., 2006; White et al., 1989). Image processing methods have been proposed to score more complex behavior states (Jhuang et al., 2010; Parkison et al., 2012; Pereira et al., 2020; Singh et al., 2019). However, sleep sub-states remain difficult to assess precisely by this means, even if some methods seem to report good approximations (McShane et al., 2012). Indeed, if the recording conditions are adapted (field of view, sampling rate, resolution, contrast, light, color video, stereo vision ...) one can measure breathing rate, heart rate, or eye movements (Al-Naji et al., 2019; Ben-Simon et al., 2009; Lauridsen et al., 2011). As these parameters are highly correlated with vigilance states, if recorded, they could bring more information. Infrared thermography is another way to assess an animal's physiological states (Ferretti et al., 2019). Thermal cameras measure the radiative light of the subject in the infrared spectrum which is directly proportional to the surface temperature (wavelength between 9000 to 14,000 nm). In addition to measuring the thermal state, this method can also measure breathing rate (Hu et al., 2018; Murthy et al., 2009) and help identify sleep states (Seba et al., 2017).

The integration of different types of measurements certainly would enhance assessments of the level of vigilance. In addition, the emergence of deep-learning methods and open-source software will allow researchers to develop new ways to assess vigilance states. However, the efficacy of such approaches necessarily depends upon the quality of the video images. To use video quantitatively, the recording settings (like shutter speed, gain, frame rate, and brightness) need to be fixed to provide stable light exposure. Also, the animal should always be at the same distance from the camera to avoid any bias due to the magnification effect. To measure physiological parameters, the region of interest (chest, face, skin, body, and limbs) needs to remain visible. These constraints of video can lead the experimenter to exclude shelters, habitat complexity, or natural substrates (plants, stones ...) that could be incompatible with video monitoring. Then, the animal may become "tethered" to a specific location by the constraints of the video system. In addition, the huge volume of the data generated and the time devoted to image analysis also could constitute a limitation when a project requires a large sample size.

Thus, although video is a good alternative to tethered, electrophysiological recordings, it depends heavily on the quality (visibility) of the video. The compromise is either to constrain the mobility of the subject into a "poor" environment to keep it in the field of view of the cameras, or to analyze only periods when the subject is in the field of view, an unsatisfactory solution that might lead to biases in the data. The combination between piezo sensors into the cage floor that can measure heart rate, breathing rate and/or activity parameters (Carreño-Muñoz et al., 2020; Yaghouby et al., 2016), video processing (McShane et al., 2012), or deep learning technics (Mathis et al., 2018) has been also used and could provide a better approximation of vigilance states by cumulating information and reducing the data extraction time.

Telemetry

Telemetry refers to technology that transfers data in real time though wireless transmission (radio, ultrasonic, or infrared). This method allows signals from classical electrodes and sensors for EEG, ECG, EOG, EMG, temperature monitoring to be measured, but without tethering the animal to the recording equipment. Telemetry requires that a device for transmitting the signals is fixed to the animal and a receiver to be connected to the recording station. This technology is possible due to the miniaturization of low-power electronics. The signals can be transmitted using radio frequencies. Each transmission technology has its strengths and weaknesses. Radio frequency technology (frequency range from around 20 kHz to around 300 GHz) is the more widespread in untethered electrophysiology. The radio frequency ranges used fall into two primary ranges, less than 1 GHz and higher. Lower frequencies need less energy to transmit a signal per meter due to their better ability to cross matter (air, biological tissue, cages, wall ...). To transmit information through radiofrequency, a transmitter converts the signal into specific frequencies, which is called modulation. The frequencies used for the modulation belong to an allocated frequency band (bandwidth) around a carrier frequency. Then all signals are usually transmitted on one carrier frequency. As a consequence, the larger the bandwidth, the larger the bit rate will be (number of data points per second in bits per second, bps). The way the modulation is done is defined by a communication protocol. As an example, Wi-Fi, Bluetooth, or radio-frequency identification (RFID) are three generic protocols widely used, with different frequency ranges and bandwidths adapted to the application. Other proprietary protocols also exist and can be used for specific applications by using dedicated transmitters/receivers. The data sent by the transmitter are then collected and decoded by the receiver. The maximum transmitter/receiver distance is determined by the transmission power which is legally regulated for each frequency band. The choice of one technology over another then comes down to the goal of the instrumentation and the constraints of the recording environment. The major criteria are: (1) the bitrate, which depends mainly on the number of channels recorded, the sampling rate and the resolution. (2) the transmission range, which is determined by the size and complexity of the animal's enclosure. (3) the power consumption which is related to the transmission range. (4) the battery size and weight, which determines the recording duration.

For research on animals, telemetric electrophysiological devices are usually implanted into the abdominal cavity (Cesarovic et al., 2011; Snelderwaard et al., 2006). These systems often use frequencies lower than 1 GHz (400, 800 MHz) because of their capacity to cross biological tissue (Axelsson et al., 2007; Knot and Lee, 2016; Lundt et al., 2015). As implantable devices need to be as small as possible, they usually record only a few channels at low-sampling rate and the transmission distance is low.

Recording the EEG with implantable devices is difficult in small animals as the sampling rate needs to be high and large batteries are required to record for days or weeks. Therefore, such implantable devices usually only record physiological parameters like heart rate, blood pressure, and body temperature, parameters that do not require such a high-sampling rate. One way to overcome to these limitations is to use near field communication like the RFID protocol (13 MHz), a passive technology that requires no battery in the transmitter, or more generally wireless communications employing inductive links. This transmission has a very low bitrate (<1 kbps) and a very low-distance range (few centimeters). Transmitting only one channel at very low-sampling rate is thus possible with this technology (Caldara et al., 2016). In summary, the use of implantable technology to measure vigilance states is possible but remains difficult due to the sampling requirements of the EEG and limitations of the technology.

Another approach is to use an external transmitter. The transmitter can be fixed over the head of laboratory animals, in a backpack, or, in humans, in a pocket. These devices generally use frequencies higher than 1 GHz, with 2.4 GHz being the most widespread for common, research and medical uses. Different communication protocols exist to transmit data in the 2.4 GHz band that vary in terms of reliability, loss of data, redundancy, error correction, etc., which directly affect power consumption and the bitrate by sending more or less data. Wi-fi contains 11 overlapping frequency bands (bandwidth 20 MHz) that theoretically allow for a bit rate ranging from 11 Mbps to 1,3 Gbps at 100 m (Wi-fi 802.11ac). Bluetooth also works in the 2.4 GHz frequency band, but is subdivided into 40 bands (bandwidth 2 MHz). This makes it less power consuming than Wi-fi, but the bitrate is also reduced; a comparison to Wi-fi, Bluetooth can be used to transmit 2–3 Mbps at 10 m (for example, two EEG and one EMG sampled at 256 Hz with a 16 bits resolution represents 12 kbps). The development of a telemetric device to monitor vigilance states is then a compromise between, autonomy, number of channels, sampling rate, type of measurement, size, and weight. Multiple transmitters have been developed with different specifications to record brain activity and physiology (Caldara et al., 2016; Harrison et al., 2011; Massot et al., 2019a; Mohseni et al., 2005; Weiergräber et al., 2005; Ye et al., 2008; Zayachkivsky et al., 2013). However, very few were designed to characterize the whole phenotype of vigilance states. Indeed, in humans, polysomnography is largely used to quantify sleep states by recording different EEG derivation associated with physiological and behavioral measures, whereas in animal research, vigilance states can be measured for long periods of time wirelessly using only a few channels (often EEG and EMG). If more measurements are needed, the autonomy of the device then becomes insufficient for long-term experiments. Combining electrophysiological measurements, brain activity, temperatures, activity from accelerometery and arousal threshold evaluation for multiple days to make polysomnography in small animals was recently proposed by Massot et al. (Massot et al., 2019a).

We saw that different tools are available to quantify vigilance states. In addition to the previously mentioned constraints, video and telemetric methods are not suitable for recording in freely moving animals in natural and unlimited environments (far from any receiver, computer). Global System for Mobile (GSM) communication protocol (700–800 MHz) could transmit data all around the world through a mobile antenna. Theorically, GSM can carry the large volume of electrophysiological data, but the power consumption requires a large battery (GSM 2G; bitrate 9.6 kbps), making it incompatible with continuous recordings in small animals. However, it could be developed for telemedicine allowing the deployment of vigilance state tele-monitoring all around the world. Very high frequency (VHF, 30–300 MHz) or global positioning system (GPS, 1,2 and 1,5 HGz, bitrate: 50 bps) could also be used, however the bitrate relative to the power consumption still remains incompatible with continuous monitoring of physiological parameters (like EEG, ECG, etc.). Given the limitations of telemetry, other methods for recording vigilance states in the wild are needed. One solution is to store the data on the subject itself.

Biologging

Biologging can be defined as biological measurements collected and stored on the subject itself. Multiple biologgers have been developed to record different parameters from animals (speed, position, angle, movement, body temperature, etc) and from the environment (temperature, pressure, altitude, position, humidity, light, etc.) (Whitford and Klimley, 2019; Williams et al., 2020). These measures are often collected from accelerometers, gyroscopes, magnetometers, thermistors, pressure sensors, or GPS. The device can be carried on the back, the head, or implanted sub-cutaneously. One major limitation of biologgers is the need to recatch the animal to remove/download the device (Buil et al., 2019; Rafiq et al., 2019). Although drop-off mechanisms have been developed, these are large and only suitable for very large animals. An alternative to animal recapture or drop-off mechanisms is to retrieve the data remotely when the animal is near a receiver (downloading stations). As this approach is in effect intermittent or delayed telemetry, it is limited by some of the same constraints as telemetry. Despite the limitations of biologging technologies, this approach is being used in more and more biological studies, including those assessing vigilance states.

Initially, biologging techniques were largely developed to track animals and quantify their activity (Rutz and Hays, 2009). Assuming that a correlation exists between the level of activity and vigilance states, such actigraphy could provide an estimate of time spent awake and asleep. It seems likely that activity sensors (mainly accelerometers, but sometimes including magnetometers and gyroscopes) placed on the head give a more accurate measure of vigilance state than those placed on the body, as awake animals often move only their head. An obvious limitation of actigraphy is that an animal can be awake while keeping the body and head still. This is particularly true for ectotherms that can bask in the sun motionless, but awake (Libourel et al., 2018). Conversely, some animals are able to remain active while sleeping with only one half of the brain (Lyamin et al., 2008, 2018; Rattenborg et al., 2016). Another limitation of actigraphy is that it usually cannot be used to distinguish between NREM and REM sleep (Rattenborg et al., 2017).

Given these limitations, it is essential to verify whether actigraphy provides an accurate estimate of vigilance states. To do this, actigraphy needs to be compared to measures of sleep and wakefulness, derived from video or electrophysiology recordings (Lesku et al., 2012; Rattenborg et al., 2017). Validation of actigraphy is often absent from publications or insufficiently demonstrated. Similarly, despite the widespread use of devices purporting to measure human sleep (e.g., accelerometry-based mobile phones applications, fitness watches, etc.), their validity remains disputed (Boe et al., 2019; Haghayegh et al., 2019).

Recently, electrophysiological biologgers have been developed and used to asses directly brain states in animals (Massot et al., 2019b; Vyssotski, 2005) and humans (Mikkelsen et al., 2019). Combined with accelerometry, electrophysiological biologgers provide direct measures of vigilance states and waking behaviors. Using this technology, important questions on sleep adaptations are starting to been addressed in the wild (Lesku et al., 2012; Rattenborg et al., 2008, 2016; Scriba et al., 2013c; Voirin et al., 2014). However, the number of studies of this sort remains limited, in part because of the difficulty to deploy the devices and retrieve the data. Indeed, to measure the EEG, muscle tone, and ocular activity, surgery is usually required. Less invasive, nonsurgical, methods exist (i.e., subcutaneous and cutaneous EEG electrodes), but the electrode remains much more susceptible to movement artifacts (Scriba et al., 2013a), and in the case of cutaneous electrodes, are not suitable for long-term, undisturbed recordings of animals (Scriba et al., 2013b). Measuring and quantifying vigilance states in the wild remains of great interest, but is limited to specific animal models as technical and practical limitations still exist.

Limitations

Our ability to assess vigilance states and, more generally, behavioral states is limited by multiple factors and compromises. The most important is the number of biological parameters required to quantify and identify the states. The more parameters we have, the more precise its description will be. However, this necessarily increases the quantity of data collected. As vigilance states, more specifically sleep states, can be identified with certainty with at least one EEG and another variable depending on the type of animal (EMG, EOG, head position, …), this directly limits the type of wireless technology that can be used to measure vigilance states. This also often requires invasive procedures, as actually no non-invasive methods are able to provide precise and stable, long-term measures of the vigilance states. This certainly constitutes one important axis for future development.

Chronic recordings lasting several days are required to fully understand an animal's sleep patterns (amount, circadian timing, architecture, and homeostatic regulation). The major constraint of performing such recordings wirelessly (telemetry or biologger) is the size of the battery that the animal needs to carry. Currently, this limitation makes it impossible to record very small animals (<30 g) wirelessly, as it is recommended that the weight of the instrumentation should not exceed 5% of the animal weight (Portugal and White, 2018), at least in the wild. The placement of the weight on the animal (head vs. back), the type of animal and its morphology also condition the type of device and the maximum weight that the animal can carry. The size of the battery is determined, in part, by the volume of data collected, which is obviously determined by the sampling rate, the number of biological parameters collected and the needed recording duration. This is even more true when videos are recorded at the same time with animal-borne cameras (Thomson and Heithaus, 2014). An alternative to reduce the weight is to remove the battery and to provide power by induction. However, this supposes to have a quite small distance between the transmitter and the energy source (few centimeters), limiting its use in a large enclosure. Each type of battery has its own energy density specific to its chemical processes that influence the size and weight of the battery. Overall, the battery remains the main limiting factor, in terms of weight and size, to quantify vigilance states wirelessly in small animals. Indeed, the size of electronic devices is actually very small compared to the battery needed to make them work for a long time.

Another point limiting the use of any devices to assess vigilance states is its impact itself on the animal's behavior. Indeed, catching an animal, conducting a surgery, fitting a backpack, or fixing any device on it could modify its normal behavior. Even if sleep is a behavior that cannot be avoided, its architecture can be modified by the stress induced. As no perfect method exists to avoid this, the impact of the stress induced by the device itself should not be ignored. Letting the animal habituate and recover for at least multiple days after the equipment phase is necessary. In the case of wild recordings, this requires either a delay when starting the recordings after the surgery or larger batteries. Comparing the behavior of instrumented and uninstrumented animals can also be informative. For example, in a highly competitive setting, male sandpipers instrumented to record their sleep were not more likely to lose their territory than males that were instrumented. Their activity patterns were also similar to uninstrumented birds (Lesku et al., 2012). The long-term impact should also be assessed when possible. In humans (or big animals), the limiting factor is generally not the weight or the invasiveness of the device, but the wires between the sensors, like electrodes, and the transmitters. This is also true for small animals when recorded wirelessly with devices placed off the head. The ergonomics of the recording device then constitute a key factor for the use of such devices to monitor behavioral states. The use of thin wires, implanted subcutaneously, when possible, to avoid entanglement is often the best solution to limit problem.

As a summary, recording vigilance states wirelessly is a balance between the precision of the vigilance state determination, the number of bioparameters to record, the duration of the recording, the invasiveness, the stability and quality of the signals acquired, the maximum weight that the subject can carry, and the autonomy.

Despite some limitations, assessing vigilance states wirelessly remains of great interest. This approach is extremely dynamic and new technologies are arising. This will allow researchers to address new questions and reach a new understanding of vigilance states.

New opportunities offered by the wireless monitoring of the vigilance states

The wireless identification and quantification of vigilance states is highly valuable for research in animal and humans across multiple scientific domains. Clinicians usually conduct polysomnography at the hospital to identify sleep pathologies. However, the advent of wireless technology allows polysomnography to be conducted at home, improving the quality of the measures and the comfort of the patient, avoiding the disturbance induced when sleeping at the hospital. Despite the fact that some improvements are still needed to obtain good and stable EEG recordings at home, this approach will undoubtedly be highly beneficial to sleep clinicians.

Removing the cable is also of interest for researchers who use animal models. Indeed, tethered vigilance state recordings highly limit the animal's movement and freedom, even when the weight of the tether is counterbalanced. In the lab, wireless technology allows animals to be housed in enriched, spatially complex (3D), naturalistic environments, directly improving the animal's welfare and, as a consequence, the sleep quality by reducing stress. In the wild, wireless technology allows vigilance states to be recorded in completely unrestrained animals. This permits investigating potential relationships between sleep and various ecological factors, such as sexual competition, parental care, or predation. Indeed, despite being an apparently simple behavior, sleep is present in all species carefully examined, and has been implicated in various important processes, such as memory consolidation, brain development, emotion regulation, immune system regulation, and brain maintenance. As such, sleep plays an important role in maintaining adaptive performance during wakefulness. Studying how vigilance states vary relative to the environment and how it affects daily animal performance, then, is of high interest; wireless biologging technology has now made this possible (Aulsebrook et al., 2016; Rattenborg et al., 2017).

Finally, evaluating vigilance states while the subject is performing a task, such as driving a car, can provide information on their level of attention. Developing wireless tools that detect drowsiness or the onset of sleep, for example, could provide feedback to prevent an accident. Domains that require a high level of attention, like driving, could therefore could benefit greatly from a wireless vigilant state recording (Bergasa et al., 2006; Guo et al., 2018).

Even if multiple limitations still need to be overcome, the wireless monitoring of vigilance states is undoubtedly opening new frontiers in research, diagnosis, and safety.

References

Aguiar, P., Mendonça, L., Galhardo, V., 2007. OpenControl: a free opensource software for video tracking and automated control of behavioral mazes. J. Neurosci. Methods 166, 66–72. https://doi.org/10.1016/j.jneumeth.2007.06.020.

Al-Naji, A., Tao, Y., Smith, I., Chahl, J., 2019. A pilot study for estimating the cardio-pulmonary signals of diverse exotic animals using a digital camera. Sensors 19, 5445. https://doi.org/10.3390/s19245445.

Aserinsky, E., Kleitman, N., 1953. Regularly occurring periods of eye motility, and concomitant phenomena, during sleep. Science 118, 273–274.

Aulsebrook, A.E., Jones, T.M., Rattenborg, N.C., Roth, T.C., Lesku, J.A., 2016. Sleep ecophysiology: integrating neuroscience and ecology. Trends Ecol. Evol. 31, 590–599. https://doi.org/10.1016/j.tree.2016.05.004.

Axelsson, M., Dang, Q., Pitsillides, K., Munns, S., Hicks, J., Kassab, G.S., 2007. A novel, fully implantable, multichannel biotelemetry system for measurement of blood flow, pressure, ECG, and temperature. J. Appl. Physiol. 102, 1220–1228. https://doi.org/10.1152/japplphysiol.00887.2006.

Ben-Simon, A., Ben-Shahar, O., Segev, R., 2009. Measuring and tracking eye movements of a behaving archer fish by real-time stereo vision. J. Neurosci. Methods 184, 235–243. https://doi.org/10.1016/j.jneumeth.2009.08.006.

Bergasa, L.M., Nuevo, J., Sotelo, M., Barea, R., Guillén, M.E., 2006. Real-time system for monitoring driver vigilance. In: IEEE Trans. Intell. Transp. Systs, 3, pp. 1303–1308. https://doi.org/10.1109/ISIE.2005.1529113.

Berger, H., 1929. Über das elektrenkephalogram des menschen. I. Arch. Psychiat. Nervenkr. 87, 527–570.

Blumberg, M.S., Lesku, J.A., Libourel, P.-A., Schmidt, M.H., Rattenborg, N.C., 2020. What is REM sleep? Curr. Biol. 1, R38–R49. https://doi.org/10.1016/j.cub.2019.11.045.

Boe, A.J., McGee Koch, L.L., O'Brien, M.K., Shawen, N., Rogers, J.A., Lieber, R.L., Reid, K.J., Zee, P.C., Jayaraman, A., 2019. Automating sleep stage classification using wireless, wearable sensors. NPJ Digit. Med. 2, 1–9. https://doi.org/10.1038/s41746-019-0210-1.

Borbély, A.A., 1982. A two process model of sleep regulation. Hum. Neurobiol. 1, 195–204.

Buil, J.M.M., Peckre, L.R., Dörge, M., Fichtel, C., Kappeler, P.M., Scherberger, H., 2019. Remotely releasable collar mechanism for medium-sized mammals: an affordable technology to avoid multiple captures. Wildl. Biol. 1–7. https://doi.org/10.2981/wlb.00581.

Buzsáki, G., Anastassiou, C.A., Koch, C., 2012. The origin of extracellular fields and currents — EEG, ECoG, LFP and spikes. Nat. Rev. Neurosci. 13, 407–420. https://doi.org/10.1038/nrn3241.

Caldara, M., Nodari, B., Re, V., Bonandrini, B., 2016. Miniaturized blood pressure telemetry system with RFID interface. Electronics 5, 51. https://doi.org/10.3390/electronics5030051.

Carreño-Muñoz, M.I., Medrano, M.C., Leinekugel, T., Bompart, M., Martins, F., Subashi, E., Aby, F., Frick, A., Landry, M., Grana, M., Leinekugel, X., 2020. Detecting fine and elaborate movements with piezo sensors, from heartbeat to the temporal organization of behavior. bioRxiv. https://doi.org/10.1101/2020.04.03.024711, 2020.04.03.024711.

Caton, R., 1875. Electrical currents of the brain. J. Nerv. Ment. Dis. 2, 610.

Cesarovic, N., Jirkof, P., Rettich, A., Arras, M., 2011. Implantation of radiotelemetry transmitters yielding data on ECG, heart rate, core body temperature and activity in free-moving laboratory mice. J. Vis. Exp. 57, 3260. https://doi.org/10.3791/3260.

Chouchou, F., Desseilles, M., 2014. Heart rate variability: a tool to explore the sleeping brain? Front. Neurosci. 8 https://doi.org/10.3389/fnins.2014.00402.

Crispim Junior, C.F., Pederiva, C.N., Bose, R.C., Garcia, V.A., Lino-de-Oliveira, C., Marino-Neto, J., 2012. ETHOWATCHER: validation of a tool for behavioral and video-tracking analysis in laboratory animals. Comput. Biol. Med. 42, 257–264. https://doi.org/10.1016/j.compbiomed.2011.12.002.

Delcourt, J., Becco, C., Vandewalle, N., Poncin, P., 2009. A video multitracking system for quantification of individual behavior in a large fish shoal: advantages and limits. Behav. Res. Methods 41, 228–235. https://doi.org/10.3758/BRM.41.1.228.

Ferretti, A., Rattenborg, N.C., Ruf, T., McWilliams, S.R., Cardinale, M., Fusani, L., 2019. Sleeping unsafely tucked in to conserve energy in a nocturnal migratory songbird. Curr. Biol. 29, 2766–2772.e4. https://doi.org/10.1016/j.cub.2019.07.028.

Galvani, L., 1791. De viribus electricitatis in motu musculari. Commentarius. Bonoiensi Sci. Artium Intituo Acad. Comment. 7, 363–418.

Guo, Z., Pan, Y., Zhao, G., Cao, S., Zhang, J., 2018. Detection of driver vigilance level using EEG signals and driving contexts. IEEE Trans. Reliab. 67, 370–380. https://doi.org/10.1109/TR.2017.2778754.

Haghayegh, S., Khoshnevis, S., Smolensky, M.H., Diller, K.R., Castriotta, R.J., 2019. Accuracy of wristband fitbit models in assessing sleep: systematic review and meta-analysis. J. Med. Internet Res. 21 https://doi.org/10.2196/16273.

Harrison, R.R., Fotowat, H., Chan, R., Kier, R.J., Olberg, R., Leonardo, A., Gabbiani, F., 2011. Wireless neural/EMG telemetry systems for small freely moving animals. IEEE Trans. Biomed. Circ. Syst. 5, 103–111. https://doi.org/10.1109/TBCAS.2011.2131140.

Hu, M., Zhai, G., Li, D., Fan, Y., Duan, H., Zhu, W., Yang, X., 2018. Combination of near-infrared and thermal imaging techniques for the remote and simultaneous measurements of breathing and heart rates under sleep situation. PLoS One 13, e0190466. https://doi.org/10.1371/journal.pone.0190466.

Jhuang, H., Garrote, E., Yu, X., Khilnani, V., Poggio, T., Steele, A.D., Serre, T., 2010. Automated home-cage behavioural phenotyping of mice. Nat. Commun. 1, 68. https://doi.org/10.1038/ncomms1064.

Jouvet, M., Michel, F., Courjon, J., 1959. [On a stage of rapid cerebral electrical activity in the course of physiological sleep]. C. R. Seances Soc. Biol. Fil. 153, 1024–1028.

Knot, H.J., Lee, D., 2016. A novel freely moving animal based blood pressure, ECG, and dual body temperature telemetry system for group housed mice in social context. Faseb. J. 30 https://doi.org/10.1096/fasebj.30.1_supplement.lb595 lb595–lb595.

Land, M., 1992. Locomotion and visual behavior of mid-water Crustaceans. J. Mar. Biol. Assoc. U K 72, 41–60. https://doi.org/10.1017/S0025315400048773.

Lauridsen, H., Hansen, K., Wang, T., Agger, P., Andersen, J.L., Knudsen, P.S., Rasmussen, A.S., Uhrenholt, L., Pedersen, M., 2011. Inside out: modern imaging techniques to reveal animal anatomy. PloS One 6, e17879. https://doi.org/10.1371/journal.pone.0017879.

Lesku, J.A., Rattenborg, N.C., Valcu, M., Vyssotski, A.L., Kuhn, S., Kuemmeth, F., Heidrich, W., Kempenaers, B., 2012. Adaptive sleep loss in polygynous pectoral sandpipers. Science 337, 1654–1658. https://doi.org/10.1126/science.1220939.

Libourel, P.-A., Barrillot, B., Arthaud, S., Massot, B., Morel, A.-L., Beuf, O., Herrel, A., Luppi, P.-H., 2018. Partial homologies between sleep states in lizards, mammals, and birds suggest a complex evolution of sleep states in amniotes. PLoS Biol. 16, e2005982. https://doi.org/10.1371/journal.pbio.2005982.

Lundt, A., Wormuth, C., Siwek, M.E., Müller, R., Ehninger, D., Henseler, C., Broich, K., Papazoglou, A., Weiergräber, M., 2015. EEG radiotelemetry in small laboratory rodents: a powerful state-of-the art approach in neuropsychiatric, neurodegenerative, and epilepsy research [WWW Document]. Neural Plast. https://doi.org/10.1155/2016/8213878.

Lyamin, O., Manger, P., Ridgway, S., Mukhametov, L., Siegel, J., 2008. Cetacean sleep: an unusual form of mammalian sleep. Neurosci. Biobehav. Rev. 32, 1451–1484. https://doi.org/10.1016/j.neubiorev.2008.05.023.

Lyamin, O.I., Kosenko, P.O., Korneva, S.M., Vyssotski, A.L., Mukhametov, L.M., Siegel, J.M., 2018. Fur seals suppress REM sleep for very long periods without subsequent rebound. Curr. Biol. 28, 2000–2005.e2. https://doi.org/10.1016/j.cub.2018.05.022.

Massot, B., Arthaud, S., Barrillot, B., Roux, J., Ungurean, G., Luppi, P.-H., Rattenborg, N.C., Libourel, P.-A., 2019a. ONEIROS, a new miniature standalone device for recording sleep electrophysiology, physiology, temperatures and behavior in the lab and field. J. Neurosci. Methods 316, 103–116. https://doi.org/10.1016/j.jneumeth.2018.08.030.

Massot, B., Rattenborg, N.C., Hedenström, A., Akesson, S., Libourel, P.-A., 2019b. An implantable, low-power instrumentation for the long term monitoring of the sleep of animals under natural conditions. In: Proceeding EMBC, Berlin, Germany. Presented at the 41st Annual International Conference of the IEEE Engineering in Medicine & Biology Society (EMBC), Berlin, Germany.

Mathis, A., Mamidanna, P., Cury, K.M., Abe, T., Murthy, V.N., Mathis, M.W., Bethge, M., 2018. DeepLabCut: markerless pose estimation of user-defined body parts with deep learning. Nat. Neurosci. 21, 1281–1289. https://doi.org/10.1038/s41593-018-0209-y.

McShane, B.B., Galante, R.J., Biber, M., Jensen, S.T., Wyner, A.J., Pack, A.I., 2012. Assessing REM sleep in mice using video data. Sleep 35, 433–442. https://doi.org/10.5665/sleep.1712.

Mikkelsen, K.B., Tabar, Y.R., Kappel, S.L., Christensen, C.B., Toft, H.O., Hemmsen, M.C., Rank, M.L., Otto, M., Kidmose, P., 2019. Accurate whole-night sleep monitoring with dry-contact ear-EEG. Sci. Rep. 9, 16824. https://doi.org/10.1038/s41598-019-53115-3.

Miller, N., Gerlai, R., 2012. Automated tracking of zebrafish shoals and the analysis of shoaling behavior. In: Kalueff, A.V., Stewart, A.M. (Eds.), Zebrafish Protocols for Neurobehavioral Research. Humana Press, Totowa, NJ, pp. 217–230. https://doi.org/10.1007/978-1-61779-597-8_16.

Mohseni, P., Najafi, K., Eliades, S.J., Wang, X., 2005. Wireless multichannel biopotential recording using an integrated FM telemetry circuit. IEEE Trans. Neural Syst. Rehabil. Eng. 13, 263–271. https://doi.org/10.1109/TNSRE.2005.853625.

Murali, N.S., Svatikova, A., Somers, V.K., 2003. Cardiovascular physiology and sleep. Front. Biosci. 8, S636–S652. https://doi.org/10.2741/1105.

Murthy, J.N., van Jaarsveld, J., Fei, J., Pavlidis, I., Harrykissoon, R.I., Lucke, J.F., Faiz, S., Castriotta, R.J., 2009. Thermal infrared imaging: a novel method to monitor airflow during polysomnography. Sleep 32, 1521–1527.

Parkison, S.A., Carlson, J.D., Chaudoin, T.R., Hoke, T.A., Schenk, A.K., Goulding, E.H., Pérez, L.C., Bonasera, S.J., 2012. A low-cost, reliable, high-throughput system for rodent behavioral phenotyping in a home cage environment. Annu. Int. Conf. IEEE Eng. Med. Biol. Soc. 2012, 2392–2395. https://doi.org/10.1109/EMBC.2012.6346445.

Parmeggiani, P.L., 2003. Thermoregulation and sleep. Front. Biosci. 8, s557–s567.

Patel, T.P., Gullotti, D.M., Hernandez, P., O'Brien, W.T., Capehart, B.P., Morrison, B., Bass, C., Eberwine, J.E., Abel, T., Meaney, D.F., 2014. An open-source toolbox for automated phenotyping of mice in behavioral tasks. Front. Behav. Neurosci. 8 https://doi.org/10.3389/fnbeh.2014.00349.

Pereira, T.D., Shaevitz, J.W., Murthy, M., 2020. Quantifying behavior to understand the brain. Nat. Neurosci. 23, 1537–1549. https://doi.org/10.1038/s41593-020-00734-z.

Piéron, H., 1913. Le problème physiologique du sommeil. Masson.

Portugal, S.J., White, C.R., 2018. Miniaturization of biologgers is not alleviating the 5% rule. Methods Ecol. Evol. 9, 1662–1666. https://doi.org/10.1111/2041-210X.13013.

Rafiq, K., Appleby, R.G., Edgar, J.P., Jordan, N.R., Dexter, C.E., Jones, D.N., Blacker, A.R.F., Cochrane, M., 2019. OpenDropOff: an open-source, low-cost drop-off unit for animal-borne devices. Methods Ecol. Evol. 10, 1517–1522. https://doi.org/10.1111/2041-210X.13231.

Rattenborg, N.C., de la Iglesia, H.O., Kempenaers, B., Lesku, J.A., Meerlo, P., Scriba, M.F., 2017. Sleep research goes wild: new methods and approaches to investigate the ecology, evolution and functions of sleep. Philos. Trans. R. Soc. Lond. B Biol. Sci. 372 https://doi.org/10.1098/rstb.2016.0251.

Rattenborg, N.C., Voirin, B., Cruz, S.M., Tisdale, R., Dell'Omo, G., Lipp, H.-P., Wikelski, M., Vyssotski, A.L., 2016. Evidence that birds sleep in mid-flight. Nat. Commun. 7, 12468. https://doi.org/10.1038/ncomms12468.

Rattenborg, N.C., Voirin, B., Vyssotski, A.L., Kays, R.W., Spoelstra, K., Kuemmeth, F., Heidrich, W., Wikelski, M., 2008. Sleeping outside the box: electroencephalographic measures of sleep in sloths inhabiting a rainforest. Biol. Lett. 4, 402–405. https://doi.org/10.1098/rsbl.2008.0203.

Roebuck, A., Monasterio, V., Gederi, E., Osipov, M., Behar, J., Malhotra, A., Penzel, T., Clifford, G.D., 2014. A review of signals used in sleep analysis. Physiol. Meas. 35, R1–R57. https://doi.org/10.1088/0967-3334/35/1/R1.

Rutz, C., Hays, G.C., 2009. New frontiers in biologging science. Biol. Lett. 5, 289–292. https://doi.org/10.1098/rsbl.2009.0089.

Salem, G.H., Dennis, J.U., Krynitsky, J., Garmendia-Cedillos, M., Swaroop, K., Malley, J.D., Pajevic, S., Abuhatzira, L., Bustin, M., Gillet, J.-P., Gottesman, M.M., Mitchell, J.B., Pohida, T.J., 2015. SCORHE: a novel and practical approach to video monitoring of laboratory mice housed in vivarium cage racks. Behav. Res. Methods 47, 235–250. https://doi.org/10.3758/s13428-014-0451-5.

Samson, A.L., Ju, L., Ah Kim, H., Zhang, S.R., Lee, J.A.A., Sturgeon, S.A., Sobey, C.G., Jackson, S.P., Schoenwaelder, S.M., 2015. MouseMove: an open source program for semi-automated analysis of movement and cognitive testing in rodents. Sci. Rep. 5, 16171. https://doi.org/10.1038/srep16171.

Schmidt, M.H., Valatx, J.L., Schmidt, H.S., Wauquier, A., Jouvet, M., 1994. Experimental evidence of penile erections during paradoxical sleep in the rat. Neuroreport 5, 561–564.

Scriba, M.F., Ducrest, A.L., Henry, I., Vyssotski, A.L., Rattenborg, N.C., Roulin, A., 2013a. Linking melanism to brain development: expression of a melanism-related gene in barn owl feather follicles covaries with sleep ontogeny. Front. Zool. 10, 42. https://doi.org/10.1186/1742-9994-10-42.

Scriba, M.F., Harmening, W.M., Mettke-Hofmann, C., Vyssotski, A.L., Roulin, A., Wagner, H., Rattenborg, N.C., 2013b. Evaluation of two minimally invasive techniques for electroencephalogram recording in wild or freely behaving animals. J. Comp. Physiol. 199, 183–189. https://doi.org/10.1007/s00359-012-0779-1.

Scriba, M.F., Roulin, A., Henry, I., Ducrest, A.L., Vyssotski, A.L., Rattenborg, N.C., 2013c. Mammalian-like rem sleep ontogeny in barn owls in the wild. In: Presented at the SFN.

Seba, A., Istrate, D., Guettari, T., Ugon, A., Pinna, A., Garda, P., 2017. Thermal-signature-based sleep analysis sensor. Informatics 4, 37. https://doi.org/10.3390/informatics 4040037.

Singh, S., Bermudez-Contreras, E., Nazari, M., Sutherland, R.J., Mohajerani, M.H., 2019. Low-cost solution for rodent home-cage behaviour monitoring. PLoS One 14, e0220751. https://doi.org/10.1371/journal.pone.0220751.

Snelderwaard, P.C., van Ginneken, V., Witte, F., Voss, H.P., Kramer, K., 2006. Surgical procedure for implanting a radiotelemetry transmitter to monitor ECG, heart rate and body temperature in small *Carassius auratus* and *Carassius auratus* gibelio under laboratory conditions. Lab. Anim. 40, 465–468. https://doi.org/10.1258/002367706778476325.

Snyder, F., Hobson, J.A., Morrison, D.F., Goldfrank, F., 1964. Changes in respiration, heart rate, and systolic blood pressure in human sleep. J. Appl. Physiol. 19, 417–422.

Thomson, J.A., Heithaus, M.R., 2014. Animal-borne video reveals seasonal activity patterns of green sea turtles and the importance of accounting for capture stress in short-term biologging. J. Exp. Mar. Biol.Ecol. Charism. Mar. Mega-Fauna 450, 15–20. https://doi.org/10.1016/j.jembe.2013.10.020.

Tobler, I., Schwierin, B., 1996. Behavioural sleep in the giraffe (*Giraffa camelopardalis*) in a zoological garden. J. Sleep Res. 5, 21–32.

Tort, A.B.L., Neto, W.P., Amaral, O.B., Kazlauckas, V., Souza, D.O., Lara, D.R., 2006. A simple webcam-based approach for the measurement of rodent locomotion and other behavioural parameters. J. Neurosci. Methods 157, 91–97. https://doi.org/10.1016/j.jneumeth.2006.04.005.

Toscani, L., Gangemi, P.F., Parigi, A., Silipo, R., Ragghianti, P., Sirabella, E., Morelli, M., Bagnoli, L., Vergassola, R., Zaccara, G., 1996. Human heart rate variability and sleep stages. Ital. J. Neurol. Sci. 17, 437–439. https://doi.org/10.1007/BF01997720.

Ungurean, G., Barrillot, B., Martinez-Gonzalez, D., Libourel, P.-A., Rattenborg, N.C., 2020. Comparative perspectives that challenge brain warming as the primary function of REM sleep. iScience 101696. https://doi.org/10.1016/j.isci.2020.101696.

Voirin, B., Scriba, M.F., Martinez-Gonzalez, D., Vyssotski, A.L., Wikelski, M., Rattenborg, N.C., 2014. Ecology and neurophysiology of sleep in two wild sloth species. Sleep 37, 753–761. https://doi.org/10.5665/sleep.3584.

Vyssotski, A.L., 2005. Miniature neurologgers for flying pigeons: multichannel EEG and action and field potentials in combination with GPS recording. J. Neurophysiol. 95, 1263–1273. https://doi.org/10.1152/jn.00879.2005.

Weiergräber, M., Henry, M., Hescheler, J., Smyth, N., Schneider, T., 2005. Electrocorticographic and deep intracerebral EEG recording in mice using a telemetry system. Brain Res. Protoc. 14, 154–164. https://doi.org/10.1016/j.brainresprot.2004.12.006.

White, W.J., Balk, M.W., Lang, C.M., 1989. Use of cage space by guineapigs. Lab. Anim. 23, 208–214. https://doi.org/10.1258/002367789780810617.

Whitford, M., Klimley, A.P., 2019. An overview of behavioral, physiological, and environmental sensors used in animal biotelemetry and biologging studies. Anim. Biotelemetry 7, 26. https://doi.org/10.1186/s40317-019-0189-z.

Williams, H.J., Taylor, L.A., Benhamou, S., Bijleveld, A.I., Clay, T.A., Grissac, S. de, Demšar, U., English, H.M., Franconi, N., Gómez-Laich, A., Griffiths, R.C., Kay, W.P., Morales, J.M., Potts, J.R., Rogerson, K.F., Rutz, C., Spelt, A., Trevail, A.M., Wilson, R.P., Börger, L., 2020. Optimizing the use of biologgers for movement ecology research. J. Anim. Ecol. 89, 186–206. https://doi.org/10.1111/1365-2656.13094.

Yaghouby, F., Donohue, K.D., O'Hara, B.F., Sunderam, S., 2016. Noninvasive dissection of mouse sleep using a piezoelectric motion sensor. J. Neurosci. Methods 259, 90–100. https://doi.org/10.1016/j.jneumeth.2015.11.004.

Ye, X., Wang, P., Liu, J., Zhang, S., Jiang, J., Wang, Q., Chen, W., Zheng, X., 2008. A portable telemetry system for brain stimulation and neuronal activity recording in freely behaving small animals. J. Neurosci. Methods 174, 186–193. https://doi.org/10.1016/j.jneumeth.2008.07.002.

Zayachkivsky, A., Lehmkuhle, M.J., Fisher, J.H., Ekstrand, J.J., Dudek, F.E., 2013. Recording EEG in immature rats with a novel miniature telemetry system. J. Neurophysiol. 109, 900–911. https://doi.org/10.1152/jn.00593.2012.

CHAPTER 10

Wearable and nonwearable sleep-tracking devices

Laronda Hollimon[1], Ellita T. Williams[1], Iredia M. Olaye[2], Jesse Moore[1], Daniel Volshteyn[3], Debbie P. Chung[1], Janna Garcia Torres[1], Girardin Jean-Louis[1] and Azizi A. Seixas[1]
[1]NYU Grossman School of Medicine, New York, NY, United States; [2]Weill Cornell Medicine of Cornell University, New York, NY, United States; [3]Cornell University, New York, NY, United States

Introduction

The current chapter provides a broad and in-depth overview of wearable and nonwearable sleep devices, specifically highlighting their (i) features (hardware and software), (ii) accuracy, reliability, and validity, and (iii) utility and applications in assessment, diagnosis, and management of sleep health parameters including sleep duration, quality, disorders, efficiency and sleep architecture and stages. Despite the heterogeneity of these wearable and nonwearable sleep devices, there are fundamental similarities in their sensor hardware, connectivity (wireless or Bluetooth), software, engagement features, and types of sleep-wake parameters measured, like sleep duration, sleep parameters (e.g., sleep architecture and stages of sleep). Conversely, wearable and nonwearable sleep devices differ across several areas which include: their diagnostic capabilities (e.g., ability to assess and diagnose sleep disorders like sleep apnea), ability to measure clinical outcomes such as brain activity via electroencephalogram (EEG), pulse oxygen, heart rate, and respiration via electrocardiogram (ECG). The heterogeneity of these devices highlights the exciting nature of the ever-growing digital health market in sleep and circadian science and medicine, which to be fully embraced in mainstream academic medicine must be continuously vetted for accuracy, reliability, validity, and utility of new devices and innovations. Our bifurcation of sleep devices into wearables and nonwearables adds value to a body of literature heavily focused on distinguishing consumer technology from clinical devices. Categorizing sleep devices within the framework of wearable and non-wearable devices democratizes sleep medicine—where consumer devices are not quickly dismissed in favor of clinical devices—as more data show

Methodological Approaches for Sleep and Vigilance Research
ISBN 978-0-323-85235-7
https://doi.org/10.1016/B978-0-323-85235-7.00004-1

new generation consumer devices having similarly high accuracy, reliability, and validity as clinical devices.

Sleep trackers and wearables

Sleep wearables and trackers are digital devices that monitor and track sleep parameters and behaviors, such as total sleep time (TST), time in bed (TIB), sleep efficiency (SE), wake after sleep onset (WASO), and stages of sleep [nonrapid eye movement (NREM), rapid eye movement (REM), deep and light sleep]. These wearables and trackers are ubiquitous and heterogenous as they can be found in smartwatches, fitness trackers, phones, and other forms of biometric devices.

What are wearables and nonwearables?

A wearable is a type of electronic device powered by microprocessors that collects and processes diverse data (e.g., biometric, biospecimen, and environmental) into meaningful trends and insights, and transfers and receives data through the internet. Wearables are generally worn, tethered, tattooed, or implanted on an individual's body, clothing, or footwear. Wearables come in many forms such as a watch, bracelet, headband, or other accessories. Conversely, nonwearable devices are typically immobile devices that perform similar tasks as wearables such as collecting, processing, receiving, and transferring biometric data (e.g., sleep data). However, unlike wearable devices, nonwearables are generally higher grade medical devices due to their superior hardware, immobile (or hard to carry around), and perform specific tasks in certain restricted contexts [e.g., polysomnography (PSG) in a sleep lab]. The superior hardware in nonwearable devices allows them to capture more voluminous and diverse data and more stable and accurate assessment of sleep data, thus making them more reliable and accurate medical devices.

In sleep/circadian science and medicine, the use of wearables and nonwearables is a hot and controversial debate in sleep medicine because some believe that the proliferation of consumer technology wearables dilutes the rigor and integrity of the science and medicine, thus leading to faulty clinical practice. Others, however, believe that wearables are game changers as they increase population access to sleep assessment/monitoring and raise awareness about sleep health at the population level. Although this debate has merit, we argue that there is a place for both wearables and nonwearables devices in sleep/circadian science and medicine. In fact, the viability of sleep/circadian science and medicine and improvement of population sleep health depend in part on whether wearables and non-wearables can co-exist and be seamlessly integrated in research and clinical care. Ideally, wearables can serve as first-line therapeutics, objectively screening for poor sleep health in the population and offering basic digital interventions to address poor sleep, and nonwearables (specifically clinical devices) can serve as thorough diagnostic tools to determine a diagnosis and appropriate medical treatment.

Features

Hardware and software of sleep devices

One area in which sleep wearables and nonwearables have revolutionized the field of sleep/circadian sciences and medicine is in their hardware and software features. The rapid rise of wearable and nonwearable sleep devices can be attributed to their evolving hardware and capabilities, as devices become smaller, inconspicuous, and portable, while expanding their functionality capturing novel types of big data with high velocity, variety, volume, and veracity. Although the hardware features of wearables and nonwearables vary, for the purposes of this book chapter, we only focus on their physical features (device chassis, electronic circuit board), sensor technology (type of sensors), connectivity (analog, Bluetooth, internet, or Wi-Fi), and type of biometric signals they measure and record.

Physical features: The physical features of wearables and nonwearables (e.g., consumer technology products) include their chassis, electronic, sensors, microprocessing systems, and biometric signal they measure. The physical chassis of wearables and nonwearables are designed and built to measure sleep and wake cycles conveniently and across several contexts. Thus, the hardware features are small, relatively inconspicuous, and portable with built-in sensor technology to conduct high capacity and throughput microprocessing that include measuring, collecting, and transferring data in real time. The first generation of sleep wearables and consumer technology devices (e.g., Fitbit, Jawbone, and Misfit) had difficulty distinguishing sleep and wake across different situations, specifically when the user was lying awake in bed with limited movement. Over the years, wearables have modified their hardware, specifically their sensor technology and algorithms, to better detect sleep and wake, especially for nuanced situations where it may be difficult to differentiate the two states.

Sensors: Today, wearable features have evolved to allow for more comprehensive data collection leading to more accurate estimates of sleep and wake. Wearables now capture motion/accelerometer data (as was the case in first generation devices), body position (estimating supine positioning as an indicator of sleep), temperature of user, environmental factors (such as light and noise), and biometric markers such as heart rate variability through heart rate monitoring and electrocardiography (ECG) or electroencephalography (EEG). Conversely, nonwearables, which do not need to be worn on-person, are also able to distally capture similar data as wearables, such as body motion, environmental factors, and heart rate through ballistocardiography generally used in mattress sleep trackers and devices. Both wearables and nonwearables triangulate all the aforementioned diverse data in their proprietary algorithms to estimate sleep and wake. The physical features of PSG devices (portable and immobile) have significant advantages, as they can measure a wide range of biometric data such as: eye movement (via electrooculography EOG), body movement, brain waves (via EEG), heart rate (via ECG), blood oxygen levels, and your respiration (See Table 10.1).

One major physical feature that distinguishes wearables from nonwearables is the integrated circuit board and microprocessor systems. The integrated circuit boards in sleep devices have heterogenous sensor technology. As these integrated circuit boards become smaller and more portable, their applications in measuring sleep has evolved. With advancements to hardware features, current wearable and nonwearable

Table 10.1 Type of sleep wearables, what they measure, and examples.

Type of sleep tracker	What is measured	Examples
Smartphone application	Body movement, noise	Sleep cycle alarm clock, MotionX 24/7, AutoSleep
Fitness wearable or watch	Body movement, heart rate, noise	Jawbone, Fitbit, Apple Watch, Withings
Mattress sleep tracker	Body movement, heart rate, noise	Beddit, Tomorrow Sleeptracker Moniter, Withings Sleep Tracking Pad
Sleep tracking headgear	Body movement, heart rate	Phillips SmartSleep Deep Sleep Headband
Nightstand	Noise, temperature, motion	Resmed S^{+}, SleepScore Max Tracker

devices are now used to diagnose patients with sleep disruptions and disorders, with comparable accuracy to PSG, the gold standard approach of assessing sleep and wake episodes (Polysomnography (sleep study)). In fact, current wearables and nonwearables are like miniaturized actigraphy (ACT) and PSG devices with sophisticated sensor technology that include leads to monitor brainwave (electroencephalography), movement in the muscles (electromyography), eye movement (electrooculography), heart rate (electrocardiography), oxygen level using pulse oximetry, and microphone to record sound.

Besides PSG devices, ACT (actimetry sensor) is used to monitor sleep. ACT is used in lab and at-home sleep studies. With the different tools available to monitor and analyze sleep in sleep studies, scientist and researchers are looking at alternatives that are less expensive, require less equipment, maximize comfort and ease of use for patients, and can be used outside the lab. Currently, there are a number of PSG and ACT devices used for at-home sleep studies. Table 10.2 below lists some of the available PSG and ACT devices on the market and the sensors available in each device. Unlike, PSG and ACT, consumer wearables are wellness devices that track and monitor different biometric outcomes and many environmental factors (light, noise, temperature, humidity, and air quality) that can affect sleep. Consumer wearables come in many forms and serve multiple purposes, from being worn on the wrist (smartwatches or fitness trackers), the head, or being incorporated into the bed mattress. New generation consumer wearables go beyond tracking sleep and now offer a wide range of digital therapeutics where they serve as sleep aids.

However, consumer wearables monitor and collate sleep data in different ways. There are three hardware features that are essential for the collection and processing of sleep data. The first feature is their ability to be portable. A mobile application will use sensors available on in smartphone such as the accelerometer (movement), gyroscope (orientation), GPS (Geo-location), and microphone (sound) to track how long the user/consumer has slept for. The second feature is their varied and unique sensor technology. They use a wide selection of sensor technologies which include but are not limited to: an accelerometer (movement), tri-wavelength sensor, gyroscope (orientation), GPS (Geo-location), microphone (sound), SpO2 sensor, optical heart rate monitor, barometric altimeter, ambient light sensor, vibration motor, thermometer, ECG sensor, and skin temperature sensor. Other wearables that are worn as headbands, or can be placed under the mattress, use other types of sensors such as sonometer, air bladder, piezo

Table 10.2 Hardware and features.

Device	Sensors
Apple Watch Series 5	• GPS/GNSS • Built-in compass • Barometric altimeter • Optical heart sensor • Electrical heart sensor • Accelerometer up to 32 g-forces • Gyroscope • Improved ambient light sensor • Digital crown with haptic feedback • Speaker • Wi-Fi (802.11 b/g/n 2.4GHz) • Bluetooth 5.0
Fitbit smartwatch (Versa 2)	• 3-axis accelerometer • Optical heart rate monitor • Altimeter • Vibration motor • Relative SpO2 sensor • NFC • Ambient light sensor • Wi-Fi antenna (802.11 b/g/n) • Microphone
Fitbit tracker (Charge 3)	• 3-axis accelerometer • Optical heart rate monitor • Altimeter • Vibration motor • Relative SpO2 sensor • Near field communication (NFC) (in special editions only)
Garmin (Fenix 6S)	• GPS • GLONASS • Galileo • Garmin Elevate wrist heart rate monitor • Barometric altimeter • Compass • Gyroscope • Accelerometer • Thermometer • Pulse Ox
Samsung Galaxy Watch Active2	• Accelerometer • Barometer • Gyro Sensor

Table 10.2 Hardware and features.—cont'd

Device	Sensors
	• HR Sensor • Light Sensor • ECG Sensor
Withings Sleep Tracking Mat	• Air Bladder • Setup LED • Metrics: o Sleep duration o Sleep cycle o Heart rate o Snoring duration
Beddit 3	• Piezo force • Capacitive touch • Humidity (USB plug) • Temperature (USB plug)
Philips Smart Sleep headband	• Sticky sensor behind the ear (type of sensor not available)
Dreem 2 headband	• EEG sensors x 6 • Pulse oximeter • Accelerometer • Sonometer
S+ by ResMed	'Records the light, noise, and temperature conditions in your room' (S Plus By ResMed Sleep Tracker)
ActiGraph GT9X Link (Phillps)	• Dynamic range (primary accelerometer) • Dynamic Range (secondary accelerometer) • Gyroscope dynamic range • Magnetometer dynamic range • Heart rate monitoring • Wear time sensor • Gyroscope/magnetometer • Proximity detection
CleveMed SleepView Monitor	• 8 channels • Heart rate • Pulse oximetry • Respiratory airflow (compatible with CPAP) • Snore • Body position • Auxiliary (Second RIP respiratory effort, Thermal airflow, or IDcheck) • Actigraphy (with web portal)

Continued

Table 10.2 Hardware and features.—cont'd

Device	Sensors
CleveMed Sapphire PSG	• 22 channels • 6 EEG channels • 2 EOG channels • 2 Chin EMG channels • 1 ECG channel • 3 EMG channels • Left leg • Right leg • 1 additional EMG or snore mic • Thoracic effort • Abdominal effort • Thermistor • Body position • Auxiliary DC Airflow (pressure based) • Snore (derived from airflow) • Pulse oximetry
SOMNOwatch plus (Actigraphy)	• 7 channels • Body position/movement (separate acc for 3 axis) • Ambient light • Patient marker with acoustic tone • Piezo tone generator built in (programmable) • Up to 8 different external sensors available
SOMNOscreen Plus	• 17 Channel Headbox with 12 EEG/EOG • Ground, reference • Continuous impedance, 2 EMG • OPTIONAL—32 Channel Headbox with 25 EEG/EOG with continuous impedance, 6 differential, 1 ECG • 2 PLM • Pressure (flow and snore) • Flow thermistor • Effort (thorax/abdomen) • SpO2, Pulse rate • Body position • Movement (sleep/wake determination) • CPAP/BiPAP—pressure • Microphone • Wireless data receiver
Nox T3 Sleep Monitor	• 2 bipolar channels; for recording of electrocardiography (ECG), electromyography (EMG), electroencephalography (EEG) or electrooculography (EOG)

Table 10.2 Hardware and features.—cont'd

Device	Sensors
	• 1 ground channel • 1 pressure/cannula channel; for recording of nasal or mask pressure • 2 respiratory effort channels; for recording of abdomen and thorax ventilatory effort signals • 3-D built-in acceleration sensor; for recording of patient's position and activity • Built-in microphone; for recording of audio and snoring • Built-in Bluetooth module; to support wireless connectivity allowing the device to record signals from compatible auxiliary devices
Nox A1 PSG System	• 13 unipolar channels; for recording of electroencephalography (EEG), electrooculography (EOG) and submental electromyography (EMG) • 1 ground channel • 4 bipolar channels; for recording of electrocardiogram (ECG), periodic limb movements (PLM), bruxism, or additional EMG • 1 pressure/cannula channel; for recording of nasal or mask pressure • respiratory effort channels; for recording of abdomen and thorax ventilatory effort signals • 3-D built-in acceleration sensor; for recording of patient's position and activity • Built-in light sensor; for recording of ambient light • Built-in microphone; for recording of audio and snoring • Built-in Bluetooth module; to support wireless connectivity allowing the device to record signals from compatible auxiliary devices

force, capacitive touch, and humidity to triangulate sleep data and environmental factors that affect sleep.

Connectivity: The third characteristic hardware feature of wearable devices is their connectivity. Wearable and nonwearable devices can be connected through wire-based connections like through a USB port or by wireless technology. By nature, wearables devices are wireless and thus need the necessary wireless technology to communicate with an application on the mobile or digital devices phone or desktop the devices. There are six types of wireless connectivity that wearable devices utilize. These are near field communication (NFC), Bluetooth low energy (BLE), Bluetooth classic, Wi-Fi, and cellular. NFC wireless technology works best with devices that require low amounts of power and can transfer small amounts of data. BLE technology is a low-cost option and requires little power. As a tradeoff, BLE has low-throughput potential of data and the range in which data are transmitted is confined to usually 100 m. Bluetooth classic wireless technology provides higher bandwidth and is the standard method of streaming audio. Wi-Fi technology is ideal for transferring large data with limited lag time. Cellular technology is best when there is no Wi-Fi and uses cellular network to transmit data. Cellular technology does not need a bridge device like a smart phone to transmit data via the cloud. However, cellular technology is no ideal for wearable devices as it consumes a lot of power and the devices are typically larger.

Types of sleep wearables and nonwearables based on biometric signals: The type of biometric signal a device captures and processes is another essential feature that distinguishes wearables and nonwearables. Signature features in these devices are built to capture and measure a wide range of sleep and biometric signals, such as brain activity, autonomic, and movement. Devices that measure sleep through brain activity are generally headgear accessories that capture EEG, frontalis muscle electromyogram (EMG), or electrooculogram (EOG) signals [e.g., Zeo, iBrain (Neuro Vigil), Dreem headband and Beddr] (Kelly et al., 2012). Brain-based sleep devices can capture sleep architecture and stages such as NREM sleep and REM sleep, as well as detect abnormal breathing (a sign of a sleep breathing disorders like sleep apnea). Devices that measure sleep through autonomic signals generally use a combination of sensors, ports, and electrodes to measure ECG, EEG, EMG, EOG, body position, ACT, and respiration and cardiac physiology. Autonomic devices can measure a wide range of biometric signals such as respiration/pulmonary physiology, heart rate, and cardiopulmonary coupling, such as respiratory-derived heart rate variability and R-wave amplitude fluctuations that signify mechanical changes in breathing which is unique during sleep. Autonomic devices typically include wire electrodes that attach to an individual's chest to measure cardiopulmonary (heart and lung) function through ECG and ACT that measures total sleep parameters. Automimic devices are particularly helpful in detecting sleep fragmentation and sleep breathing disorders. Movement-based devices determine sleep/wake cycles through an actimeter sensor technology. Majority of consumer grade wearables use actimeter and accelerometer sensor technology to determine sleep parameters. Mattress or bed-based sleep devices are typically devices placed on an individual's mattress to detect a wide range of sleep parameters (TST and sleep staging are the most notable), heart rate, respiration rate, body movement, and sometimes snoring through a pressure-sensing pad. These types of devices, such as EarlySense, Air Cushion, and Withings, have high-correspondence reliability with PSG in measuring NREM and wake, but mixed reliability and sensitivity measuring REM sleep, where Linen Sensor having high correspondence with PSG and air cushion having low correspondence.

Software: engagement and algorithm

In addition to hardware features, the software features embedded in wearables and nonwearables is another characteristic feature that distinguishes wearables, nonwearables, and medical-grade sleep machines. Software can be categorized into user experience, interface that keeps the user engaged, and algorithms used to transform and score raw biometric signals into meaningful endpoints to provide insight to user about their sleep. The engagement features of consumer grade wearables and nonwearables are highly sophisticated aiming to provide the user interesting content and experiences to increase use of the device. Conversely, the engagement features of medical-grade sleep machines and devices on the other hand are less developed and generally providing the user minimal data. Differences in engagement across these classes of devices are primarily due to their function. For example, since the primary goal of medical-grade devices is to aid healthcare provider in diagnosis or monitoring of patients, their user interface is generally minimal and tailored for the healthcare provider.

The engagement software for consumer-grade wearables and nonwearable devices aims to keep the user engaged by showing their process, meeting their goals, or suggesting how to get better sleep. These metrics are shown to users through a variety of data visualizations as charts, graphs, nudges, and messages to promote user engagement and behavioral changes, such as making sleep health a day-to-day priority. These devices measure several sleep parameters such as sleep stages (light sleep, deep sleep, and REM), sleep quality, and other proprietary sleep outcomes, specific to the device.

The second software feature is based on the device's algorithms. Algorithms, which consist of mathematical formulas and machine learning (ML) technology, serve as the connective tissue between hardware (sensors technology) and engagement software such as user-experience and interface. They help to process raw data and score these raw data into meaningful and quantifiable outcomes. The typical algorithmic workflow includes capturing specific biometric signals through a variety of sensing channels and processing those signals into a cloud computing framework to estimate sleep parameters. Although algorithms are generally proprietary, there are a few notable open access sleep algorithms that have been reliably used to process wear time and accelerometer data; these include: Choi, Hecht, and Troian (Knaier et al., 2019; Syed et al., 2020).

Limitations of wearable and nonwearable software features: Wearables use neural network models to detect and classify wake, light NREM (classified as light sleep in consumer technology devices), deep NREM (slow wave sleep or stage N3), and REM. However, wearables are unable to reliably measure sleep staging discretely, which is in direct contrast with nonwearables that use PSG algorithms and scoring. Wearable algorithms have difficulty detecting sleep staging, such as combining stage N1 and stage N2 sleep and classifying it as light sleep without clinical or anatomical grounding and heavily weigh sleep quality in the percentage of time an individual spends in stage N3. Wearable algorithms are based on another unfounded notion that majority of sleep quality is based on the amount of stage N3 sleep an individual receives. This is contradictory to mainstream views of sleep staging where in fact the majority of sleep staging occurs in stage N2 and not stage N3. Another limitation to wearable algorithms is that they are unable to detect subtle transitions from stage N1 to stage N2. Stage N1 is characterized by slowing of brain waves, while stage N2 is characterized by sleep spindles and K-complexes, which are difficult to detect in wearables. The limitation in detecting discrete sleep staging and sleep architecture prevents wearables from detecting abnormalities in sleep architecture such as sleep fragmentation.

Reliability, validity, and accuracy

Wearables and other consumer sleep technologies are emerging as a popular alternative to PSG (the gold-standard) and research-grade ACT for sleep and circadian rhythm assessment (Smith et al., 2018; Ancoli-Israel et al., 2003) because of their consumer-facing appeal; most notably consumer accessibility and automated sleep summaries (Zee et al., 2014). The consumer accessibility of wearables is an important quality for sleep researchers and clinicians because it can ameliorate the barrier to engage potential participants into sleep research (Kahawage et al., 2020). By drawing on the familiarity that potential participants may already have with their wearable device, sleep researchers and clinicians are better positioned to do work that is supported by public opinion. Furthermore, access to real-time sleep data from wearable devices is important because clinicians and investigators may be able to address questions about sleep and circadian rhythm, efficiently, at low cost, and perhaps bypass the need for a PSG-based sleep study (de Zambotti et al., 2016). As the gold-standard for sleep assessment, PSG requires a sleep laboratory and sleep laboratory infrastructure that is time-intensive, labor-intensive, and expensive when compared to wearables. Therefore, reconciling the use of wearables against PSG in sleep and circadian research in terms of validity, reliability, and accuracy is critical for the field.

Polysomnography: the gold standard

PSG is gold standard to assess sleep and sleep disorders by providing detailed physiologic output through several channels that allow clinicians and researchers to document the stages of sleep, extent of sleep-disordered breathing, and abnormal sleep behaviors. These channels are EEG, EOG, EMG, ECG/EKG, oral/nasal sensors to assess breathing, and abdominal or thoracic belts to assess chest movement. Through PSG investigators can see sleep parameters like TST, WASO, SE, TIB, and time of lights on/off, which may all be relevant to the sleep study. In addition to diagnosing sleep disorders, the comprehensive nature of PSG-based sleep studies makes it ideal for assessing sleep concerns (Buysse, 2014). The exhaustive and definite quality of PSG provides granularity to sleep and circadian rhythm research that allows investigators to address's multiple layers of the research question. For example, the amount of time a person spends in N3 is a definite metric that PSG will show and can inform the investigator on the extent of deep sleep a person may be getting.

Actigraphy for sleep and circadian assessment

PSG-based sleep studies typically occur in a laboratory and can go on from one to five nights. The short range of data collection of PSG prevents it from assessing sleep and circadian rhythm disorders that manifest or change in intensity over a course of weeks or even months. As a result, ACT, the research-grade accelerometry-based technology, can be used to capture longitudinal data and does not require a laboratory or technician to score sleep parameters. ACT is validated alternative to PSG for sleep and circadian rhythm assessment (Ancoli-Israel et al., 2003) that also provides parameters like WASO, SOL, TST, and SE. The consecutive, longitudinal capture of this information with ACT gives context to underlying sleep disorders and abnormalities. This context that allows clinicians and researchers to make recommendations informed by human behavior and sleep-hygiene patterns (Bei et al., 2016), but complementing ACT with either another objective or subjective measure is still important (Sadeh, 2011).

Validity, reliability, and accuracy requirements of wearables

The two main factors that determine whether an ACT-based device can be used as an alternative to PSG are the device's approval by the FDA for sleep research and the device meeting the 95% threshold for sensitivity and 45% threshold for specificity when compared to PSG (Cheung et al., 2020; Marino et al., 2013; Depner et al., 2020). These factors are requirements for ACT to be used in place of PSG for sleep or circadian research and, by extension, are likely requirements a consumer wearable device may need to meet to be used in sleep research. Without FDA approval, wearables are considered as part of the wellness classification as opposed to the medical device classification (Depner et al., 2020). However, conducting rigorous reliability and validity testing of a wearable device against ACT and PSG may be a more tangible way to use the wearable in a sleep research study than it would be to obtain FDA approval as a medical device. The Digital Health Software Precertification by the FDA, however, is an example of the ongoing initiatives to make it easier for wearables to become medical devices (Depner et al., 2020).

Consumer accessibility of wearable devices

Despite the ambulatory appeal that both the wearables and the research-grade ACT may have, wearables may still be favorable over ACT because wearables are less expensive and have a broader consumer accessibility. In this case of consumer accessibility, the barrier is not going to a laboratory and having a sleep technician as with PSG. Conversely, one of the main barriers that is overcome with using a wearable over ACT is the ready-made access investigators and participants have to the sleep data. Accessing these sleep data does not require a software interface and can be exported to the participant's smart phone via a smart phone application in real time (Sawka and Friedl, 2018). Another barrier that is overcome by using wearables over ACT is the wearable's ability to engage the participant in the sleep research process. Participants can troubleshoot these devices, monitor aspects of data collection, and have seamless access to technical support from the device's manufacturer. Research-grade ACT may not be (1) as familiar to potential participants as their own wearable device may be and (2) require direction from the investigator and the ACT manufacturer to troubleshoot. Both of these factors can hinder the research process by limiting the network of available help (one technical department of an ACT company vs. an established, multiplatform industry with wearable devices) and by stifling the dynamic of participant-investigator partnership.

Triangulation of gold standard, actigraphy, and wearables

Robust validity and reliability of wearables against ACT and PSG allow for a triangulation of sleep data that shows the limitations in the wearable device. This is important because the type of sleep parameter being assessed might not be able to be captured by the wearable device and it is important to know the extent of this limitation. For example, the sleep parameter TIB may be relevant for assessing insomnia and unlike PSG, neither ACT nor wearables can capture without the participant's input. Investigators studying this sleep parameter may therefore consider how else to assess TIB (sleep diary, self-report) or consider a different sleep parameter. Furthermore, ACT and wearables overestimate sleep (low specificity) and underestimate wake (low sensitivity) and serve as another point for investigators to evaluate when assessing sleep parameters like SOL and WASO (corresponds with underestimation sleep) (Marino et al., 2013; Taibi et al., 2013; Toon et al., 2016) and SE and TST (corresponds with overestimation) (Taibi et al., 2013; Toon et al., 2016; Sivertsen et al., 2006).

Limitation of wearables

Summary outputs of sleep data by wearables is a limitation that does not allow for the nuanced triangulation between PSG and ACT for two main reasons. First, summary outcomes of wearable devices may lack the granularity needed to determine the intensity of diversion from sleep parameter commonly reported in PSG or ACT; this may be the case when whole night data of wearable output (de Zambotti et al., 2016). For example, minute-by-minute assessment of TST and WASO allowed investigators to optimize the threshold of the wearable device (Arc) to train it against the

WASO and TST of the PSG (Cheung et al., 2020). Second, comparing the sleep staging that wearable devices give to that of PSG, even those with EEG capability, is a limitation because sleep staging requires the skill of a technician to evaluate the sleep microstructure (i.e., K-complexes, slow wave activity) that informs the scoring of sleep macrostructure (N1, N2, N3, and REM). It is not clear that algorithms of wearable devices can offer this nuance, especially in the scant collaborative efforts between industry and sleep/circadian rhythm scientists (Malhotra et al., 2013). Moreover, wearables, once validated for sensitivity and specificity, may be better suited for nighttime sleep (Depner et al., 2020) because the automatic daytime sleep scoring is limited in differentiating the array of daytime activities (Cook et al., 2018).

Recommended validation practices

Therefore, validation of wearable devices for use in sleep or circadian rhythm research can begin with assessing the wearables' main outcomes against the main outcomes of PSG (de Zambotti et al., 2016; Depner et al., 2020). When validating the wearable, replication of the results and standardization of the data analysis (Bland-Altman plots, epoch-by-epoch analysis) needs to be prioritized because this level of rigor will allow other groups to replicate and reproduce the results. It is also important to shape the methodology of the validation study based on the validation results of the wearable that has been previously published. Incorporating results from previously published validation studies on the wearable can ensure that methodological limitations those groups identified have been accounted for and addressed in the current validation study. Once validated, the device can be used to address a specific scientific question to replicate and extend the findings from the initial validation study.

Limitations of polysomnography

Although PSG is the gold standard concerning measuring sleep data and assessing sleep disorders with its detailed physiologic outputs, there is one flaw in PSG where wearables excel: sleeping in the comfort of your own home. PSG studies are conducted almost exclusively within the lab in a hospital. According to the well-documented "first-night effect" (FNE), our brains sleep less deeply the first few nights in a new environment. FNE is an accepted pitfall of PSG as this phenomenon can lead to decreased TST, lower sleep efficiencies, reduction in REM sleep, and longer REM latencies on the first night of testing (Agnew et al., 1966). Even your first night sleeping in a hotel, you do not get as much quality sleep as you normally would simply because your body and mind have not adapted to the fact that they are sleeping in a new environment. Therefore, as most PSG studies only take one night worth of data, these data can be a misrepresentation of the problems a patient may be having in a typical night sleeping at home. Wearables, on the other hand, are not influenced by the FNE. Since wearables can monitor a patient's sleep from the comfort of their own bedrooms, they can avoid the slight inaccuracies that FNE presents in a sleep study using PSG.

Overcoming the "black box" and future directions

The proprietary algorithm or "black box" of consumer wearable devices can be a challenge to sleep and circadian rhythm scientist that are validating the wearable against PSG because the algorithm prevents scientists from knowing how sleep and wakefulness are scored. In addition, the algorithm dictates the output of the wearable's sleep data which may be summarized and therefore not pliable to the rigorous design and statistical needs of sleep and circadian rhythm scientist (raw data) (de Zambotti et al., 2016). While inability to know the algorithm and access the data in a more granular form are challenges, these cannot be the sole reason to discard the use of wearable technology for research. Likewise, algorithms for PSG and ACT software/firmware may also be unknown and requires updates as well and the field has become familiar with these realities (Depner et al., 2020). Collaborative efforts between industry and sleep/circadian rhythm scientists can help wearables be better incorporated into rigorous research endeavors while influencing the public's narrative about sleep health with partnership from industry (big data, analytics) and research (expertise, context) (Depner et al., 2020).

We therefore argue that observed differences in reliability and validity in estimating sleep across wearables, nonwearables, and PSG are in large part due to the varying physical features and the data that they collect; therefore, the data collected by each should be contextualized. To hold wearables and nonwearables to the standard of PSG is impossible, and to proport to be a replacement to PSG is an overshot of expectations. There is a place for wearables, nonwearables, and PSG devices in sleep medicine and science; it ultimately depends on their utility and application.

Utility and applications of wearables and nonwearables

Wearables and nonwearables have revolutionized sleep and circadian medicine as they are used to assess and track sleep health parameters, diagnose sleep disorders, and deliver interventions to improve sleep health. We argue that their inclusion in the healthcare continuum and workflow will improve access to sleep care services, thus increasing the number of people screened, assessed, diagnosed, and treated for a sleep issue. Adding wearables and nonwearables will also benefit the healthcare ecosystem as it is likely that patients will be more adherent to assessment and treatment regimens, issues that plague healthcare providers and payers.

Assessing sleep health parameters

As indicated above, wearables and nonwearables assess and track various sleep health parameters, such as TST, sleep staging, sleep quality, and sleep disorders. The different sleep stages are NREM stages 1–3, and REM (Polysomnography (sleep study)) sleep. Consumer wearables break down sleep as light sleep, deep sleep, and REM sleep (What should I know). Consumer wearable mobile applications will also show the user what time they went to sleep, the number of hours the user has slept for the day, and the average number of hours slept for the week and the month (Track your sleep). This information can be detected automatically or by manual user input, depending on the application in question. Many applications also calculate a "sleep quality" percentage on several proprietary factors, but predominantly the number of movements during the night and moments when you are fully awake. One application, Sleep Cycle, also compares your sleep quality when sleeping through different weather conditions, air pressure, and even by your activity (steps or workouts) throughout the day. One advantage of wearables (such as an Apple Watch or Fitbit) is that since they stay on a patient's wrist the entire day, they also track all the activity you do throughout the day and can inform a user if more activity throughout the day will correlate to better quality sleep. Another factor measured by most wearables is a regularity score, or how consistent your sleep schedule is weekly. Consistency, or going to bed and waking up at approximately the same time every day, is another determining factor in quality sleep. Showing a wearable's user the regularity score will encourage them to improve their consistency.

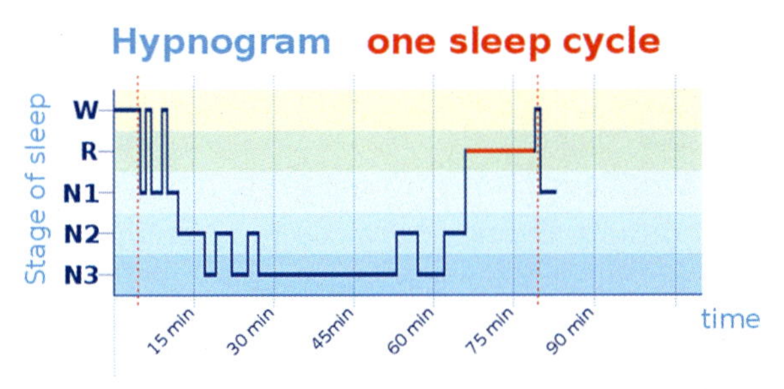

Figure 10.1 Fitbit sleep stages of a typical night's sleep.

Most wearables model a user's sleep by a graph with sleep stages on the y-axis, and time on the x-axis, as shown below in Fig. 10.1. The sleep cycles through the different stages of sleep multiple times throughout one night.

Diagnosis of diseases and health conditions

A sleep diagnostic evaluation is initiated if a patient presents to a healthcare provider with sleep disorder symptoms (daytime sleepiness, snoring, and choking or gasping during sleep) and have sleep disorder risk factors (obesity, male gender, and advanced age) (Kline et al., 2021). Although there are evaluation tools such as algorithms and risk scores, the gold standards for sleep apnea diagnostic tests are in-laboratory PSG and unattended home sleep apnea testing (HSAT). Diagnostic in-laboratory PSG can monitor patient's movement (accelerometer), brainwaves (electroencephalography), heart rate (electrocardiography), eye movement, oxygen levels (pulse oximetry), and noise (microphone). The PSG device gathers the greatest amount of information on the patient and their sleep as compared to other methods. The HSAT, at-home sleep study consist of less equipment and only monitoring the patient's breathing and oxygen level for obstructive sleep apnea (OSA). The patient's doctor will have to determine if the patient has any sleep disorders based on the sleep study results. The diagnosis of OSA is based upon the presence or absence of related symptoms and the results of the diagnostic tests.

A physician may recommend a patient undergoes a PSG if he/she suspects sleep apnea or another sleep-related breathing disorder, periodic limb movement disorder, narcolepsy, REM sleep behavior disorder, unexplained chronic insomnia, and many more. The sleep study may not only be using the data from the PSG but also may require a sleep technician to supervise the study when suspecting disorders like REM sleep behavior disorder may be present, which is characterized by sudden body movements, vocalizations, and vivid dreams during REM sleep. Consumer wearables and nonwearables are not medical devices (Polysomnography (sleep study)), but they have features that can help make a user aware to certain conditions that a physician should investigate for further diagnoses. These devices have features of detecting sleep apnea (Polysomnography (sleep study)), heart rate irregularities (Polysomnography (sleep study); What should I know; Haselton, 2017; Use fall), and even falls (Use fall) off the bed.

Using wearables to deliver interventions

In the previous studies, wearables were also used outside of tracking sleep. One major intervention wearables are capable of delivering are reminders to stop prolonged sedentary behavior. Prolonged sedentary behavior, which is defined as maintaining sedentary behavior for more than 30 min, is associated with many adverse health outcomes such as heart disease and diabetes. Wearables can detect prolonged periods without activity (not including sleeping) and send a reminder to stand up or go for a quick walk.

Wearables were also used to deliver interventions to patients with depression, anxiety, and psychotic disorders (Aledavood et al., 2019) using smartphone mobile applications and ML to predict mood (Cho et al., 2019). A smartphone mobile application in conjunction with wearables collected data on 55 patients with mood disorders for 2 years, and by the end, the researchers created an algorithm that had a mood state prediction of about 70%. It is expected that the aforementioned research can be used in the future to improve the prognosis of patients with mood disorders. There are many principal ways in which consumer wearables can be used in research and clinical settings outside of just sleep interventions.

The use of wearables and nonwearables across the lifespan

Sleep wearables can be used on any population of people, given that they work the same way, disregarding age, gender, etc. The populations who use sleep wearables most often are adults who are healthy (Yoon et al., 2018).

These adults use consumer wearables as a wellness device to track and monitor other activity like fitness, diet, and check their messages and emails (Inc, 2019; Piwek et al., 2016). In the clinical setting, consumer wearables are also being used in helping teenagers monitor their sleeping habits (George et al., 2019). While adults are the most common user, there is a push for adolescents to use wearables to improve their sleep quality. More than half of all teens ages 15 and older sleep less than 7 h per night and about 85% get less than the recommended 8–10 h of sleep per night (Sleepy Teens, 2015). This shows the necessity and urgency of creating incentives or encouraging teenagers to use sleep wearables to create more awareness for this issue and improve their sleep quality. If teenagers develop bad habits while young, these teenagers will continue these bad habits into adulthood and eventually pass them onto their children. Over the last decade, the use of wearables regarding fitness and wellness tracking has escalated exponentially and will likely continue to do so as their role in sleep health becomes more prominent. There is a diverse population using sleep wearables in a medical setting and everyday life, many of which are pursuing a healthier lifestyle.

Conclusion

The proliferation of wearable and nonwearable sleep devices has revolutionized sleep and circadian sciences and medicine. In this book chapter, we describe the hardware and software of these devices, reliability and validity of devices, and utility of devices. We argue that though there is a debate as to whether these devices should be integrated in sleep medicine, we stress that if they are integrated carefully, they can improve limitations in the delivery of sleep care.

Acknowledgment

The authors acknowledge the support of several funding agencies and efforts of study staff, key personnel, and participants who all contributed to make the study successful.

Funding sources

The authors report no conflicts of interest. This research was supported by funding from the National Institutes of Health: K01HL135452, K07AG052685, R01HL152453, R01MD007716, R01HL142066. The funding sources had no role in the design, conduct, or analysis of the study, or in the decision to submit the manuscript for publication.

References

Agnew, H.W., Webb, W.B., Williams, R.L., 1966. The first night effect: an EEG study of sleep. Psychophysiology 2 (3), 263–266. https://doi.org/10.1111/j.1469-8986.1966.tb02650.x.

Aledavood, T., Torous, J., Triana Hoyos, A.M., Naslund, J.A., Onnela, J.P., Keshavan, M., 2019. Smartphone-Based Tracking of Sleep in Depression, Anxiety, and Psychotic Disorders, vol. 21. Current Medicine Group LLC 1. https://doi.org/10.1007/s11920-019-1043-y.

Ancoli-Israel, S., Cole, R., Alessi, C., Chambers, M., Moorcroft, W., Pollak, C.P., 2003. The role of actigraphy in the study of sleep and circadian rhythms. Sleep 26 (3), 342–392. https://doi.org/10.1093/sleep/26.3.342.

Bei, B., Wiley, J.F., Trinder, J., Manber, R., 2016. Beyond the mean: a systematic review on the correlates of daily intraindividual variability of sleep/wake patterns. Sleep Med. Rev. 28, 108–124. https://doi.org/10.1016/j.smrv.2015.06.003.

Buysse, D.J., 2014. Sleep health: can we define it? Does it matter? Sleep 37 (1), 9–17. https://doi.org/10.5665/sleep.3298.

Cheung, J., Leary, E.B., Lu, H., Zeitzer, J.M., Mignot, E., 9 September 2020. PSG validation of minute-to-minute scoring for sleep and wake periods in a consumer wearable device. PLoS One 15. https://doi.org/10.1371/journal.pone.0238464.

Cho, C.H., Lee, T., Kim, M.G., In, H.P., Kim, L., Lee, H.J., 2019. Mood prediction of patients with mood disorders by machine learning using passive digital phenotypes based on the circadian rhythm: prospective observational cohort study. J. Med. Internet Res. 21 (4). https://doi.org/10.2196/11029.

Cook, J.D., Prairie, M.L., Plante, D.T., 2018. Ability of the multisensory jawbone UP3 to quantify and classify sleep in patients with suspected central disorders of hypersomnolence: a comparison against polysomnography and actigraphy. J. Clin. Sleep Med. 14 (5), 841–848. https://doi.org/10.5664/jcsm.7120.

de Zambotti, M., Godino, J.G., Baker, F.C., Cheung, J., Patrick, K., Colrain, I.M., 2016. The boom in wearable technology: cause for alarm or just what is needed to better understand sleep? Sleep 39 (9), 1761–1762. https://doi.org/10.5665/sleep.6108.

Depner, C.M., Cheng, P.C., Devine, J.K., et al., 2020. Wearable technologies for developing sleep and circadian biomarkers: a summary of workshop discussions. Sleep 43 (2). https://doi.org/10.1093/sleep/zsz254.

George, M.J., Rivenbark, J.G., Russell, M.A., Ng'eno, L., Hoyle, R.H., Odgers, C.L., 2019. Evaluating the use of commercially available wearable wristbands to capture adolescents' daily sleep duration. J. Res. Adolesc. 29 (3), 613–626. https://doi.org/10.1111/jora.12467.

Haselton, T., 2017. Here's Why People Keep Buying Apple Products. CNBC. Published May 1. https://www.cnbc.com/2017/05/01/why-people-keep-buying-apple-products.html. (Accessed 8 July 2020).

Inc, G., 2019. One in Five U.S. Adults Use Health Apps, Wearable Trackers. Gallup.com. Published December 11. https://news.gallup.com/poll/269096/one-five-adults-health-apps-wearable-trackers.aspx. (Accessed 8 July 2020).

Kahawage, P., Jumabhoy, R., Hamill, K., de Zambotti, M., Drummond, S.P.A., 2020. Validity, potential clinical utility, and comparison of consumer and research-grade activity trackers in Insomnia Disorder I: in-lab validation against polysomnography. J. Sleep Res. 29 (1). https://doi.org/10.1111/jsr.12931.

Kelly, J.M., Strecker, R.E., Bianchi, M.T., 2012. Recent developments in home sleep-monitoring devices. ISRN Neurol. 2012, 10. https://doi.org/10.5402/2012/768794, 768794.

Kline, L.R., Collop, N., Finlay, G., 2021. Clinical Presentation and Diagnosis of Obstructive Sleep Apnea in Adults. Uptodate Com Internet. Published online 2017.

Knaier, R., Höchsmann, C., Infanger, D., Hinrichs, T., Schmidt-Trucksäss, A., 2019. Validation of automatic wear-time detection algorithms in a free-living setting of wrist-worn and hip-worn ActiGraph GT3X+. BMC Publ. Health 19 (1), 244. https://doi.org/10.1186/s12889-019-6568-9.

Malhotra, A., Younes, M., Kuna, S.T., et al., 2013. Performance of an automated polysomnography scoring system versus computer-assisted manual scoring. Sleep 36 (4), 573–582. https://doi.org/10.5665/sleep.2548.

Marino, M., Li, Y., Rueschman, M.N., et al., 2013. Measuring sleep: accuracy, sensitivity, and specificity of wrist actigraphy compared to polysomnography. Sleep 36 (11), 1747–1755. https://doi.org/10.5665/sleep.3142.

Piwek, L., Ellis, D.A., Andrews, S., Joinson, A., 2016. The rise of consumer health wearables: promises and barriers. PLoS Med. 13 (2). https://doi.org/10.1371/journal.pmed.1001953.

Polysomnography (sleep study) – Mayo Clinic. https://www.mayoclinic.org/tests-procedures/polysomnography/about/pac-20394877. (Accessed 10 December 2020).

S Plus By ResMed Sleep Tracker: The smarter sleep solution. S Plus by ResMed. https://splus.resmed.com/. (Accessed 8 July 2020).

Sadeh, A., 2011. The role and validity of actigraphy in sleep medicine: an update. Sleep Med. Rev. 15 (4), 259–267. https://doi.org/10.1016/j.smrv.2010.10.001.

Sawka, M.N., Friedl, K.E., 2018. Emerging wearable physiological monitoring technologies and decision aids for health and performance. J. Appl. Physiol. 124 (2), 430–431. https://doi.org/10.1152/japplphysiol.00964.2017.

Sivertsen, B., Omvik, S., Havik, O.E., et al., 2006. A comparison of actigraphy and polysomnography in older adults treated for chronic primary insomnia. Sleep 29 (10), 1353–1358. https://doi.org/10.1093/sleep/29.10.1353.

2015. Sleepy Teens – New Study Says Teens are Even More Sleep-Deprived than We Thought. ChildrensMD. Published February 25. https://childrensmd.org/browse-by-age-group/sleepy-teens-new-study-says-teens-even-sleep-deprived-thought/. (Accessed 13 February 2021).

Smith, M.T., McCrae, C.S., Cheung, J., et al., 2018. Use of actigraphy for the evaluation of sleep disorders and circadian rhythm sleep-wake disorders: an American academy of sleep medicine systematic review, meta-analysis, and GRADE Assessment. J. Clin. Sleep Med. 14 (07), 1209–1230. https://doi.org/10.5664/jcsm.7228.

Syed, S., Morseth, B., Hopstock, L.A., Horsch, A., 2020. Evaluating the performance of raw and epoch non-wear algorithms using multiple accelerometers and electrocardiogram recordings. Sci. Rep. 10 (1), 5866. https://doi.org/10.1038/s41598-020-62821-2.

Taibi, D.M., Landis, C.A., Vitiello, M.V., 2013. Concordance of polysomnographic and actigraphic measurement of sleep and wake in older women with insomnia. J. Clin. Sleep Med. 09 (03), 217–225. https://doi.org/10.5664/jcsm.2482.

Toon, E., Davey, M.J., Hollis, S.L., Nixon, G.M., Horne, R.S.C., Biggs, S.N., 2016. Comparison of commercial wrist-based and smartphone accelerometers, actigraphy, and PSG in a clinical cohort of children and adolescents. J. Clin. Sleep Med. 12 (03), 343–350. https://doi.org/10.5664/jcsm.5580.

Track your sleep on Apple Watch and use Sleep on iPhone. Apple Support. https://support.apple.com/en-us/HT211685. (Accessed 10 December 2020).

Use fall detection with Apple Watch. Apple Support. https://support.apple.com/en-us/HT208944. (Accessed 10 December 2020).

What should I know about Fitbit sleep stages? https://help.fitbit.com/articles/en_US/Help_article/2163.htm. (Accessed 7 December 2020).

Yoon, H., Hwang, S.H., Choi, J.W., Lee, Y.J., Jeong, D.U., Park, K.S., 2018. Slow-wave sleep estimation for healthy subjects and OSA patients using R-R intervals. IEEE J. Biomed. Health Inform. 22 (1), 119–128. https://doi.org/10.1109/JBHI.2017.2712861.

Zee, P.C., Badr, M.S., Kushida, C., et al., 2014. Strategic opportunities in sleep and circadian research: report of the joint task force of the sleep research society and American academy of sleep medicine. Sleep 37 (2), 219–227. https://doi.org/10.5665/sleep.3384.

CHAPTER 11

Clinical trial: imaging techniques in sleep studies

Luigi Ferini-Strambi[1,2], Andrea Galbiati[1,2], Maria Salsone[2,3]
[1]Vita-Salute San Raffaele University, Milan, Italy; [2]IRCCS San Raffaele Scientific Institute, Department of Clinical Neurosciences, Neurology-Sleep Disorder Center, Milan, Italy; [3]Institute of Molecular Bioimaging and Physiology, National Research Council, Segrate, Italy

Imaging in normal human sleep

Sleep is a fascinating and enigmatic phenomenon occurring, with varying patterns, in all known animal species. Human sleep is typically considered a global phenomenon affecting the whole brain uniformly and simultaneously (Fuller et al., 2011), despite being characterized by reduced sensory activity, responsiveness to stimuli and conscious awareness (Tagliazucchi et al., 2013). To reconcile these two apparently contradictory aspects, sleep is now understood as a complex state that is "by, for, and of the brain" rather than simply a state of reduced wakefulness (Hobson, 2005). There is convergence that sleep may present characteristic signature of neuroelectrical and metabolic activity across specific brain structures/areas as well as the brain activity dynamically changes during the sleep. Indeed, sleep is a spontaneous process since the sleep stages are defined according to the occurrence and amount of specific phasic activities (Rechtschaffen and Kales, 1968). During nonrapid eye movement (NREM) sleep, brain activity is organized by spontaneous coalescent cerebral rhythms: spindles and slow waves (Steriade and McCarley, 2005) while during rapid eye movement (REM) sleep, the phasic activity is characterized by the presence of ponto-geniculo-occipital (PGO) waves (Mouret et al., 1963). These different types of sleep affect the brain activity differently and to play distinct roles in a variety of cognitive functions (Takamitsu et al., 2014). Finally, sleep is a dynamic process as seen by its interaction with the current environment: perception of external stimulation is decreased but not abolished during sleep (Dang-Vu et al., 2010).

Some fundamental questions should be addressed when discussing the normal human sleep. Firstly, how the human brain sleeps; secondly, how sleep and brain interact and which is the nature of this interaction. That the sleep can influence the brain and vice versa, brain can modulate the sleep.

Methodological Approaches for Sleep and Vigilance Research
ISBN 978-0-323-85235-7
https://doi.org/10.1016/B978-0-323-85235-7.00012-0

For instance, how cortical connectivity and excitability change during the sleep and still, what are the potential sleep-induced changes in the brain. This is possibile considering that, although consciousness abruptly ceases when falling asleep, the cortical activity which is a sine qua non for conscious experience does not fade away across all sleep stages (Kaufmann and Wehrle, 2006). Thus, the human brain seems not "sleep" during the sleep, but it works, learns, and consolidates the memory. For all these reasons, it should be also investigated the clinical implications/consequences of this dynamic and precious interaction occurring every night of our life. The answers to all these questions may provide fundamental insights not only to understand the sleep research but also the brain function as whole. Finally, considering the lack of communication with sleeping subjects, we need to rigorous recordings to scientifically identify and characterize the human sleep.

The most salient answers to these interesting questions resulted from applying novel techniques from field of the neuroscience research. From the original intuition of von Economo in 1930 "sleep as a problem of localisation" (von Economo, 1930), researchers have used a variety of approaches to study the relationship between sleep and brain activity. Novel emerging techniques have been developed for investigating this intriguing interaction and with time, these approaches have become increasingly sophisticated. Before the invention of electroencephalography (EEG), researchers studied sleep by examining its behavioral characteristics (e.g., arousal threshold) (Duyn, 2012). Nevertheless, it was soon realized that there were limitations to the characterization of sleep based on arousal threshold and the level of EEG slow-wave activity alone (Duyn, 2012). Neuroimaging techniques have refined to provide a noninvasive investigation allowing for the identification of subtle changes in cerebral blood flow, metabolism and connectivity as well as the neural structures involved in the regulation of normal sleep-wake cycle in humans. Indeed, regional differences in sleep EEG dynamics indicate that sleep-related brain activity involves local brain processes with sleep stage specific activity patterns of neuronal populations (Kaufmann and Wehrle, 2006) being segregated within specific cortical and subcortical areas in relation to the sleep stages (Dang-Vu et al., 2007). In the wide scenario of neuroimaging field, functional brain imaging such as positron emission tomography (PET) first, and functional MRI (fMRI) later, has been used in humans to noninvasively investigate the neural mechanisms underlying the generation of sleep. Briefly, PET shows the distribution of compounds labeled with positron-emitting isotopes and is characterized by a relatively low spatial as well as temporal resolution compared with fMRI (Kaufmann and Wehrle, 2006). Indeed, PET imaging allows the assessment of brain activity over a period of time such as a specific stage of sleep, but it cannot capture the changes in brain activity of a short event such as NREM-sleep oscillations. Conversely, fMRI, in particular, the blood oxygen level-dependent (BOLD) measures the variations in brain perfusion related to neural activity and presents relatively high spatial and temporal resolution (Bandettini, 2009). Furthermore, several studies have been carried out combining neuroimaging and EEG to explore regional specific brain activity during the sleep. These studies refined the description of brain function beyond the stages of sleep and provide new insights into the mechanisms of spontaneous brain activity in humans (Kaufmann and Wehrle, 2006). Finally, recent developments in whole-brain neuroimaging support the examination of more sophisticated features of brain networks through functional connectivity and structural connectivity analyses, the detection of task-related and resting-state functional networks and the development of mechanistic computational models (Discovery, 2019). Taken together, these evidences put in the spotlight the precious contribute of neuroimaging to provide reliable data on the mechanisms and regulation of the normal human sleep.

This chapter has the ambition to critically review the existing literature focused on the neuroimaging application to the field of sleep research. Since global and regional patterns of brain activity could vary across the different sleep stages from wakefulness, NREM sleep to REM sleep, here we discuss the major functional differences within the sleep stages from a neuroimaging perspective. We also highlight the neuronal correlates of phasic activity charactering NREM/REM stages and the sleep-dependent modifications of the brain activity.

Imaging in nonrapid eye movement sleep

Neuronal correlates of nonrapid eye movement sleep

NREM sleep has long been considered as a state of brain quiescence, partly because a drop of brain activity with brain oxygen metabolism, cerebral blood flow, and glucose metabolism were shown to be decreased when compared to wakefulness and REM sleep (Maquet, 2010). Quantitatively, this decrease has been estimated at around 40% during slow wave sleep (SWS) compared to wakefulness (Dang-Vu et al., 2010; Maquet et al., 1990). Decreases in brain activity have been consistently found in the several structures and cortical areas including the medial prefrontal cortex (MPFC), in agreement with a homeostatic need for brain energy recovery (Dang-Vu et al., 2010).

Among the most important contributions of the functional neuroimaging such as PET imaging to the field of sleep research has been the identification of how global and regional brain activity dynamically changes during the normal human sleep. Consistent are the works of Maquet's group, that for first have used functional neuroimaging to investigate human sleep by presenting functional brain maps in different sleep stages (Dang-Vu et al., 2005, 2010; Maquet, 2000, 2010; Desseilles et al., 2011). Neuroimaging data show that NREM sleep is characterized by a global decrease in cerebral blood flow, and a decrease in regional cerebral blood flow (rCBF) in the dorsal pons, mesencephalon, thalami, basal ganglia, basal forebrain and anterior hypothalamus, prefrontal cortex, anterior cingulate cortex, and precuneus (Dang-Vu et al., 2007). All these brain structures include neuronal populations involved in arousal and awakening, as well as areas which are among the most active ones during wakefulness (Dang-Vu et al., 2010; Maquet, 2000). Some considerations: first, the deactivation in the brainstem and thalamus is not surprising and is in agreement with NREM sleep generation mechanisms in mammals, in which a decreased firing rate in brainstem structures causes a hyperpolarization of thalamic neurons and a cascade of events inducing the formation of NREM sleep rhythms (e.g., spindles, K-complexes, delta, and slow oscillations) (Dang-Vu et al., 2007). Second, decreased activity during NREM sleep were also located the basal ganglia. A possible explanation is that decreasing activity in the striatum could be related to a lower propensity to arousal: afferents arising from the striatum may entrain the disinhibition of the pedunculopontine tegmental nucleus (PPT) and result in cortical activation, thus promoting wakefulness (Mena-Segovia & GiordanoM, 2003). Another area where the rCBF is significantly decreased is the precuneus (Dang-Vu et al., 2007; Maquet et al., 1997; Braun et al., 1997; Andersson et al., 1998), a region particularly active during wakefulness. This may reflect more the waning of waking-dependent processes rather than sleep-promoting mechanisms (Borbely, 2001). Third, the pattern of deactivation appears to be heterogeneous distributed throughout the cortex. When compared to wakefulness, the primary cortices were the less deactivated respect to the associative cortical (Maquet et al., 1997). Moreover, among the associative cortices, the least active areas in NREM sleep were observed in the frontal [in particular, in the dorsolateral prefrontal (DLPF) and orbital prefrontal cortex], parietal- and less consistently in the temporal and insular lobes (Dang-Vu et al., 2007; Maquet et al., 1997; Braun et al., 1997; Andersson et al., 1998) (Table 11.1). Although the reasons for this cortical distribution remain until unclear, one possible explanation is that since association cortices are the most active cerebral areas during wakefulness and because sleep deep is homeostatically related to prior waking activity at the regional level (Borbely, 2001), these cortices might be more profoundly

Table 11.1 Table listing the main neuronal correlates and the relative brain activity during nonrapid eye movement and rapid eye movement sleep stages.

Human sleep-stages	Brain activity	Brain structures	Imaging
NREM sleep	Decreased activity Decreased activity Decreased connectivity Increased connectivity	Associative cortices (primary cortices (were less deactivated) Thalamic and hypothalamic regions, the cingulate cortex, the right insula and adjacent regions of the temporal lobe, the inferior parietal lobule and the inferior/middle frontal gyri. Thalami to heteromodal regions (e.g., medial frontal gyrus and posterior cingulate/precuneus), Default-mode network (DMN)	PET (Maquet et al., 1997; Braun et al., 1997; Andersson et al., 1998) EEG-fMRI (Kaufmann and Wehrle, 2006) fMRI-FC (Picchioni et al., 2014) fMRI-FC (Koike et al., 2011; Mason et al., 2007)
NREM-phasic activity *Spindles* *Slow waves*	Increased activity Increased connectivity Increased activity	Thalami, the anterior cingulate cortex, the left anterior insula and the superior temporal gyrus Subiculum to frontal, lateral temporal, and motor cortical regions and to the insula Medial prefrontal cortex (MPFC)	fMRI (Schabus et al., 2007)-BOLD fMRI-FC (Andrade et al., 2011) EEG-fMRI (Dang-Vu et al., 2008)

Continued

Table 11.1 Table listing the main neuronal correlates and the relative brain activity during nonrapid eye movement and rapid eye movement sleep stages.—cont'd

Human sleep-stages	Brain activity	Brain structures	Imaging
REM sleep	Increased activity Increased activity Reduced activity	Posterior visual cortex Temporo-occipital areas included inferior temporal cortex and fusiform gyrus Inferior and middle frontal gyrus (DLPF) precuneus, posterior cingulate cortex and part of the parietal cortex (temporo-parietal region, inferior parietal lobule)	PET (Maquet et al., 1996; Braun et al., 1998) fMRI (Wehrle et al., 2007) MEG (Ioannides et al., 2009)
REM-phasic activity Ponto geniculo occipital waves	Increased regional cerebral blood flow Increased activity in relationship to rapid eye movements Increased activity in association with associated with rapid eye movements	Occipital cortex and the lateral geniculate bodies of the thalamus Posterior thalamus and occipital cortex Pons, thalamus, and primary visual cortex	PET (Peigneux et al., 2001) EEG/fMRI (Wehrle et al., 2007) fMRI (Miyauchi et al., 2009)

DMN, default-mode network; *fMRI-BOLD*, functional MRI- blood oxygen level-dependent; *fMRI-FC*, functional MRI-Functional connectivity; *MEG*, magnetoencephalography; *MPFC*, medial prefrontal cortex; *NREM*, nonrapid eye movement; *PET*, positron emission tomography; *rCBF*, regional cerebral blood flow; *REM*, rapid eye movement.

influenced by NREM sleep than primary cortices (Discovery, 2019). Taken together, PET studies provide precious insights on which brain areas are active or not during the sleep. This neuroimaging approach, however, shows as the major limitation the lack of information on how the different brain areas are connected and dialogged during the NREM sleep.

In fMRI, spontaneous signal fluctuations have recently been used to study functional cerebral connectivity (Fox and Raichle, 2007). Functional connectivity analysis is a noninvasive tool able to capture and decode how brain regions construct a spontaneous network by detecting the correlation of BOLD signals among different brain regions. In the last years, the main target of fMRI studies has been to investigate the corticocortical connectivity. Connectivity within and between specific areas/structures in particular thalamus and neocortex underlies several fundamental processes, including relay of sensory signals during wakefulness and increased sensory awareness thresholds during sleep (Picchioni et al., 2014; Jones, 1991; Llinas and Steriade, 2006). Cumulating evidences demonstrated that during SWS, the corticocortical connectivity is disrupted consistently with the decreased activity observed using EEG (Picchioni et al., 2014; Massimini et al., 2005). Furthermore, there is a decrease in neocortical integration as the neurophysiologic basis of decreased consciousness during sleep (Picchioni et al., 2014; Tononi, 2005). To support this, Picchioni et al. sought to answer the following question: does thalamocortical connectivity decrease during NREM sleep in humans? In particular, these authors investigated whether thalamocortical signaling between the thalamus and the neocortex was decreased from wakefulness to NREM sleep (Picchioni et al., 2014). They found statistical significance of the difference between correlations obtained during wakefulness and during SWS. Results of this study provide evidence that neocortical regions displaying decreased thalamic connectivity were all heteromodal regions (e.g., medial frontal gyrus and posterior cingulate/precuneus), whereas there was a complete absence of neocortical regions displaying increased thalamic connectivity during NREM sleep in humans (Picchioni et al., 2014) (Table 11.1). Changes in thalamocortical connectivity may act as a universal "control switch" for changes in consciousness that are observed in coma, general anesthesia, and natural sleep (Picchioni et al., 2014).

In addition, other authors simultaneously measured spontaneous EEG and fMRI during the night's first sleep cycles and report here, for the first time, NREM sleep stage and regional specific alterations of the BOLD response (Kaufmann and Wehrle, 2006). They described a specific pattern of decreased brain activity during sleep and suggest that this pattern must be synchronized for establishing and maintaining sleep. According their results, a decreased activity may occur in specific regions only during NREM sleep stages, rather than wakefulness. NREM sleep stage 2 usually linked to the loss of self-conscious awareness was associated with signal decreases comprising thalamic and hypothalamic regions, the cingulate cortex, the right insula and adjacent regions of the temporal lobe, the inferior parietal lobule, and the inferior/middle frontal gyri (Table 11.1). This is not surprising since hypothalamic region known to be of particular importance in the regulation of the sleep-wake cycle shows specific temporally correlated network activity with the cortex while the system is in the sleeping state, but not during wakefulness. Hypothalamus is involved in a reciprocal network of wake-promoting nuclei in the brainstem as well as the lateral hypothalamus itself and sleep-promoting neurons inside the hypothalamic ventrolateral preoptic nucleus (Kaufmann and Wehrle, 2006). In contrast to the decreased thalamic and hypothalamic activity, the connectivity pattern in default-mode network (DMN), during deep NREM sleep is increased when compared to REM sleep. DMN is a spontaneous brain network in the resting state related to high-level brain functions such important component of consciousness, i.e., mind-wandering and unconstrained activity (Koike et al., 2011; Mason et al., 2007) (Table 11.1). Thus, exploring the DMN during the sleep state might be really of interest. Some authors reported that brain activity of individual brain regions and functional interactions between pairs of regions significantly were increased in the DMN during SWS and decreased during REM sleep. In contrast, the network activity of the frontoparietal and sensory-motor networks showed the opposite pattern (Watanabe et al., 2014). These findings suggest that the brain activity may be dynamically modulated even in a sleep stage and that the pattern of modulation depends on the type of the large-scale brain networks (Watanabe et al., 2014). Despite marked progress has been performed on the macrostructure/architecture of human sleep, some questions remain until to address to characterize the microstructural NREM events.

Neuronal correlates of phasic nonrapid eye movement sleep-oscillations

It is well-documented that brain activity during a specific stage of sleep is not constant and homogeneous over time, but is structured by spontaneous, transient, and recurrent neural processes (Maquet, 2010). Of interest, the spontaneous brain oscillations of NREM sleep have been widely investigated both in animal and human models. Neuroimaging studies have investigated the neural correlates of phasic events characterizing the architecture of NREM sleep stages such as spindles and slow waves.

Spindles

The key structure for the generation of NREM sleep oscillations, in particular, spindles is the thalamus: within the thalamus, "pacemakers" of spindle oscillations are located in thalamic reticular neurons (Pare et al., 1987;

Steriade et al., 1987). On the other hand, neocortex is essential for the induction, synchronization, and termination of spindles, despite these can be generated within the thalamus also in the absence of the cerebral cortex (Steriade and McCarley, 2005). A consistent line of fMRI results demonstrated that spindles were associated with increased brain responses in the lateral and posterior aspects of the thalamus, as well as in paralimbic (anterior cingulate cortex, insula) and neocortical (superior temporal gyrus) areas. This finding constitutes the first neuroimaging report showing that NREM sleep is characterized by transient increases in brain activity organized by specific neural events. It also confirms the active involvement of thalamic structures in the generation of spindles and the participation of specific cortical areas in their modulation in humans (Dang-Vu et al., 2010). Spindles, however, may exist in the two potential subtypes: the most spindles in humans recorded in central and parietal regions and display a frequency of 13–14 Hz ("fast" spindles), others are prominent on frontal derivations with a frequency below 13 Hz ("slow" spindles) (Jankel and Niedermeyer, 1985; Jobert et al., 1992). Thus, we can expect that these two different sleep spindle categories may be modulated by segregated neural systems and associated with activation of distinct cortical networks.

Both slow and fast spindles have been shown to be related to increases in BOLD signal in the thalami, the anterior cingulate cortex, the left anterior insula, and the superior temporal gyrus (Schabus et al., 2007) (Table 11.1). However, brain activity associated with slow-spindles displayed a distribution very close to the patterns associated with unspecified spindles whereas fast-spindles activity appears associated with larger activations in several cortical areas, including precentral and postcentral gyri, MPFC, and hippocampus (Dang-Vu et al., 2010). The recruitment of partially segregated cortical networks for the two types of spindles suggests that they may have different functions, with fast spindles being involved in sensorimotor and mnemonic information processing (Dang-Vu et al., 2010).

Not only thalamus is related to the spindles brain activity. Recent evidences revealed important reorganization of the spontaneous hippocampal connectivity during NREM sleep (Andrade et al., 2011). Especially during NREM2, the hippocampus was more strongly connected with the temporal, insular, occipital, and cingulate cortices than during wakefulness or slow-wave sleep. Spindle-related activity overlapped with the hippocampal connectivity map (contrasting sleep stage 2 with wakefulness). The connectivity pattern of the subiculum to frontal, lateral temporal, and motor cortical regions and to the insula showed a strong interaction with

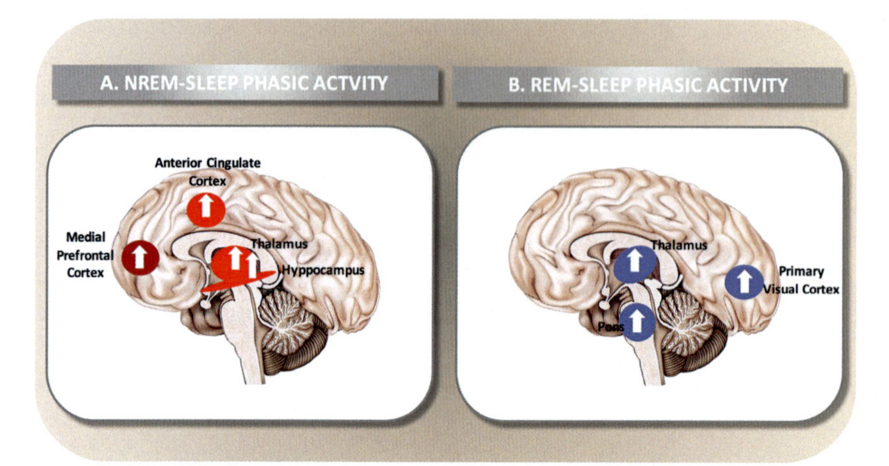

Figure 11.1 ***Brain cortices/structures mainly activated during the normal human sleep.*** Schematic representation of the brain activations during nonrapid eye movement (A) and rapid eye movement (B) sleep phasic activity. Colors: red (spindles activity); burgundy (slow waves activities); blue [ponto-geniculo-occipital (PGO) waves-like activities].

occurrence of sleep spindles (Andrade et al., 2011) (Table 11.1). Increased connectivity between hippocampus and neocortical regions in NREM2 suggests an increased capacity for possible global information transfer, while connectivity in slow-wave sleep is reflecting a functional system optimal for segregated information reprocessing (Andrade et al., 2011). These altered functional patterns likely reflect differences in information processing, ae relevant to differentiating sleep stage-specific contributions to neural plasticity as proposed in sleep-dependent memory consolidation (Andrade et al., 2011). Fig. 11.1A shows a schematic representation of brain structures/cortex (red color) mainly activated during NREM-sleep phasic activity (spindles).

Slow waves

Marked progress has been performed in understanding of mechanisms underlying the normal human sleep, by the identification of brain structures involved in the generation and/or modulation of slow waves. Significant increases in human brain activity have been reported in association with slow waves in several cortical areas using a combined EEG-fMRI approach (Dang-Vu et al., 2008). In particular, the main area associated with slow waves is the MPFC thus confirming the frontal predominance of slow wave activity during human NREM sleep (Dang-Vu et al., 2008) (Table 11.1). This activation of the MPFC is in line with findings from topographical scalp EEG studies (Massimini et al., 2004; Happe et al., 2002), and it is expected since it is involved in the memory processing occurring during sleep. Changes in brain activity have been also detected in others structures including the inferior and medial frontal gyrus, parahippocampal gyrus, precuneus, posterior cingulate cortex, pontomesencephalic tegmentum, and cerebellum. A decrease activity of brainstem nuclei located in the pontomesencephalic tegmentum, nuclei implicated is necessary in to generate NREM sleep oscillations (Steriade and McCarley, 2005; Steriade et al., 1990). In contrast to the classical view of brainstem nuclei promoting vigilance and wakefulness, these neuroimaging data suggest that several pontine structures might be active during NREM sleep concomitant with slow waves, thereby modulating cortical activity even during the deepest stages of sleep (Dang-Vu et al., 2010). Finally, as occurs for spindles also the slow waves may be present in two distinct categories: high and medium amplitude. The high slow waves activated brainstem and mesio-temporal areas, while medium-amplitude slow waves preferentially activated inferior and medial frontal areas. The preferential activation of mesio-temporal areas (with high-amplitude slow waves) suggests that the amplitude of the wave may be a crucial factor in the recruitment of brain structures potentially involved in the processing of memory traces during the sleep (Dang-Vu et al., 2010). Fig. 11.1A, shows a schematic representation of the cortex (burgundy color) mainly activated during NREM-sleep phasic activity (slow waves).

Imaging in rapid eye movement sleep

Neuronal correlates of rapid eye movement sleep

There is evidence that, the pattern of brain activity during REM sleep as assessed by functional neuroimaging studies is markedly different from patterns detected in NREM sleep. In contrast to NREM, REM sleep is characterized by sustained neuronal activity, high-cerebral energy requirements, and cerebral blood flow. Cerebral cortex becomes globally deactivated as the sleep stages progress and then reactivates to the waking level during REM sleep (Kubota et al., 2011). Further evidence of activation of the cerebral cortex during REM sleep is the desynchronization of the EEG, one of the criteria of REM sleep (Kubota et al., 2011). When compared to wakefulness and/or NREM sleep, the pattern of cerebral activity during REM sleep is not only different but also peculiar of this sleep stage. Indeed, it is characterized by simultaneously regional activations and deactivations: the activations were mainly located in pontine tegmentum, thalamus, amygdala, hippocampus, anterior cingulate cortex, temporo-occipital areas, basal forebrain, cerebellum, and caudate nucleus, while the deactivations in the DLPF cortex, posterior cingulate gyrus, the pre-cuneus, and inferior parietal cortex (Dang-Vu et al., 2007). Of interest, a contrast between the activation of posterior cortical areas and deactivation/quiescence of the associative frontal and parietal cortices appears to emerge. With respect to the cortical involvement during REM sleep, several PET studies have revealed an activation in the posterior visual cortex during REM sleep (Maquet et al., 1996; Braun et al., 1998) and a significant positive correlation with REM density (Braun et al., 1998; Peigneux et al., 2001) (Table 11.1). Moreover, a recent fMRI study (Wehrle et al., 2007) showed an activation of the temporo-occipital areas included inferior temporal cortex and fusiform gyrus, areas which are considered as extras-triate cortices belonging to the ventral visual stream (Braun et al., 1997). In addition, the functional interactions between posterior cortical areas appeared to be different during REM sleep as compared to wakefulness (Braun et al., 1998): extrastriate cortex activation was significantly correlated with primary visual cortex (striate cortex) deactivation during REM sleep, while their activities are usually positively correlated during wakefulness. Regional deactivations during REM sleep have been reported in the DLPF (inferior and middle frontal gyrus), precuneus, posterior cingulate cortex, and part of the parietal cortex (temporo-parietal region, inferior parietal lobule) (Braun et al., 1997; Maquet et al., 1996; Maquet et al., 2005). Neuroimaging data concerning the DLPF activity during REM sleep are, however, debated. Indeed, a line of neuroimaging studies demonstrate that DLPF showed a significant increase during REM sleep, rather than a deactivation as compared to the wakeful state (Ioannides et al., 2009) (Table 11.1). Results from a recent spectroscopy study investigating the hemodynamic changes of the DLPF throughout the REM sleep, also indicate that the appearance of the first REM that occurred just after onset of the REM sleep coincides with the activation/reactivation of the DLPF (Kubota et al., 2011). Although the reasons for cortical deactivations are still unclear, one possible explanation is linked to the amygdalar inputs: cortical areas less active during REM sleep received only few inputs from the amygdala, whereas those more active received rich amygdalar inputs (Marini et al., 1992), thus suggesting that the amygdala can modulate the cortical activity during REM sleep. This is in line with the demonstration that functional interactions between the amygdala and the occipitotemporal cortices were different in the context of REM sleep than in NREM sleep or wakefulness (Desseilles et al., 2006). Thus the amygdalocortical network during REM sleep might contribute to the selective processing of emotionally relevant memories (Maquet et al., 1996). In addition, a meta-analysis of Maquet and coworkers refines the description of the functional neuroanatomy of normal human sleep and characterize the hypoactivity in frontal and parietal areas during REM sleep. It involves the inferior and middle frontal gyrus as well as the posterior part of the inferior parietal lobule. Interestingly enough, the superior frontal gyrus, the medial frontal areas, the intraparietal sulcus, and the superior parietal cortex are not less active in REM sleep than during wakefulness (Maquet et al., 2005). This peculiar distribution of regional brain activity during REM sleep might correlate with some features of cognition, as reflected in dream reports (Maquet et al., 2005). Especially, this regional metabolic pattern provides new insights on the possible neural bases of dream features such as self and characters' mind representations, the poor episodic recall, the lower ability of external stimuli to break the dream narrative, and the difficulty to organize one's oneiric behavior toward a well identified and persistent goal (Maquet et al., 2005).

In contrast to the distribution of the activity within the cortex, the pattern of segregated activity in the subcortical structures as well as in limbic areas is easily explained by the known neurophysiological mechanisms which generate REM sleep and the dreams in animals (Maquet et al., 2005). These involve cholinergic processes arising from brainstem structures located in the PPT and laterodorsal tegmentum activating the cortex via the thalamus and basal forebrain (Maquet et al., 2005; Ioannides et al., 2009; Amaral and Price, 1984; Maquet and Phillips, 1998; Datta, 1995, 1997; Marini et al., 1992). Thus, is not surprising that pontine tegmentum, thalamic nuclei and basal forebrain, as well as cerebellum reflect input from the brainstem. Activations during REM sleep were also observed in the caudate nucleus although the role of this structures in REM sleep physiology remains speculative (Braun et al., 1997). Finally, the amygdala seems to modulate other key features of REM sleep. For instance, the large variability in heart rate during REM sleep could be explained by a prominent influence of the amygdaloid complexes (Desseilles et al., 2006). Both amygdala and hippocampal formation are also critical for memory systems (Bechara et al., 1995) and may, thus, participate in the processing of memory traces during REM sleep, as reviewed elsewhere (Maquet et al., 2003; Maquet, 2001; Rauchs et al., 2005; Dang-Vu et al., 2006).

REM sleep is also the sleep stage during which dreams are prominent. Recent advances in neuroimaging techniques have allowed researchers to link dream features to well-known patterns of brain activation to identify EEG markers of dreaming and dream recall, and even to predict dreaming and dream content in real time (Siclari et al., 2020). These studies have also demonstrated that some brain areas are consistently and locally activated during dreams, irrespective of sleep stages (Siclari et al., 2020). In the view, the functional brain mapping activating during REM sleep some brain structure rather than others might therefore also be interpreted in light of dreaming properties (Dang et al., 2007, 2009; Schwartz and Maquet, 2002). The main interactions that could explain this phenomenon is related to some evidences. First, the perceptual aspects of dreams are always visual components, whereas auditory components are present in 40%–60% of dreams, movement and tactile sensations in 15%–30%, and finally smell and taste in less than 1% (Dang-Vu et al., 2007). Thus it is possible to speculate that these could be related to the activation of posterior (occipital and temporal) cortices typically occurring during REM sleep. Indeed, patients with occipitotemporal lesions may report a cessation of visual dreams imagery (Solms MThe, 1997). Second, the dream content is also characterized by the prominence of emotions, and especially negative emotions such as fear and anxiety (Dang-Vu et al., 2007; Strauch et al., 1996; Nielsen et al., 1991). Thus, emotional features in dreams would be related to the activation of amygdaloid complexes, involved in the modulation of responses to threatening stimuli or stressful situations during wakefulness and hightly activated during REM sleep (Dang-Vu et al., 2007). Finally, the relative hypoactivation of the prefrontal cortex, would explain the alteration in logical reasoning, working memory, episodic memory, and executive functions that manifest themselves in dream reports from REM sleep awakenings (Maquet, 2000; Maquet et al., 1996; Maquet et al., 2005; Hobson et al., 1998; Hobson et al., 2003).

Neuronal correlates of phasic rapid eye movement sleep-activity

In animal models, the phasic activity during REM sleep is characterized by the occurrence of PGO waves. These are prominent phasic bioelectrical potentials, strictly related to REMs occurring in isolation or in bursts during the transition from NREM to REM sleep or during REM sleep itself (Dang-Vu et al., 2010; Datta, 1999; Callaway et al., 1987). The neural evidence for the existence of PGO in the brain is related to the REM-related activation of the primary visual cortex without visual input from the retina thus constituting a bridge between REM sleep and dreaming (Datta, 2000). PGO waves are most recorded in many areas of animal brain in particular, pons, the lateral geniculate bodies, and the occipital cortex (Dang-Vu et al., 2007, 2010). PGO are functionally significant for the brain plasticity and for the processes of learning and memory consolidation (Mavanji and Datta, 2003; Datta et al., 2004; Miyauchi et al., 2009).

In humans, functional neuroimaging data suggest that the REMs observed during REM sleep could be generated by mechanisms similar to PGO waves in animals (Dang-Vu et al., 2010). In a PET imaging study, Peigneux et al. found correlations during REM sleep between the density of REMs and rCBF in the occipital cortex and the lateral geniculate bodies of the thalamus (Peigneux et al., 2001) (Table 11.1). In a simultaneous EEG/fMRI recording, Wehrle et al. show distinct magnetic resonance imaging signal increases in the posterior thalamus and occipital cortex in close temporal relationship to REMs during human REM sleep (Wehrle et al., 2007) (Table 11.1). They found a positive correlation between BOLD signal in these areas and the density of REMs in the normal human volunteers. Moreover, Miyauchi and co-workers conducted an event-related fMRI study to assess the brain activations time-locked to the onset of REMs in healthy subjects (Miyauchi et al., 2009). They found an increase in brain activity associated with REMs in the pons, thalamus, and primary visual cortex. Furthermore, the time-course analysis of BOLD responses indicated that the activation of the pontine tegmentum, ventroposterior thalamus, and primary visual cortex started before the occurrence of REM (Miyauchi et al., 2009) (Table 11.1). Considering that these areas are those in which PGO waves are recorded in animals, it is possible to speculate that processes similar to PGO waves are responsible for the generation of REMs in humans. Taken together, these neuroimaging findings, consistent with cell recordings in animal experiments confirms that PGO-like activities may exist in humans and contribute to shape the functional brain mapping of REM sleep (Dang-Vu et al., 2010). On the other hand, the activation of other areas such as the limbic areas including the parahippocampal gyrus and amygdala simultaneously with REMs suggests that REMs and/or their generating mechanism are not merely an epiphenomenon of PGO waves, but may be linked to the triggering activation of these areas (Miyauchi et al., 2009). Additional activations were located in the putamen, anterior cingulate cortex suggesting that these are structures potentially involved in the modulation of phasic human REM sleep activity (Dang-Vu et al., 2010). These evidences also emphasize the need to use combined approaches to detect PGO-like activities in human for studying neuronal networks underlying sleep regulation. Fig. 11.1B shows a schematic representation of brain structures/cortex (blue color) mainly activated during REM-sleep phasic activity (PGO-like activities).

Conclusions and future directions

In conclusion, precious has been the contribute of neuroimaging in the last decades to identify and decode the mechanisms underlying to the physiology of human sleep. PET/SPECT are tools highly specialized to characterize the network of deactivated/deactivated brain areas during NREM sleep as well as activated/deactivated structures during REM sleep. However, these techniques assessing the brain activity across a long period of time such as a specific sleep stage are not able to capture the changes in brain activity of short and fast events such as the oscillations of NREM sleep including spindles and slow waves. The development of combined approaches such as EEG/fMRI has made possible the identification of phasic events occurring during sleep stages. Functional neuroimaging studies allow to demonstrate that NREM sleep cannot be reduced to a state of quiescence, characterized by global and regional decrease of brain activity. Rather, it appears to be an active state during which the increase of activity in specific brain regions appears to be synchronized to equally specific NREM sleep phasic oscillations. REM sleep stage is characterized by an increased activity of the brain. A contrast between the activation of posterior visual areas and deactivation/quiescence of the associative frontal and parietal cortices is of interest. This apparent discrepancy could be the neural correlate to explain the fascinating phenomenon of dreaming. As future directions, additional studies performed during the nocturnal sleep rather than in wakefulness are needed to detect the dynamic changes of brain activity occurring across the different sleep stages.

Conflict of interests

The authors declare not to have commercial or financial relationships that could be represented as a potential conflict of interest.

References

Amaral, D.G., Price, J.L., 1984. Amygdalo-cortical projections in the monkey (*Macaca fascicularis*). J. Comp. Neurol. 230 (4), 465–496.

Andersson, J.L., Onoe, H., Hetta, J., et al., 1998. Brain networks affected by synchronized sleep visualized by positron emission tomography. J. Cerebr. Blood Flow Metabol. 18 (7), 701–715.

Andrade, K.C., Spoormaker, V.I., Dresler, M., et al., 2011. Sleep spindles and hippocampal functional connectivity in human NREM sleep. J. Neurosci. 31 (28), 10331–10339, 13.

Bandettini, P.A., 2009. What's new in neuroimaging methods? Ann. NY Acad. Sci. 1156, 260–293.

Bechara, A., Tranel, D., Damasio, H., Adolphs, R., Rockland, C., Damasio, A.R., 1995. Double dissociation of conditioning and declarative knowledge relative to the amygdala and hippocampus in humans. Science 269 (5227), 1115–1118.

Borbely, A.A., 2001. From slow waves to sleep homeostasis: new perspectives. Arch. Ital. Biol. 139 (1–2), 53–61.

Braun, A.R., Balkin, T.J., Wesenten, N.J., et al., 1997. Regional cerebral blood flow throughout the sleep-wake cycle. An H2(15) O PET study. Brain 120 (Pt 7), 1173–1197.

Braun, A.R., Balkin, T.J., Wesensten, N.J., et al., 1998. Dissociated pattern of activity in visual cortices and their projections during human rapid eye movement sleep. Science 279 (5347), 91–95.

Callaway, C.W., Lydic, R., Baghdoyan, H.A., Hobson, J.A., 1987. Ponto-geniculo-occipital waves: spontaneous visual system activity during rapid eye movement sleep. Cell. Mol. Neurobiol. 7, 105–149.

Dang, V.T.T., Schabus, M., Desseilles, M., Schwartz, S., Maquet, P., 2007. Neuroimaging of REM sleep and dreaming. In: McNamara, P., Barrett, D. (Eds.), The New Science of Dreaming. Praeger Publishers, Westport, pp. 95–113.

Dang, V.T.T., Schabus, M., Cologan, V., Maquet, P., 2009. Sleep: implications for theories of dreaming and consciousness. In: Banks, W.P. (Ed.), Encyclopedia of Consciousness. Elsevier, Oxford, pp. 357–373.

Dang-Vu, T.T., Desseilles, M., Laureys, S., et al., 2005. Cerebral correlates of delta waves during non-REM sleep revisited. Neuroimage 28 (1), 14–21.

Dang-Vu, T.T., Desseilles, M., Peigneux, P., Maquet, P., 2006. A role for sleep in brain plasticity. Pediatr. Rehabil. 9 (2), 98–118.

Dang-Vu, T.T., Desseilles, M., Petit, et al., 2007. Neuroimaging in sleep medicine. Sleep Med. 8 (4), 349–372.

Dang-Vu, T.T., Schabus, M., Desseilles, M., et al., 2008. Spontaneous neural activity during human slow wave sleep. Proc. Natl. Acad. Sci. U. S. A. 105, 15160–15165.

Dang-Vu, T.T., Desseilles, M., Schabus, M., et al., 2010. Functional neuroimaging insights into the physiology of human sleep, 33 (12), 1589–1603.

Datta, S., Mavanji, V., Ulloor, J., Patterson, E.H., 2004. Activation of phasic pontine-wave generator prevents rapid eye movement sleep deprivation-induced learning impairment in the rat: a mechanism for sleep-dependent plastic¬ity. J. Neurosci. 24, 1416–1427.

Datta, S., 1995. Neuronal activity in the peribrachial area: relationship to behavioral state control. Neurosci. Biobehav. Rev. 19, 67–84.

Datta, S., 1997. Cellular basis of pontine ponto-geniculo-occipital wave generation and modulation. Cell. Mol. Neurobiol. 17, 341–365.

Datta, S., 1999. PGO wave generation: mechanism and functional significance. In: Mallick, B.N., Inoue, S. (Eds.), Rapid Eye Movement Sleep. Narosa Publishing House, New Delhi, pp. 91–106.

Datta, S., 2000. Avoidance task training potentiates phasic pontine-wave density in the rat: a mechanism for sleep-dependent plasticity. J. Neurosci. 20, 8607–8613.
Desseilles, M., Dang Vu, T., Laureys, S., et al., 2006. A prominent role for amygdaloid complexes in the variability in heart rate (VHR) during rapid eye movement (REM) sleep relative to wakefulness. Neuroimage 32 (3), 1008–1015.
Desseilles, M., Dang-Vu, T.T., Maquet, P., 2011. Chapter 6. Functional neuroimaging in sleep, sleep deprivation, and sleep disorders. Handb. Clin. Neurol. 98, 71–94.
Discovery of key whole-brain transitions and dynamics during human wakefulness and non-REM sleep. Nat. Commun. 10 (1), 2019, 1035.
Duyn, J.H., 2012. EEG-fMRI methods for the study of brain networks during sleep. Front. Neurol. 2 (3), 100.
Fox, M.D., Raichle, M.E., 2007. Spontaneous fluctuations in brain activity observed with functional magnetic resonance imaging. Nat. Rev. Neurosci. 8, 700–711.
Fuller, P.M., Fuller, P., Sherman, D., et al., 2011. Reassessment of the structural basis of the ascending arousal system. J. Comp. Neurol. 519, 933–956.
Happe, S., Anderer, P., Gruber, G., Klosch, G., Saletu, B., Zeitlhofer, J., 2002. Scalp topography of the spontaneous K-complex and of delta-waves in human sleep. Brain Topogr. 15, 43–49.
Hobson, J.A., Pace-Schott, E.F., Stickgold, R., Kahn, D., 1998. To dream or not to dream? Relevant data from new neuroimaging and electrophysiological studies. Curr. Opin. Neurobiol. 8, 239–244.
Hobson, J., Pace-Schott, E., Stickgold, R., 2003. Dreaming and the brain: toward a cognitive neuroscience of conscious states. In: Pace-Schott, E., Solms, M., Blagrove, M., Harnad, S. (Eds.), Sleep and Dreaming. Cambridge University Press, Cambridge, pp. 1–50.
Hobson, J.A., 2005. Sleep is of the brain, by the brain and for the brain. Nature 437, 1254–1256.
Ioannides, A.A., Kostopoulos, G.K., Liu, L., Fenwick, P.B., 2009. MEG identifies dorsal medial brain activations during sleep. Neuroimage 44, 455–468.
Jankel, W.R., Niedermeyer, E., 1985. Sleep spindles. J. Clin. Neurophysiol. 2, 1–35.
Jobert, M., Poiseau, E., Jahnig, P., Schulz, H., Kubicki, S., 1992. Topographical analysis of sleep spindle activity. Neuropsychobiology 26, 210–217.
Jones, E.G., 1991. The anatomy of sensory relay functions in the thalamus. Prog. Brain Res. 87, 29–52.
Kaufmann, C., Wehrle, R., 2006. Brain activation and hypothalamic functional connectivity during human non-rapid eye movement sleep: an EEG/fMRI study. Brain 129, 655–667.
Koike, T., Kan, S., Misaki, M., Miyauchi, S., 2011. Connectivity pattern changes in default-mode network with deep non-REM and REM sleep. Neurosci. Res. 69 (4), 322–330.
Kubota, Y., Takasu, N.N., Horita, S., et al., 2011. Dorsolateral prefrontal cortical oxygenation during REM sleep in humans. Brain Res. 10 (1389), 83–92.
Llinas, R.R., Steriade, M., 2006. Bursting of thalamic neurons and states of vigilance. J. Neurophysiol. 95, 3297–3308.
Maquet, P., Phillips, C., 1998. Functional brain imaging of human sleep. J. Sleep Res. 7 (Suppl. 1), 42–47.
Maquet, P., Dive, D., Salmon, E., et al., 1990. Cerebral glucose utilization during sleep-wake cycle in man determined by positron emission tomography and [18F]2-fluoro-2-deoxy-D-glucose method. Brain Res. 513, 136–143.
Maquet, P., Degueldre, C., Delfiore, G., Aerts, J., Peters, J.M., Luxen, A., et al., 1997. Functional neuroanatomy of human slow wave sleep. J. Neurosci. 17 (8), 2807–2812.
Maquet, P., Smith, C., Stickgold, R., 2003. Sleep and Brain Plasticity. Oxford University Press, Oxford.

Maquet, P., Ruby, P., Maudoux, A., et al., 2005. Human cognition during REM sleep and the activity profile within frontal and parietal cortices: a reappraisal of functional neuroimaging data. Prog. Brain Res. 150, 219–227.
Maquet, P., Peters, J.M., Aerts, J., et al., 1996. Functional neuroanatomy of human rapid-eye movement sleep and dreaming. Nature 383, 163–166.
Maquet, P., 2000. Functional neuroimaging of normal human sleep by positron emission tomography. J. Sleep Res. 9, 207–231.
Maquet, P., 2001. The role of sleep in learning and memory. Science 294 (5544), 1048–1052.
Maquet, P., 2010. Understanding non rapid eye movement sleep through neuroimaging. World J. Biol. Psychiatr. 11 (Suppl 1), 9–15.
Marini, G., Gritti, I., Mancia, M., 1992. Enhancement of tonic and phasic events of rapid eye movement sleep following bilateral ibotenic acid injections into centralis lateralis thalamic nucleus of cats. Neuroscience 48, 877–888.
Mason, M.F., Norton, M.I., Van Horn, J.D., et al., 2007. Wandering minds: the default network and stimulus-independent thought. Science 315, 393–395.
Massimini, M., Huber, R., Ferrarelli, F., Hill, S., Tononi, G., 2004. The sleep slow oscillation as a traveling wave. J. Neurosci. 24, 6862–6870.
Massimini, M., Ferrarelli, F., Huber, R., Esser, S.K., Singh, H., Tononi, G., 2005. Breakdown of cortical effective connectivity during sleep. Science 309, 2228–2232.
Mavanji, V., Datta, S., 2003. Activation of the phasic pontine-wave generator enhances improvement of learning performance: a mechanism for sleep-dependent plasticity. Eur. J. Neurosci. 17, 359–370.
Mena-Segovia, J., GiordanoM, 2003. Striatal dopaminergic stimulation produces c-Fos expression in the PPT and an increase in wakefulness. Brain Res. 986 (1–2), 30–38.
Miyauchi, S., Misaki, M., Kan, S., et al., 2009. Human brain activity time-locked to rapid eye movements during REM sleep. Exp. Brain Res. 192 (4), 657–667.
Mouret, J., Jeannerod, M., Jouvet, M., 1963. L'activite électrique du systeme visuel au cours de la phase paradoxale du sommeil chez le chat. J. Physiol. 55, 305–306.
Nielsen, T.A., Deslauriers, D., Baylor, G.W., 1991. Emotions in dream and waking event reports. Dreaming 1, 287–300.
Pare, D., Steriade, M., Deschenes, M., Oakson, G., 1987. Physiological characteristics of anterior thalamic nuclei, a group devoid of inputs from reticular thalamic nucleus. J. Neurophysiol. 57, 1669–1685.
Peigneux, P., Laureys, S., Fuchs, S., et al., 2001. Generation of rapid eye movements during paradoxical sleep in humans. Neuroimage 14, 749–758.
Picchioni, D., Pixa, M.L., Fukunaga, M., et al., February 1, 2014. Decreased connectivity between the thalamus and the neocortex during human nonrapid eye movement sleep. Sleep 37 (2), 387–397.
Rauchs, G., Desgranges, B., Foret, J., Eustache, F., 2005. The relationships between memory systems and sleep stages. J. Sleep Res. 14 (2), 123–140.
Rechtschaffen, A., Kales, A., 1968. A Manual of Standardized Terminology, Techniques and Scoring System for Sleep Stages of Human Subjects. University of California. Brain Information Service/Brain Research Institute, Los Angeles.
Schabus, M., Dang-Vu, T.T., Albouy, G., et al., 2007. Hemodynamic cerebral correlates of sleep spindles during human non-rapid eye movement sleep. Proc. Natl. Acad. Sci. U. S. A. 104 (32), 13164–13169.
Schwartz, S., Maquet, P., 2002. Sleep imaging and the neuro-psychological assessment of dreams. Trends Cognit. Sci. 6, 23–30.
Siclari, F., Valli, K., Arnulf, I., 2020. Dreams and nightmares in healthy adults and in patients with sleep and neurological disorders. Lancet Neurol. 19, 849–859.

Solms, MThe Neuropsychology of Dreams, 1997. Lawrence Erlbaum Associates Inc., Mahwah.
Steriade, M., McCarley, R.W., 2005. Brain Control of Wakefulness and Sleep. Springer, New York.
Steriade, M., Domich, L., Oakson, G., Deschenes, M., 1987. The deafferented reticular thalamic nucleus generates spindle rhythmicity. J. Neurophysiol. 57, 260–273.
Steriade, M., Datta, S., Pare, D., Oakson, G., Curro Dossi, R.C., 1990. Neuronal activities in brain-stem cholinergic nuclei related to tonic activation processes in thalamocortical systems. J. Neurosci. 10 (8), 2541–2559.
Strauch, I., Meier, B., Foulkes, D., 1996. In Search of Dreams: Results of Experimental Dream Research. State University of New York Press, Albany.
Tagliazucchi, E., Behrens, M., Laufs, H., 2013. Sleep neuroimaging and models of consciousness. Front. Psychol. 4, 256.
Takamitsu, W., Shigeyuki, J., Takahiko, T., et al., 2014. Network-dependent modulation of brain activity during sleep. Neuroimage 98, 1–10.
Tononi, G., 2005. Consciousness, information integration, and the brain. Prog. Brain Res. 150, 109–126.
von Economo, C., 1930. Sleep as a problem of localization. J. Nerv. Ment. Dis. 71, 249–259.
Watanabe, T., Kan, S., Koike, et al., September, 2014. Network-dependent modulation of brain activity during sleep. Neuroimage 98, 1–10.
Wehrle, R., Kaufmann, C., Wetter, T.C., et al., 2007. Functional microstates within human REM sleep: wrst evidence from fMRI of a thalamocortical network specifc for phasic REM periods. Eur. J. Neurosci. 25, 863–871.

CHAPTER 12

Objective questionnaires for assessment sleep quality

David L. Streiner
Department of Psychiatry and Behavioural Neurosciences, McMaster University, Hamilton, ON, Canada

Introduction

Both clinical investigations and research into sleep rely on various ways of assessing its depth and quality, as well as factors that may interfere with it or be affected by it. Ibáñez et al. (2018) have developed a taxonomy of assessment methods used in sleep. These evaluations can be either medically assisted (e.g., polysomnography, continuous positive airways pressure (CPAP) titration test) or self-assessed, with the latter divided into the three categories of questionnaires, sleep diaries, and hardware devices, such as video cameras and accelerometers. In this chapter, we will focus primarily on questionnaires, although many of the issues that will be discussed, such as reliability and validity, are also relevant to the other assessment approaches. The chapter can be read from two perspectives: as a "procedures manual" for those who wish to develop their own scale to measure some aspect of sleep, and as guide for users of existing scales to determine if they meet the criteria necessary to perform adequately.

To begin, though, we need to discuss some issues of terminology. In this chapter, we will use the term *scale* to describe an instrument, either completed by the patients themselves or administered by others, consisting of a number of items that measure a single, unidimensional attribute, such as restless leg syndrome or sleep apnea. There are two implications from this. First, because all of the items tap one domain, the items tend to be highly correlated with one another. For the same reason, not every aspect of the phenomenon needs to be included in the scale; whatever may be missed by omitting a given item will be detected by the other, correlated items on the scale. That is, the items in a scale are only a sample of all possible questions that can be asked. In contrast, an *index* is more akin to a checklist, consisting of unrelated items. Perhaps the most widely known one is the Apgar Scale (Apgar, 1953), which, despite its name, would be considered to be an index using this terminology. Its five items (skin color, heart rate, muscle tone,

Methodological Approaches for Sleep and Vigilance Research
ISBN 978-0-323-85235-7
https://doi.org/10.1016/B978-0-323-85235-7.00007-7

reflexes, and respiration) may be correlated in healthy neonates, but not in those with a medical condition. An infant with cerebral palsy, for example, may have excellent skin color, respiration, and heart rate, but poor muscle tone and reflexes; whereas one with aortic stenosis would score poorly on skin color and heart rate, but well on the other items. Unlike a scale, the items on an index are the entire universe of items; if one is changed or omitted, then the nature of the index itself changes. Finally, an *inventory* is an instrument consisting of a number of individual scales, which may or may not be correlated among themselves; an example would be the Pittsburgh Sleep Quality Index (Buysse et al., 1989), which consists of 19 items grouped into seven components or sub-scales (subjective sleep quality, sleep latency, sleep duration, habitual sleep efficiency, sleep disturbances, use of sleeping medication, and daytime dysfunction). These definitions are not universally used, but will help in our discussions about constructing various instruments.

Steps in scale construction

Is a new scale required?

Before embarking on the long, and often tedious, process of developing a new scale, it is necessary to ask whether a new scale is really required. Many scales already exist (see, for example, Kurtis et al., 2018; Lomeli et al., 2008; Smith and Wegener, 2003) and the question is whether yet another one will add anything of value. Of the over 10,000 scales listed in *The Directory of Unpublished Experimental Measures* (Goldman et al., 1996), the vast majority have been used by only a small number of people, usually the scale's developer and perhaps his or her students. What this means is that it is very likely that an appropriate scale already exists and can be uncovered with a bit of effort. Appendix B in *Health Measurement Scales: A Practical Guide to their Development and Use* (Streiner et al., 2015) lists scores of published and on-line sources of published and unpublished instruments. It is strongly recommended that these resources be consulted before starting to devise a new test.

It should also be borne in mind that devising a new scale is not something that can be done over one or two weekends and a few bottles of beer (although some indeed seem as if they were constructed in this way). The process is a long one, consisting of devising the items, checking them for comprehension and lack of bias, assessing the reliability of the scale as a whole, and then establishing its validity. This can take many months or even years. Unless the primary goal is to derive a new scale, as opposed to simply constructing a tool to be used in a research project, the researcher's time may be better spent in other ways.

Devising the items

Perhaps the most important step in constructing an instrument—a scale, an index, or an inventory—is writing the items. If they are not constructed well, then the tool cannot measure what it is designed to, and if important areas are omitted, they cannot be later recaptured. Unless the process of devising items is done well, all of the later steps will be fruitless. There are many potential sources of items. The most widely used is to borrow from other scales. There are a number of reasons for this. First, there are only a limited number of ways that one can ask how long it takes to fall asleep or whether the respondent uses hypnotics. Second, it saves time, in that the items from an earlier instrument likely have been checked regarding their interpretability and freedom from the errors we will discuss shortly. However, there are some cautions that need to be observed. Previous scales may be copyrighted, and there is a fine line between "borrowing" one or two items and outright plagiarism. Moreover, some terms may become dated or their meaning may have changed over time. For example, "fagged out" can mean tired, but the term "fag" has accumulated other meanings over the years and is seen by some as offensive.

But if a new scale is being developed, it is likely because the existing ones are seen as flawed or incomplete in some way, so new items must be devised. Another source of items is clinical observation: there may be insights that you have you gleaned about sleep, its disturbances, and sequelae from your clinical practice that you do not see reflected in the existing instruments. Similarly, there may be new findings from research that need to be incorporated into a scale or index. For example, recent research has shown that poor sleep may make people more prone to type 2 diabetes and cardiovascular disease (Gibson et al., 2009), so a new scale about the consequences of sleep disturbance should inquire into these areas.

Another source of items, all too frequently overlooked, are the patients themselves. They are the ones with first-hand knowledge of what it means to have a sleep disturbance, what they feel may contribute to it, and what effects it may have on their lives. Most especially when devising a scale regarding a novel aspect of sleep problems, it is extremely helpful to get the perspective of the patients. One very useful way of eliciting information is to have focus groups, consisting of five to eight people with similar problems (e.g., Ronen et al., 1999). The number of focus groups is dictated by achieving "saturation," that is, no additional information is learned from the last one. Most often, no more than three or four such groups are required. It should be noted, though, that this is a time-consuming process because each group session must be recorded, transcribed, and then the themes extracted from the written record, often using software developed for qualitative research, such as NVivo (n.d.), which was previously called NUD*IST.

When constructing a scale, the developer should have a rough idea of the number of items it will contain, and then write three to four times that number. Later steps will weed out items that may not be properly interpreted, may be misunderstood, or suffer in some other way.

Checking the items

It may appear at first glance that responding to an item on a questionnaire is a simple and straightforward process: the respondent reads the item and then gives an answer. On closer examination, though, it is actually a fairly complex cognitive task that can be thought of as consisting of five steps (Schwarz and Oyserman, 2001): understanding the question, recalling the relevant information, inferring and estimating the response, mapping the answer onto the response alternatives, and editing the answer. Problems can arise at each stage of the process.

1. *Understanding the question*: There are many issues that may interfere with a person's ability to understand the item as it is written. The first is *reading level*. Although a high-school graduate in North America has completed 12 years of education, the average person reads only at a Grade 6 level. Thus, questions should be phrased as simply as possible, and be as short as possible. Items that have 10 to 20 characters have much higher validity coefficients than those with 70–80 characters (Holden et al., 1985). A second problem is the use of *jargon*. To most lay people, a "stool sample" is what is seen in a furniture showroom; and "shock" is what happens when you put your finger into an electrical outlet. Words such as "nutrition," "digestion," "orally," and "tissue," which most physicians would assume their patients would know, were erroneously defined by more than 50% of respondents (Samora et al., 1961). A third issue is *ambiguity*. Words like "recently" and "often" mean different things to different people. If a specific

time frame is required, it should be specified. Similarly, amounts should be defined explicitly, rather than relying on terms such as "frequently" or "rarely." Sometimes, items are *double barreled*, asking two questions at once, such as "I feel tired and worn out." Some people may say yes if they feel tired but not worn out, or vice versa; while others may say no because they are not both. This introduces error into the responses, so that items that contain the words "and" or "or" should be avoided. Finally, *negatively worded* items should not be used. They are not interpreted as the opposite of their positively worded counterparts; and factor analyses of scales with both positively and negatively worded items show that the negative ones cluster together, irrespective of their content (Lindwall et al., 2012). Words beginning with "in," "im," or "un" should be avoided.

2. *Recalling the relevant information*: Scale constructors overestimate people's ability to recall events. For example, in one study of women with suspected ovarian cancer, patients remembered having had an average of 3.4 health care visits, whereas the medical records showed that they had more than twice as many (Hess et al., 2012). In a different study, 104 people kept health diaries for three months. Immediately afterward, only 51% of health events were recalled, and this dropped to 39% after three months (Cohen and Java, 1995). One problem is that questions usually ask people to recall events within a specific time frame (e.g., "How many times in the past six months ..."), but memory is not organized like a calendar. It is more akin to an inverted pyramid, with extended periods at the top ("When I worked at this company" or "Before I was married"). Underneath are summarized events, reflecting repeated behaviors ("I was sick a lot then" or "I worked out a lot at that time"), and specific events are at the bottom. Asking people how often they had sleep difficulties over the past six months does not correspond to how those memories are actually stored. As Schwarz and Oyserman (2001) state, "Many recall questions would never be asked if researchers first tried to answer them themselves" (p. 141).
3. *Inference and estimation*: Because people have difficulty remembering specific events and when they occurred, they make an estimate, based on certain inferences. If asked how often in the past month they had trouble falling asleep, they may remember that there were two times during the last week. They would then try to determine if this was worse or better than usual, adjust their answer; and multiply by four.

Realize that this is not an accurate recall, but rather a reconstruction, and therefore prone to error. A similar phenomenon occurs when a person is asked, "How are you compared to six months ago?" People do not remember how they were this far back, and resort to various implicit theories of change. If, for example, they had been prescribed medication, they may feel that the medication had to have been effective, so that, of course, they are doing better. Yet again, this is a reconstruction, not a recall of a previous state. The further back in time (beyond one week in most situations), the more people will estimate and reconstruct.

4. *Mapping the answer onto the response alternatives*: Having remembered (or constructed) an answer, the person must now translate that response into one of the answers supplied for the item. Should three sleepless nights be called "often" or "sometimes," for example? How much pain corresponds to the number eight on a visual analog scale? Realize that people do not ordinarily think in terms of scales, and that we are forcing people to transform their subjective response into a different, and perhaps unfamiliar, format. As with any translation, something will get lost.
5. *Editing the response*: There are many biases that can come between what a person feels or remembers and what he or she puts down on the page. One is called *satisficing* (Krosnick, 1991), which means giving a response that satisfies the minimum demands of the instrument (i.e., completing the item) but one that is not the optimal answer. Satisficing can be achieved in a number of ways: by choosing the "Neutral" or "Do Not Know" alternative; answering "True" to all questions; giving the same response to all of the items, regardless of their content; and so forth. This is more likely to occur if the respondent does not understand the item, the questionnaire is seen as too long, or is of no interest to the person. Another prevalent bias is *social desirability*, which refers to consciously or unconsciously modifying the response to put the person in a good light. Thus, people tend to under-report undesirable activities, such as smoking or watching TV; and to over-report socially desirable ones, such as eating properly and exercising.

It is always a good idea to use *cognitive interviewing* with any questionnaire, whether it is a new one being developed, or an existing one that you are contemplating using in your research or clinical practice. This may take many forms, such as asking the respondents to reframe the item in their own words, to think aloud how they arrived at their answer, or to explain why they gave one response rather than another. These techniques are described in more detail in Collins (2003) and Streiner et al. (2015).

This brief introduction can only scratch the surface regarding how to write items and the possible pitfalls. To read further, see Streiner et al. (2015), Sudman and Bradburn (2004), or Aday and Cornelius (2006).

Content validity

After having written the items, the next task is to determine whether the entire domain of interest is covered (*content coverage*) and that there are no irrelevant items (*content relevance*). Collectively, content coverage and relevance are referred to as *content validity*. This step is useful, both for those constructing new scales, and for users of existing ones, to determine if they are truly covering what is of interest. Relying solely on the name of the scale can be misleading because not all scales cover all aspects of the domain, irrespective of what they are called.

The first step is to create a grid, with each row being one of the items that have been written or are on the existing scale; and the columns the areas that should be covered. For example, the columns could reflect sleep quality, sleep latency, daytime functioning, the use of sleep aids, and so forth, realizing that not all scales need necessarily cover all areas. An example is seen in Table 12.1.

Each item is read, and a check mark placed in the appropriate column. This can reveal a number of different things. For example, note that Item 5 does not have a check mark in any column, indicating that its content is not relevant and it should be eliminated. Second, Item 4 has a mark in two columns. This should be avoided because it makes it more difficult to

Table 12.1 A content validity grid.

	Area				
Item	**Sleep quality**	**Sleep latency**	**Daytime functioning**	...	**Sleep aids**
1		√			
2	√				
3			√		
4	√	√			
5					
⋮					
K					√

interpret which facet of the domain each subscale represents, and artificially builds in a correlation between the facets, which may paint a false picture regarding the relationships among them. Reading down the columns, we can determine whether each facet has a sufficient number of items. At a minimum, each should have at least three items; more when devising a new scale since, as mentioned above, some will later be eliminated. If there are not enough items, then it's back to the drawing board.

A soupçon of test theory

Before we delve into the heart of psychometrics, which deals primarily with the reliability and validity of an instrument, it is necessary to discuss a bit of the underlying conceptual theory. In what is called classical test theory, we assume that the value that we observe (X_O), be it from a paper-and-pencil questionnaire, a laboratory instrument, a manometer, or any other device that yields a number, is composed of two parts: the true score (X_T) and some error (X_E). That is,

$$X_{Observed} = X_{True} + X_{Error} \tag{12.1}$$

The term "true score" is a bit of a misnomer. It means the score we would see if the person were tested an infinite number of times, or if there were no error. It does not mean that the score is necessarily accurate or unchanging. Because of social desirability bias, for example, a person may understate how much alcohol he or she consumes, and will continue to understate it on repeated questioning. Similarly, the score may change due to the effects of some intervention, aging, or a host of other factors. Thus, X_T is a hypothetical value that we try to estimate by reducing error.

It is assumed that the error score has a mean of zero and some standard deviation. That means that it will sometimes increase the value of X_O and sometimes decrease it, but over the long run, the errors will cancel each other out. This means that longer tests tend to have less error associated with them than shorter ones; and why having a number of observers is better than having just one.

Reliability

Definitions of reliability

There are a number of ways to define reliability, each of which looks at it from a different, and complementary, perspective. Conceptually, we can think of reliability as a measure of "the extent to which measurements of individuals obtained under different circumstances yield similar results" (Streiner et al., 2015, p. 10). For example, if a person were assessed at Time 1 and then later at Time 2 (and assuming the person has not changed in the interim), to what degree do the two values agree? Similarly, if a student or patient were evaluated by two different evaluators, to what degree do they agree with one another? For this reason, some people use the terms "reproducibility" or "agreement" as synonyms for reliability. However, as we shall see, these terms capture only part of what reliability actually measures and they are not accurate descriptors of the concept.

Another way of defining reliability is that it is the ratio of the variance between people to the total variance:

$$\text{Reliability} = \frac{\text{Variability Between Subjects}}{\text{Total Variability}} \tag{12.2}$$

and since the total variability is simply the variability between subject plus the error, we can also write this as:

$$\text{Reliability} = \frac{\text{Variability Between Subjects}}{\text{Variability Between Subjects} + \text{Error}} \tag{12.3}$$

and by using σ^2 as the symbol for variance (or variability), with the subscripts S, T, and E indicating subjects, total, and error, we can say:

$$\text{Reliability} = \frac{\sigma_S^2}{\sigma_T^2} = \frac{\sigma_S^2}{\sigma_{S+E}^2} \tag{12.4}$$

That means that we can think of reliability as the proportion of the total variance that is due to differences among people. Because σ_S^2 has to be less than σ_{S+E}^2, reliability is a number between 0.0 (no reliability at all) and 1.0 (perfect reliability).

These equations point to a third way of defining reliability: the ability of an instrument to differentiate among people. This simple statement has a number of important implications. First, if a scale cannot distinguish among people, its reliability is zero. Thus we have the seemingly paradoxical situation that if everyone has the same score on a scale (as is often seen in evaluations of medical students or residents, where everyone is above average), the scale is useless because its reliability is zero. There will be perfect agreement among raters, and perfect reproducibility across assessment times, but no reliability, further reinforcing the idea that the only acceptable synonym for "reliability" is "reliability."

The second implication is that the reliability of an instrument depends on the group with which it is used. If the group is very homogeneous (e.g., patients with sleep onset difficulties), the reliability will tend to be low because the between-subject variance will be small. That same scale, used in a more heterogeneous group, such as the general population, will likely have a higher reliability because of the increased variance among the people. The third, and perhaps most important, implication that follows from this is that we cannot talk about *the* reliability of a test. The reliability will change, depending on the group being measured. Reliability is not a fixed, immutable property of the scale, but rather an interaction between the scale and the characteristics of the group. Just because a scale has been shown to be reliable in one setting does not guarantee that it will reliable in another setting. If a scale is used in a population different from the one used when it was being developed, the reliability needs to be reassessed.

Types of reliability

When defining reliability, we alluded to different types of it; here we will expand on those. If a scale is self-administered, or if a test is performed on a patient (e.g., a blood level or a CPAP titration test), then it is important to determine if two or more assessments yield the same result. If they do not, it is impossible to determine which, if either, is correct. This is referred to as *test-retest reliability*. The interval between the test and the retest depends on the nature of what is being assessed. If it is a physical measurement, then the interval can be as short as a minute or two, akin to taking a person's blood pressure three or four times. For scales and indices, though, there are other considerations. On the one hand, the interval should be long enough so that the respondents do not remember their previous answers and simply put those down; this would be a test of memory, rather than reliability. On the other hand, it should not be so long that the person's state changes during the interval. For most questionnaires, this middle ground is around two weeks. It can be longer if what is being measured is a chronic state or condition, such as anxiety or depression, and shorter if it is a more transient condition, such as pain.

Other measurement techniques rely on a rater evaluating a patient or student (e.g., the Hamilton Depression Rating Scale; Hamilton, 1960) or interpreting the results of an electroencephalography to determine the duration of various stages of sleep. As with test-retest reliability, the two or more raters must come up with similar scores for the rating to be trustworthy. This is called, for obvious reasons, *inter-rater reliability*. The important caveat is that the ratings must be done independently from one another.

The third type of reliability, *internal consistency*, focuses on the scale rather than the use of it. As we mentioned in the beginning of the chapter, scales tap a single phenomenon, and all of the items should be correlated with one another to some degree. If they are not, then it is not clear what the scale measures. For this reason, internal consistency is not calculated for indices, as the items are not expected to be correlated. Because internal consistency can be determined from only a single administration of the scale, it is the most widely used measure of reliability (Sijtsma, 2009), although not necessarily the best one.

Indices of reliability

Test-retest and inter-rater reliability are based on the correlation between the two or more administrations of the instrument. In the past, this meant using the Pearson correlation coefficient (r). However, for a number of reasons, this has now been superseded by the intraclass correlation (ICC; Shrout and Fleiss, 1979). One reason is that Pearson's r is not sensitive to systematic bias between ratings. For example, if Rater 1 always rates patients two points higher on a scale than Rater 2, Pearson's r will be 1.00, whereas the ICC will be lower. In some cases, reliability is not affected by this bias. This holds if the units used by the scale are arbitrary, which is true for almost all scales. For example, the score on the Insomnia Severity Index (Morin, 1993) is a number between 0 and 28. We can add five points to each value, or multiply them by 10 and, as long as we change the cut-points accordingly, we have not lost or gained any information. As long as both ratings rank order the people in the same order, any systematic difference can be accounted for. In other cases, though, the units are meaningful, as with blood pressure, temperature, VO_2 max, and so forth. If one reading places the person within the normal range, and the other rating in the pathological or impaired range, then this cannot be ignored. In such instances, it is imperative that the ICC be used to account for the bias. However, even when the units are arbitrary, the ICC is still preferred because it is more conservative than Pearson's r. A second reason ICC is superior to Pearson's r is that the latter can handle only two ratings; it cannot be used when there are three or more raters or assessment times. The ICC, on the other hand, can easily handle these conditions. Different versions of the ICC can also deal with situations where different raters evaluate different people. Indeed, there are many versions of the ICC, depending on whether the interest is the reliability of a single rating or the average of a number of them; we want to look at absolute agreement (whether the scores are identical) or consistency (if the rank ordering of the scores is important but the exact values are systematically different); and so forth. These versions are described by McGraw and Wong (1996) and Streiner et al. (2015) (Chapter 8).

For internal consistency, the usual measure is Cronbach's α (alpha; Cronbach, 1951). Like Pearson's r and the ICC, it can range between 0.0 and 1.0. Despite its ubiquity, it has a number of shortcomings. The primary one is that it is affected not only by the internal consistency of a scale but also by its length. It will be high if an instrument has more than about 15 items, even if the scale is heterogeneous (Cortina, 1993). Furthermore, α can severely underestimate the reliability of a scale unless all of the items have the same mean and standard deviation (Sijtsma, 2009; Trizano-Hermosilla and Alvarado, 2016). Some of the alternatives, such as McDonald's ω (omega; Zinbarg et al., 2005) are not easily calculated and do not appear in the output of most statistical software packages, so simpler alternatives, such as the mean interitem correlation or the mean item-total correlation should be reported in addition to α (Clark and Watson, 1995; Streiner, 2003b).

Standards for reliability

As mentioned previously, the reliability coefficients can range between 0.0, indicating a total lack of reliability, and a maximum of 1.0, reflecting perfect reliability. How high an instrument's reliability needs to be depends on how it will be used. Nunnally's (1978) recommendations, which are still generally accepted, are that tools used in new or developing areas of research should have reliability coefficients of at least 0.7; and in more established fields of research, the minimum would be 0.8. When used in clinical settings, where the results from an instrument influence what a clinician does or does not do with an individual patient, the reliability should be at least 0.9, and 0.95 would be desirable. This may sound like a high bar to cross over, but it is worthwhile to put this in context. Imagine that we measured 100 individuals and rank ordered them on the basis of their scores. If Person A were ranked 25th from the top and Person B 50th from the top, and the reliability of the scale was 0.8, there would be a 20% chance that a retest would reverse the rankings. The probability of reversal would be below 2.5% only when the reliability of the tool is at least 0.95 (Thorndike and Hagen, 1969).

The criteria for Cronbach's α are the same as for the other types of reliability, but with two caveats. First, as mentioned previously, α may be high simply because the scale has more than about 15 items, so with longer tests, a high value should be viewed with some degree of skepticism. Second, unlike the coefficients for test-retest and inter-rater reliability, it is possible that α can be too high. If it is above 0.90, this may indicate that the scale has redundant items that can be eliminated without any loss of accuracy (Streiner, 2003b). If the mean inter-item correlation is reported, it should lie within the range of 0.15 to 0.20 for scales that tap broad constructs; and between 0.45 and 0.50 for more narrowly focused scales (Clark and Watson, 1995).

It should also be noted that measuring internal consistency for indices does not make sense. As we have mentioned, an index, by design, consists of unrelated items in a checklist-like format, so the items would not necessarily be correlated (Streiner, 2003a). For example, the Social Readjustment Rating Scale (SRRS; Holmes and Rahe, 1967), which is a measure of psychosocial stress, consists of 43 events that are presumed to be stressful, such as assuming a mortgage, getting a moving vehicle citation, or the death of a spouse. There is no reason to believe that experiencing any one of those would be related with experiencing any other. Thus, measures of internal consistency should be used only with scales, not with indices.

In a similar vein, if one is examining an inventory—a tool consisting of a number of scales—or a scale with sub-scales, it is not logical to report internal consistency for all of the items together. If each scale is tapping a different domain, then it makes no sense to look for homogeneity among all of the items. Coefficient α may be high, but this would be due solely to the fact that there are many items. Conversely, if the inventory as a whole is homogeneous, then that means that there should not be separate scales. So again, reserve internal consistency only for unidimensional scales.

The standard error of measurement

When we defined reliability, we said that the observed score is always affected to some degree by error. This error can arise from a variety of different sources. The scale itself may be imperfect, and may contain items that are misunderstood by the respondents. This means that they may interpret the item one way at time 1 and differently at time 2. We also know that the person's motivation may influence how they respond, and this may change from one occasion to the next. The respondent may inadvertently circle the wrong response, answering with a three when they meant a four; or the person scoring the scale may add up the values incorrectly. This means that two administrations of the scale to the same individual will yield somewhat different results.

The question is how much the scores will vary from one time to the next, and this is captured by a statistic called the *standard error of measurement*, abbreviated as SEM or SE_M (and not to be confused with the standard error the mean, which is abbreviated in the same way; context should make it clear which one is meant). The formula for the SEM is:

$$\text{SEM} = \text{SD}\sqrt{1 - r_{xx}} \tag{12.5}$$

where SD is the standard deviation of the scale and r_{xx} is its reliability. This equation tells us that the smaller the SD and the higher the scale's reliability, the smaller SEM will be. From this, we can construct a confidence interval (CI) around the observed score. To derive the 95% CI, we would multiply the SEM by 1.96, and it would be:

$$95\%\ \text{CI} = \text{Score} \pm (1.96 \times \text{SEM}) \tag{12.6}$$

For example, if the person had a score of 50, and the SEM of the scale were 5, then the 95% CI would be $50 \pm (1.96 \times 5)$ or [40.2, 59.8], meaning that there is a 95% probability that their true score was between roughly 40 and 60.

Validity

Establishing the reliability of an instrument is a necessary, but not sufficient, step in scale development; it is also necessary to establish if it is valid. Older articles and text books (i.e., those written before 1960) would have defined validity as telling us whether the test is measuring what we think it's measuring (e.g., Cattell, 1946). They then would have gone on to discuss the three different "types" of validity—content validity, criterion validity, and construct validity—and perhaps to subdivide those even further into concurrent criterion validity versus predictive criterion validity, discriminant validity versus convergent validity, and so on. Endless hours have been spent (or wasted) debating whether a particular article was an example of convergent validity or criterion validity.

Around the middle of the 20th century, two themes emerged: one changing the definition of validity, and a second, related one, unifying the "types" of validity into just one. We will discuss each of those in turn.

The change in the conceptualization of what validity is emerged in the 1970s, following the work of Lee J. Cronbach (1971). Rather than focusing on the *test*, the emphasis was now placed on the *scores* and the conclusions that can legitimately be made about people based on those scores. As Landy (1986) phrased it, "Validation processes are not so much directed toward the integrity of tests as they are directed toward the inferences that can be made about the attributes of people who have produced those test scores" (p. 1186). Earlier we saw that reliability is not a fixed property of a test, but an interaction between it and the people taking it. This revised conceptualization of validity means that the same caveat applies to validity. It, too, is not fixed but depends on the instrument, the sample being evaluated, and the circumstances under which the scale is given. For example, the Sleep Disorders Questionnaire (Douglass et al., 1994) contains items such as, "I get too little sleep at night," "I have trouble getting to sleep at night," and "I wake up often during the night." While this scale may be valid for those who work regular hours during the day, it may not be valid for shift workers or those who work at night jobs. Furthermore, it, and other scales, may not be valid for people who are claiming a disability allowance because of sleep-related problems and who may therefore wish to exaggerate their difficulties. These factors do not invalidate the scales as a whole, but serve as a reminder that the validity depends on the population and the context.

Another implication from this revised definition of validity is that validity is not a yes/no conclusion, but rather one that lies along a continuum (Zumbo, 2009). Determining whether or not a scale is valid for a specific group involves synthesizing sometimes conflicting evidence from multiple sources. This means that two people, reviewing the same set of articles, may come to different conclusions, especially if they are using the scale with different populations. It also means that the job of validating an instrument can never be said to be over; there is always more that can be learned about how it performs for different groups in different circumstances.

The second change in the conceptualization of validity is that we now speak of only one "type" of validity—construct validity. We can talk about criterion *validation*—a method of establishing construct validity—but not criterion *validity*—a type; and the same is true for all other ways of validating a scale. That then raises the issue of defining what a construct is. A construct is an unseen, hypothesized abstract idea that explains why certain observed variables tend to occur together. For example, we never directly see depression. What we do see is that people who are pessimistic about the future also tend to have sleep difficulties, disturbances in their appetite (either over-eating or not feeling hungry), decreased libido, thoughts of suicide, psychomotor retardation, and so on. We hypothesize that these behaviors occur together because they are all the observable, outward manifestations of this unseen phenomenon we label "depression." Other phenomena, like intelligence, and many diagnoses, such as irritable bowel syndrome and schizophrenia, can similarly be described as constructs.

We can postulate a number of hypotheses about each construct. If, for instance, we were interested in anxiety, these hypotheses could include, (1) people who are anxious tend to perform poorly on tasks in stressful situations; (2) anxiety should be greater before a high-stakes examination than afterward; (3) symptoms of anxiety should decrease following the use of anxiolytic medication; and (4) patients attending an anxiety disorders clinic should evidence more anxiety than those in a fracture clinic; and so forth. These hypotheses are derived from our theoretical and clinical knowledge of anxiety, and they allow us to design studies to establish the validity of a scale. For example, if our new scale of anxiety is valid, then (1) people who score high on it should do less well on timed tasks than people who score low; (2) scores on the scale should be lower after a high-stakes exam than prior to it; (3) higher before taking medication than afterward; and (4) those in an anxiety clinic should score higher than those in a fracture clinic. We could also theorize that our new scale should be highly correlated with other existing scales of anxiety, and it should not be correlated with measures of intelligence.

In an extremely influential article, Cronbach and Meehl (1955) lay out the three steps necessary to establish construct validity:

1. Explicitly spell out a set of theoretical concepts and how they are related.
2. Develop a scale to measure the hypothetical construct.
3. Test the relationships among the constructs and their observable manifestations.

Thus, establishing the construct validity of a scale consists of hypothesis testing, and the more hypotheses that are tested, the better. There are a number of consequences of this. First, there are an unlimited number of hypotheses that can be postulated for any construct, meaning that there are limitless ways of establishing construct validity for a scale purporting to measure the construct. Second, as noted previously, one can never say that validity testing is complete; there is always more that can be learned about how a scale performs with different populations and in different circumstances.

Third, there are only a limited number of types of studies to establish reliability—giving the scale twice to the same group of people (for test-retest reliability) or having two or more raters evaluate the same individuals (inter-rater reliability). And from this, there are very few statistics that are used in reliability testing: the ICC coefficient and Cronbach's alpha. In contrast to this, there are many different designs and statistics that can be used for validating a scale, depending on the nature of the study. For example, determining if the new instrument agrees with an existing one may involve giving both to a group of people and using a correlation coefficient; seeing if scores change following an intervention would call for a before-after study, analyzed with a paired *t*-test; if it's hypothesized that various groups would score differently implies a two- or multigroup study, and the use of an independent *t*-test or an analysis of variance; and so on.

If the results of the study indicate that the scale is performing as expected, then that adds to the validity of the scale and the credibility of the hypotheses. However, if the study yields negative results, that may indicate that (a) the hypotheses are correct but the scale is not performing as it should; (b) the scale may be fine but the hypotheses are wrong; or (c) both the scale and the hypotheses are flawed. In the early stages of development, when we have little information about the scale, it is difficult to determine where the fault lies. As more evidence accumulates regarding the scale's validity, though, we would have more confidence in it and we would be more inclined to question the hypotheses.

Summary

Devising a new scale is a multistage process. First, the items have to be written, based on other scales, theory, research, and talking to patients. Then, the items need to be checked from a number of perspectives. Are they written well, free from jargon, and comprehensible by the target population? Do they cover all aspects of the domain of interest, while at the same time do not include irrelevant items (content coverage and relevance)? Once the final set of items has been chosen, it is mandatory to determine if the scale is reliable. Is it answered the same way on two different occasions (test-retest reliability), or completed similarly by two different observers (inter-rater reliability)? This may need to be checked even for existing scales if the population is different from the original sample (e.g., those with sleep apnea as opposed to people working shifts). Finally, the validity of the scale needs to be established. This is more difficult than for reliability, as validation is an on-going process, learning more and more about what the scores mean for different groups of people and in different situations.

References

Aday, L.A., Cornelius, L.J., 2006. Designing and Conducting Health Surveys: A Comprehensive Guide, third ed. Wiley, New York.

Apgar, V., 1953. A proposal for new method of evaluation of the newborn infant. Curr. Res. Anesth. Analg. 32 (4), 260–267. https://doi.org/10.1213/ANE.0b013e31829bdc5c.

Buysse, D.J., Reynolds III, C.F., Monk, T.H., Berman, S.R., Kupfer, D.J., 1989. The Pittsburgh Sleep Quality Index: a new instrument for psychiatric practice and research. Psychiatr. Res. 28 (2), 193–213. https://doi.org/10.1016/0165-1781(89)90047-4.

Cattell, R.B., 1946. Description and Measurement of Personality. World Book Company, New York.

Clark, L.A., Watson, D., 1995. Constructing validity: basic issues in objective scale development. Psychol. Assess. 7 (3), 309–319. https://doi.org/10.1037/1040-3590.7.3.309.

Cohen, G., Java, R., 1995. Memory for medical history: accuracy of recall. Appl. Cognit. Psychol. 9 (4), 273–288. https://doi.org/10.1002/acp.2350090402.

Collins, D., 2003. Pretesting survey instruments: an overview of cognitive methods. Qual. Life Res. 12 (3), 229–238. https://doi.org/10.1023/a:1023254226592.

Cortina, J.M., 1993. What is coefficient alpha? An examination of theory and applications. J. Appl. Psychol. 78 (1), 98–104. https://doi.org/10.1037/0021-9010.78.1.98.

Cronbach, L.J., 1951. Coefficient alpha and the internal structure of tests. Psychometrika 16 (3), 297–334. https://doi.org/10.1007/BF02310555.

Cronbach, L.J., 1971. Test validation. In: Thorndike, R.L. (Ed.), Educational Measurement. American Council on Education, Washington, DC, pp. 221–237.

Cronbach, L.J., Meehl, P.E., 1955. Construct validity in psychological tests. Psychol. Bull. 52 (4), 281–302. https://doi.org/10.1037/h0040957.

Douglass, A.B., Bornstein, R., Nino-Murcia, G., Keenan, S., Miles, L., Zarcone Jr., V.P., Guilleminault, C., Dement, W.C., 1994. The sleep disorders questionnaire. I: creation and multivariate structure of SDQ. Sleep 17 (2), 160–167. https://doi.org/10.1093/sleep/17.2.160.

Gibson, E.M., Williams 3rd, W.P., Kriegsfeld, L.J., 2009. Aging in the circadian system: considerations for health, disease prevention and longevity. Exp. Gerontol. 44 (1–2), 51–56. https://doi.org/10.1016/j.exger.2008.05.007.

Goldman, B.A., Saunders, J.L., Busch, J.C., 1996. Directory of Unpublished Experimental Measures, Volumes 1–3. American Psychological Association, Washington, D.C.

Hamilton, M., 1960. A rating scale for depression. J. Neurol. Neurosurg. Psychiatry 23 (1), 56–62. https://doi.org/10.1136/jnnp.23.1.56.

Hess, L.M., Method, M.W., Stehman, F.B., Weathers, T.D., Gupta, P., Schilder, J.M., 2012. Patient recall of health care events and time to diagnose a suspected ovarian cancer. Clin. Ovarian Other Gynecol. Cancer 5 (1), 17–23. https://doi.org/10.1016/j.cogc.2012.04.001.

Holden, R.R., Fekken, G.C., Jackson, D.N., 1985. Structured personality test item characteristics and validity. J. Res. Pers. 19 (4), 386–394. https://doi.org/10.1016/0092-6566(85)90007-8.

Holmes, T.H., Rahe, R.H., 1967. The social readjustment rating scale. J. Psychosom. Res. 11 (2), 213–218. https://doi.org/10.1016/0022-3999(67)90010-4.

Ibáñez, V., Silva, J., Cauli, O., 2018. A survey on sleep assessment methods. PeerJ 6, e4849. https://doi.org/10.7717/peerj.4849.doi.org/10.1016/j.sleep.2017.08.026.

Krosnick, J.A., 1991. Response strategies for coping with the cognitive demands of attitude measures in surveys. Appl. Cognit. Psychol. 5 (3), 213–216. https://doi.org/10.1002/acp.2350050305.

Kurtis, M.M., Balestrino, R., Rodriguez-Blazquez, C., Forjaz, M.J., Martinez-Martin, P., 2018. A review of scales to evaluate sleep disturbances in movement disorders. Front. Neurol. 9, 369. https://doi.org/10.3389/fneur.2018.00369.

Landy, F.J., 1986. Stamp collecting versus science: validation as hypothesis testing. Am. Psychol. 41 (11), 1183–1192. https://doi.org/10.1037/0003-066X.41.11.1183.

Lindwall, M., Barkoukis, V., Grano, C., Lucidi, F., Raudsepp, L., Liukkonen, J., Thøgersen-Ntoumani, C., 2012. Method effects: the problem with negatively versus positively keyed items. J. Pers. Assess. 94 (2), 196–204. https://doi.org/10.1080/00223891.2011.645936.

Lomeli, H.A., Pérez-Olmos, I., Talero-Gutiérrez, C., Moreno, C.B., González-Reyes, R., Palacios, L., de la Peña, F., Muñoz-Delgado, J., 2008. Sleep evaluation scales and questionnaires: a review. Actas Esp. Psiquiatr. 36 (1), 50–59.

McGraw, K.O., Wong, S.P., 1996. Forming inferences about some intraclass correlation coefficients. Psychol. Methods 1 (1), 30–46. https://doi.org/10.1037/1082-989X.1.1.30.

Morin, C.M., 1993. Insomnia: Psychological Assessment and Management. Guilford Press, New York.

NVivo, n.d. Available at: https://www.qsrinternational.com/nvivo-qualitative-data-analysis-software/home. Accessed 9 July 2020.

Nunnally, J.C., 1978. Psychometric Theory, second ed. McGraw-Hill, New York.

Ronen, G.M., Rosenbaum, P.L., Law, M., Streiner, D.L., 1999. Health-related quality of life in childhood epilepsy: the results of children's participation in identifying the components. Dev. Med. Child Neurol. 41 (8), 554–559. https://doi.org/10.1111/j.1469-8749.1999.tb00655.x.

Samora, J., Saunders, L., Larson, R.F., 1961. Medical vocabulary knowledge among hospital patients. J. Health Hum. Behav. 2 (2), 83–92. https://doi.org/10.2307/2948804.

Schwarz, N., Oyserman, D., 2001. Asking questions about behavior: cognition, communication, and questionnaire construction. Am. J. Educ. 22 (2), 127–160. https://doi.org/10.1177/109821400102200202.

Shrout, P.E., Fleiss, J.L., 1979. Intraclass correlations: uses in assessing rater reliability. Psychol. Bull. 86 (2), 420–428. https://doi.org/10.1037//0033-2909.86.2.420.

Sijtsma, K., 2009. On the use, the misuse, and the very limited usefulness of Cronbach's alpha. Psychometrika 74 (1), 107–120. https://doi.org/10.1007/s11336-008-9101-0.

Smith, M.T., Wegener, S.T., 2003. Measures of sleep: the Insomnia severity index, medical outcomes study (MOS) sleep scale, Pittsburgh sleep diary (PSD), and Pittsburgh sleep quality index (PSQI). Arthritis Rheum. 49 (5S), S184–S196. https://doi.org/10.1002/art.11409.

Streiner, D.L., 2003a. Being inconsistent about consistency: when coefficient alpha does and doesn't matter. J. Pers. Assess. 80 (3), 217–222. https://doi.org/10.1207/S15327752JPA8003_01.

Streiner, D.L., 2003b. Starting at the beginning: an introduction to coefficient alpha. J. Pers. Assess. 80 (1), 99–103. https://doi.org/10.1207/S15327752JPA8001_18.

Streiner, D.L., Norman, G.R., Cairney, J., 2015. Health Measurement Scales: A Practical Guide to Their Development and Use, fifth ed. Oxford University Press, Oxford.

Sudman, S., Bradburn, N.M., 2004. Asking Questions. Rev. Ed. Jossey-Bass, San Francisco, CA.

Thorndike, R.L., Hagen, E., 1969. Measurement and Evaluation in Education and Psychology. Wiley, New York.

Trizano-Hermosilla, I., Alvarado, J.M., 2016. Best alternatives to Cronbach's alpha reliability in realistic conditions: congeneric and asymmetrical measurements. Front. Psychol. 7. https://doi.org/10.3389/fpsyg.2016.00769. Article ID 769.

Zinbarg, R.E., Revelle, W., Yovel, I., Li, W., 2005. Cronbach's α, Revelle's β, and Mcdonald's ω_H: their relations with each other and two alternative conceptualizations of reliability. Psychometrika 70 (1), 123–133. https://doi.org/10.1007/s11336-003-0974-7.

Zumbo, B.D., 2009. Validity as contextualized as pragmatic explanation and its implications for validation practice. In: Lissitz, R.W. (Ed.), The Concept of Validity: Revisions, New Directions and Applications. Information Age, Charlotte, NC, pp. 65–82.

CHAPTER 13

Clinical psychoinformatics: a novel approach to behavioral states and mental health care driven by machine learning

Tetsuya Yamamoto[1,2], Junichiro Yoshimoto[3], Jocelyne Alcaraz-Silva[4], Eric Murillo-Rodríguez[2,4], Claudio Imperatori[2,5], Sérgio Machado[2,6,7] and Henning Budde[2,8]

[1]Graduate School of Technology, Industrial and Social Sciences Tokushima University, Tokushima, Tokushima, Japan; [2]Intercontinental Neuroscience Research Group, Tokushima, Tokushima, Japan; [3]Graduate School of Science and Technology, NAIST, Ikoma, Nara, Japan; [4]Laboratorio de Neurociencias Moleculares e Integrativas, Escuela de Medicina, División Ciencias de la Salud, Universidad Anáhuac Mayab, Mérida, Yucatán, México; [5]Cognitive and Clinical Psychology Laboratory, Department of Human Science, European University of Rome, Rome, Italy; [6]Department of Sports Methods and Techniques, Federal University of Santa Maria, Santa Maria, Brazil; [7]Laboratory of Physical Activity Neuroscience, Neurodiveristy Institute, Queimados, Rio de Janeiro, Brazil; [8]Faculty of Human Sciences, Medical School Hamburg, Hamburg, Germany

Introduction

To implement effective clinical psychological interventions such as cognitive-behavioral therapy (CBT), it is essential to identify problematic cognitive, behavioral, and physiological responses and to analyze the functional relationships among variables that influence them (functional analysis). By identifying controllable factors among variables based on functional analysis, it is possible to formulate an effective intervention plan to address the issue (O'Brien et al., 2016). This type of individualized assessment, represented by functional analysis, is generally performed by a skilled therapist through a process of detailed information gathering, hypothesis generation, and repeated testing. However, for variables that are not apparent in the client's consciousness (e.g., behavioral habits or physiological responses of which the client is unaware of) or that are difficult for a therapist to observe (e.g., the client's internal experiences), it often takes a great deal of time to identify the functional relationship. In addition, even for intervention methods that are assumed to be effective as a result of assessment, a period of trial and error is necessary for effective implementation and evaluation of their effectiveness. Commonly, a certain

Methodological Approaches for Sleep and Vigilance Research
ISBN 978-0-323-85235-7
https://doi.org/10.1016/B978-0-323-85235-7.00013-2

period of time is required to formulate and adjust the intervention methods appropriately for individual client needs. Therefore, it is important to develop a method to quickly understand the problem structures that maintain or exacerbate clients' difficulties and to propose the most appropriate intervention method for clients at an early stage. In this chapter, we propose the use of "machine learning (ML)" as an approach that contributes to the selection and expansion of such individualized assessment and intervention methods.

ML is a field of artificial intelligence (AI) that has recently attracted attention and is widely defined as a computational strategy that automatically determines models and parameters to reach optimal solutions to problems. This approach makes it possible to apply predictive models focused on individual states by learning functional relationships from multidimensional data sets. Therefore, the potential applications of ML for CBT and clinical psychology research, which focus on the behavior contingency specific to individuals, are appreciable. However, there are few studies that have discussed the applicability of ML to clinical psychology, and there have been no studies that assess how ML can be applied to CBT.

In this chapter, we first provide an overview of the methods, characteristics, and significance of the ML approach. Furthermore, we summarize the main research topics where ML approaches have been applied in the field of mental health and introduce some examples of applications that may contribute to clinical psychology and cognitive behavioral therapy research. Finally, we discuss the potential for future applications of ML approaches while noting their limitations.

Overview of machine learning

Definition

The history of ML dates back to the late 1950s and was derived from AI. According to Arthur Samuel, a primary pioneer of this research field, ML was defined as a "Field of study that gives computers the ability to learn without being explicitly programmed" (Samuel, 1959). Herein, the ability to learn can be interpreted as the ability to change the input–output transformation by the computer based on experienced events. From the point of view of practical data analysis, it can be regarded as a set of computer programs that aim to discover regularity hidden behind a given dataset and support human prediction and decision making based on this regularity.

Three major categories

Based on the issue to be solved, ML is classified into three schemes: supervised learning, unsupervised learning, and reinforcement learning.

Supervised learning aims to discover the regularity between input and output variables in a situation where paired instances of the inputs and their desired outputs are given and then predict an appropriate output for an unseen input. For example, imagine that we have data measured from nine people, where the datum for each person consisted of his/her body length, body weight, and sex, as shown in Fig. 13.1A. Then, suppose that we are asked to predict the sex of a person whose body length and body weight are shown by the open circle. What is the reasonable answer to this question? Most of us will draw the boundary shown by the dotted line in our mind, and consequently, the answer would be male if the data lay above the line (or female otherwise). Another example is a drug administration test in which the blood concentrations of the drug for different time points are shown in Fig. 13.1B. For this case, we can identify regularity in that the data follow the parabola demonstrated by the dotted line. Consequently, we can predict that the blood concentration at the time point of the gray line will be near the open circle. The programs for mimicking the inference

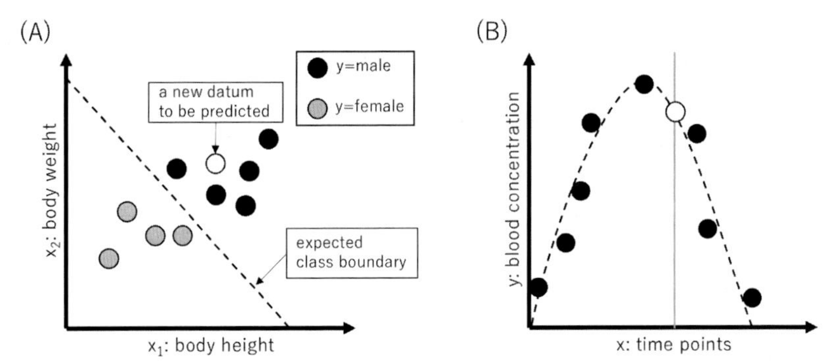

Figure 13.1 Examples of supervised learning. (A) An example of classification, in which the goal is to predict sex based on a given pair of data (body height and body weight). Black and gray *circles* denote data obtained from males and females, respectively. The open *circle* is a data point obtained from the person predicted in the task. (B) An example of regression, in which the goal is to predict the blood concentration of the drug at a specific time point based on the previous observations at different time points after administration. Black *circles* denote the data points in the previous records and the gray line is the time point at which blood concentration is to be predicted. The dotted line is an expected function that explains the regularity between the time course of the blood concentration. The open circle is a prediction result based on the expected regularity.

process, such as the aforementioned examples, are supervised learning methods (or more technically, supervised learning algorithms). Supervised learning is sometimes distinguished into two sub-categories: classification and regression. The former is the case where the desired outputs are restricted to discrete (or categorical) variables, as shown in Fig. 13.1A; the latter is the case where the desired outputs are continuous variables, as in Fig. 13.1B.

Unsupervised learning aims to extract useful information in a situation where only inputs are given without their desired outputs. For example, imagine a situation as shown in Fig. 13.2A, which is the same situation as Fig. 13.1A, except that all sex information is removed. Still, by projecting the data onto the dotted line, we can compare the data in terms of a one-dimensional (1D) view corresponding to the body size. In the example shown in Fig. 13.2B, we can see three "clusters" of data surrounded by dotted ellipses, each of which can be regarded as a group having a similar property. Both examples in Fig. 13.2 are categorized into unsupervised learning. Note that the terminology of "unsupervised" here does not stand for "random" or "meaningless." In the case of Fig. 13.2A, the data are projected onto the 1D subspace to maximize the variance (i.e., the information content) of data after the projection. In the case of Fig. 13.2B, the data are partitioned to minimize the variance within the same group. That is, even in unsupervised learning, the input-output transformation is determined based on a well-defined and meaningful criterion.

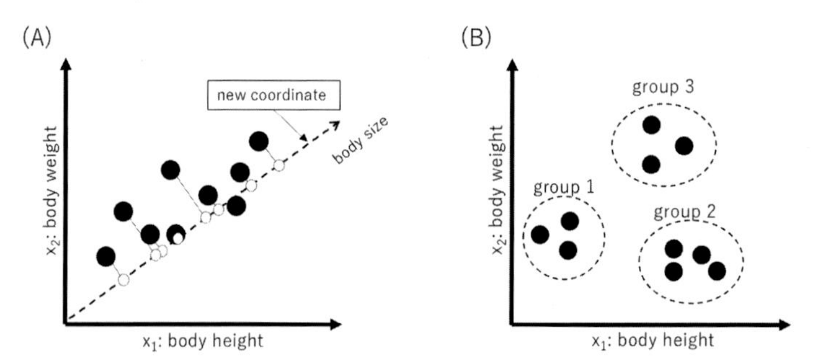

Figure 13.2 Examples of unsupervised learning. (A) An example of dimension reduction. The original data (denoted by black *circles*) are the same as Fig. 13.1A. The open circles are the projections of the original data onto the new coordinate (denoted by the dotted *line*). The variability of the data is preserved even after the projections. (B) An example of clustering. Ten data points (denoted by black *circles)* are partitioned into three groups (denoted by the dotted *ellipses*) based on "data clusters." The groups are characterized by small body (group 1), high-light body (group 2), and high-heavy body (group 3), respectively, in a data-driven manner.

Unsupervised learning is sometimes distinguished into two sub-categories: feature extraction and clustering. As in Fig. 13.2A, feature extraction is used to transform original inputs into different representations, and if the resulting representations have lower dimensionality than the originals, it is also referred to as dimension reduction. Clustering involves partitioning a given set of data into several groups without any explicitly desired outputs.

Reinforcement learning is occasionally classified into a sub-category of unsupervised learning because no desired outputs are explicitly provided. However, there is feedback referred to as "reward," which indicates whether the outcome achieved by the output is good or bad, available for each time step. In this sense, reinforcement learning is generally regarded as an independent scheme of supervised and unsupervised learning. In addition, there are two distinct features: one is a nature of "trial-and-error," indicating that the training data are not passively given but should be actively sampled through appropriate outputs; and the other is "delayed-reward," indicating that output for each time step may affect the magnitude of rewards given at its subsequent time steps.

The three schemes of ML introduced here refer to categories, but not specific algorithms: that is, there are many algorithms proposed for each scheme. Representative algorithms for supervised and unsupervised learning are summarized in Table 13.1. While we can identify many algorithms for reinforcement learning, such as the Q-learning method (Watkins and Dayan, 1992) and the SARSA method (Rummery & Niranjan, 1994); they have been omitted from Table 13.1, as suitable approximation tricks are also required in many cases [See (Sutton and Barto, 2018) for the principles and applications for reinforcement learning.].

Difference from standard statistics

In ML, we use the terminology that often appear in standard textbooks in statistics, such as least-square regression, logistic regression, and clustering. Findings demonstrated by multivariate statistics are often applied to the development of algorithms and theoretical analysis in ML. In this sense, ML can be regarded as a new branch of applied statistics. A minor difference, if any, is that ML aims at rational prediction and decision making by discovering the regularity behind a given dataset, whereas statistics is the field that addresses how to summarize data of interest efficiently and intelligibly, as observed in descriptive statistics, and develops theories to quantitatively support/reject expected interpretations for data. In addition,

Table 13.1 Representative methods for machine learning and their features.

Category	Problem setting	Method/algorithm	Features	Literature
Supervised learning	Classification	Discriminant analysis	The method assumes that the data belonging to the same class is distributed according to a normal distribution with the same mean and standard deviation.	Lachenbruch and Goldstein (1979)
		Logistic regression	The method assumes that any pair of classes is linearly separable where the class boundary is given by a line over the two-dimensional input space (or a hyper plane over the high-dimensional input space).	Cox (1958)
		Naïve bayes classifier	The method assumes that all coordinates of the input space are statistically independent of each other. If the data distribution of each coordinate is normal, the naïve bayes method can be a special case of the discriminant analysis.	Jiang et al. (2007)
	Regression	(Ordinary) least square regression	The method aims to minimize the sum of squared errors between the desired and predicted outputs that was given by the linear transformation from the inputs. Even if the prediction is given by the linear transformation of a predesigned set of nonlinear functions of inputs, such as multinomial regression and fourier series regression, it can be regarded as this method.	Fomby et al. (1984)
		Partial least square regression	In the method, the inputs and the outputs are projected onto their respective subspaces and the ordinary least square regression is applied to the subspaces. The method works better than the ordinal least square method especially when the number of samples is smaller than the dimensionality of the inputs.	Rosipal and Krämer (2006)

		Multilayer perceptron (multilayer feedforward neural network)	The method realizes the regression using layered artificial neural networks (ANNs) which are developed by analogy with basic properties of neurons. In particular, deep neural networks and deep learning refer to this type of methods with layered ANNs forming four or more layers.	Schmidhuber (2015)
	Both classification and regression	K-nearest neighbor method	The output is determined by majority voting or averaging over K data points neighboring the data point to be predicted, where K is a predefined parameter. The method is based on the principle that the similarity in the input space should inherit that in the output space.	Peterson (2009)
		Support vector machine (SVM)	The class boundary (for classification problems) is given by a hyper plane that satisfies the margin maximization principle, which claims that the hyper plane should maximize the margin (i.e., the minimum distance from data points belonging to the same class). Originally, the method was developed for classification problems and referred to as the support vector classification (SVC). Its extension for regression problems is referred to as the support vector regression (SVR).	Cortes and Vapnik (1995) Drucker et al. (1997) Cervantes et al. (2020)
		Decision tree	The output is determined by a set of if-then rules over the input space. The name of "tree" is derived from a tree structure representing the if-then rules.	Safavian and Landgrebe (1991)
		Random forest	The method is a variation of so-called ensemble learning, in which classification or regression models are made up of a large number of small decision trees based on bootstrap aggregating (a.k.a. bagging). The final output is determined by averaging over the outputs of the small decision trees.	Breiman (2001) Biau (2012)

Continued

Table 13.1 Representative methods for machine learning and their features.—cont'd

Category	Problem setting	Method/algorithm	Features	Literature
		Gradient boosting	The method is another variation of ensemble learning, in which regression models are made up of multiple "week learners" based on boosting sampling. The final output is determined by weighted average over the outputs of the week learners, where decision trees are often used for the week learners.	Friedman (2001) Natekin and Knoll (2013)
Unsupervised learning	Clustering	Hierarchical clustering	The method seeks for data clusters in a hierarchical manner, where each data point is treated as a different cluster at the initial stage and merges two clusters into one cluster successively, until all data points are merged into a single cluster.	Murtagh and Contreras (2012)
		K-means clustering	The method iteratively updates data clusters until the following procedures converge: (1) initialize K representative points, each of them represents the centroid of the corresponding cluster and K is a pre-defined parameter; (2) assign each data point to one cluster having the nearest representative points; (3) update the representative points by the centroid of data points assigned to the corresponding clusters and go to 2.	MacQueen (1967) Blömer et al. (2016)
		Gaussian mixture model clustering	The method finds data clusters, in which we assume that data points are generated by one of K distinct normal distributions and assign each data to the optimal cluster based on bayesian maximum a posteriori criterion.	Melnykov and Maitra (2010)

Dimension reduction and feature extraction	Principal component analysis (PCA)	The method projects each data point onto a lower-dimensional subspace that maximizes the variance of the resulting data points. If the data is assumed to be normally distributed, the projection also maximizes Shannon's information entropy.	Shlens (2014)
	Multidimensional scaling (MDS)	The method projects each data point onto a lower-dimensional subspace, in which pair-wise distance between data points in the original space are preserved in the subspace as much as possible.	Saeed et al. (2018)
	Independent component analysis (ICA)	The method decomposes each data point into different signals, which are assumed to be non-Gaussian distributed and statistically independent of each other. The method is typically applied to blind source separation problems.	Bell and Sejnowski (1995) Hyvarinen (1999)
	Non-negative matrix factorization (NMF)	The method factorizes the original data matrix into the product of two matrices with lower dimensionality (one is used for bases in the feature space and the other for the representations). Herein, all matrices must have no negative elements.	Lee and Seung (1999) Wang and Zhang (2013)
	Auto encoder	The method trains a multilayer perceptron to achieve identity mapping from inputs to outputs. The representation in an intermediate layer is used as the processed data after feature extraction.	Hinton (2006) Bank et al. (2020)

In the column of literature, we list original and/or review papers describing detailed principles of the corresponding methods. Besides, the literature of (Bishop, 2006; Hastie et al., 2009; Murphy, 2012) systematically introduce most methods for machine learning.

a distinctive requirement for ML is that the developed computer programs should finish all processes within an allotted time depending on the scenario. For this reason, approximation and heuristics can be used in ML more frequently than typical statistical analyses which regard strictness as important.

Why now?

Recently, ML has attracted significant attention in various fields as a tool for data analysis. There seem to be several explanations for this trend.

The first and dominant reason is that various software packages (or libraries) for ML are now freely available. For example, scikit-learn[1] and caret[2] (alternatively, mlr[3]) are representative packages for Python and R, respectively. In addition, Weka,[4] developed in JAVA, is a free stand-alone ML software that can be accessed through a graphical user interface, so that even users who do not have programming skills can perform simple data analyses based on ML.

A paradigm shift in science is also a considerable factor. Traditionally, the hypothesis-driven approach was the major trend among scientists, in which a plausible hypothesis is formulated by surveying and extending past research, and then tested through experiments and investigations. On the other side, rapid advance in sensing technology and information communication infrastructure which allows the collection of massive and diverse datasets, which are also maintained/shared via cloud services and reused for meta-analysis: We are currently in the so-called Big Data Era, which also drives researchers to the data-driven approach, in which attempts are made to extract useful information from big data and produce a highly plausible, but novel, hypotheses. The aims match the basic properties of ML characterized by the discovery of regularity for prediction and decision-making.

From a technical viewpoint, there have been several strategies proposed to solve "$p \gg n$" problems where n is the number of samples and p is the dimensionality of the inputs (e.g., the number of attributes of each sample) (Johnstone and Titterington, 2009). For these problems, significant results cannot be obtained due to multiple comparisons or indefiniteness of solution, if naive statistical analysis methods such as two-sample t-test and

[1] https://scikit-learn.org.
[2] https://cran.r-project.org/web/packages/caret/index.html.
[3] https://cran.r-project.org/web/packages/mlr/index.html.
[4] https://www.cs.waikato.ac.nz/ml/weka/.

regression analysis are applied. To solve these problems, automatic variable selection methods are available in ML. The representative methods are L1-regularization and Bayesian sparse modeling, assuming an automatic relevance determination prior distribution (Neal, 1996; Tibshirani, 1996, 2011; Yamashita et al., 2008). Using these methods, a small subset of the inputs can be identified to achieve a much better prediction than the chance level, even if the complete identification of all contributing inputs is impossible. This property is highly beneficial to data analysis in the field of mental health, in which the number of samples is generally limited due to the constraint of data collection from human volunteers.

Clinical applications of machine learning approaches in the mental health field

In the field of mental health, the major research themes using ML approaches include (1) diagnosis of mental disorders, (2) detection of symptoms and risks, and (3) prediction of prognosis and treatment effects.

Diagnosis of mental disorders

In the fields of clinical psychology and psychiatry, early diagnosis of mental disorders leads to the formulation of effective treatment and intervention methods, and the prediction of illness course and prognosis. To contribute to such diagnoses, ML research has focused on the reproducibility of diagnoses, especially by using structural and functional brain images.

The main psychiatric disorders that have been studied so far include depression (Fu et al., 2008), schizophrenia (Skåtun et al., 2017), and Alzheimer's disease (Dimitriadis et al., 2018). In recent years, ML has been used to study a variety of diseases, including anxiety (Lueken et al., 2015) and anorexia (Lavagnino et al., 2015). For example, by applying a supervised learning method (support vector machine) to functional brain activity data of 19 unmedicated depressed patients under the presentation of sad expressions, it was possible to classify patients with acute depression and healthy participants with 86% accuracy (Fu et al., 2008). Similarly, a supervised learning method (normalized discriminant analysis) was applied to the resting brain activity data of 182 patients on the schizophrenia spectrum collected at three scan sites, and they were classified with high accuracy (up to 80%) in different scanners and samples (Skåtun et al., 2017). Although the accuracy of ML using neuroimaging varies depending on the type of imaging data, the results of a recent meta-analysis revealed an overall sensitivity of 77% and specificity of 78% among patients with depression (Kambeitz et al., 2017) and 80.3% among patients with schizophrenia (Kambeitz et al., 2015). In addition to neuroimaging data, diagnostic classification with a certain degree of accuracy has been achieved by applying ML methods to video data of facial expressions recorded during clinical interviews of patients with schizophrenia (Tron et al., 2016), vocal data from patients with Alzheimer's disease during cognitive tasks (König et al., 2015), and conversational records during interviews of young people with suicidal ideation (Pestian et al., 2016).

Furthermore, supervised learning methods (support vector machines) can be used to classify children with attention-deficit/hyperactivity disorder and bipolar disorder using activity indices measured from actigraphs (Faedda et al., 2016), thus contributing to the differentiation of symptoms.

Detection of symptoms and risks

Early detection of symptoms and identification of risks is important for preventive education and early intervention to prevent disease onset and severity.

ML methods have been applied to neuroimaging data to identify specific brain activity patterns indicative of Alzheimer's disease (Doan et al., 2017), early prediction of the onset of psychosis (Koutsouleris et al., 2012), and detection of vulnerabilities that lead to depression recurrence in terms of functional connectivity (Yamamoto, 2018). Other attempts have been made to utilize various types of information, such as the detection of suicidal ideation using counseling transcripts (Oseguera et al., 2017), the detection of schizophrenia symptoms from texts written by patients (Strous et al., 2009), and the detection of mental states associated with depression and schizophrenia using speech data (Kliper et al., 2016).

In research applying ML methods to data that can be collected in daily life, an attempt has been made to detect the factors that cause headaches using daily cognitive, behavioral, and physiological information (hereafter referred to as "lifelog") of individuals (Yamamoto and Yoshimoto, 2018). In this study, a smart wristband and a smartphone were used for data collection and an unsupervised learning method (nonparametric Bayesian co-clustering) was applied to personal lifelog data measured for approximately 80 days to comprehensively visualize the interaction structure of factors associated with headaches (Fig. 13.3). In addition to suggesting the usefulness of lifelogging, this study demonstrated that it is possible to build a person-specific predictive model from data for only a single individual.

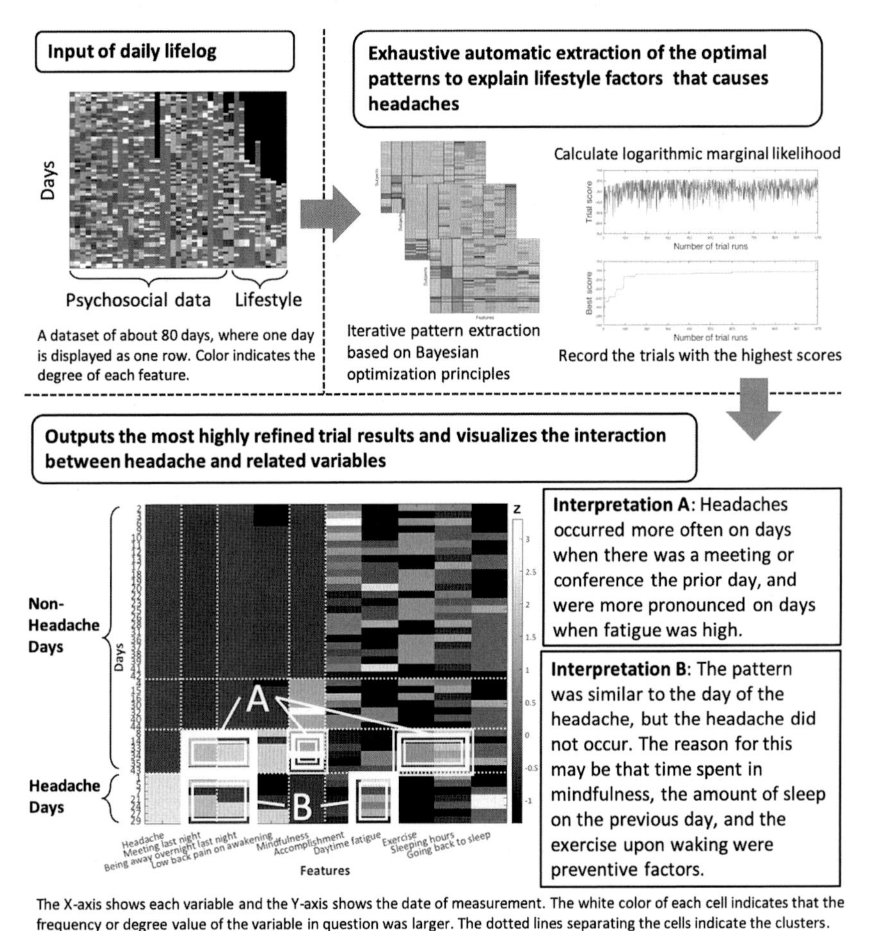

Figure 13.3 Overview and example interpretation of nonparametric Bayesian co-clustering to visualize the interaction of variables related to headache.

In addition to the detection of individuals' symptoms and risks, it is also possible to comprehensively extract the complex interaction structure of variables in population data and visualize various states and subtypes in the population. For example, a similar unsupervised learning method was applied to the data of a large-scale online survey on the mental health of Japanese citizens conducted under a declared state of emergency during the COVID-19 pandemic (Yamamoto et al., 2020). The results indicate that the background of the severely psychologically stressed group entailed an interaction of various factors, such as a high level of loneliness, deteriorating interpersonal relationships with people close to them, insomnia and anxiety related to the novel coronavirus infection, deterioration of family finances, and difficulties in work and study.

Prediction of prognosis and treatment effects

In clinical psychology and psychiatry, the ability to predict an individual's prognosis and the effectiveness of treatment methods is highly important for determining the course of symptoms, preventive interventions, such as psychoeducation, and selection of optimal psychological and pharmacological treatments.

ML has been applied to predict various conditions, such as the natural history of chronic depression over a period of 2 years (Schmaal et al., 2015), the effect of 6 weeks of antipsychotic treatment for patients with schizophrenia (Bak et al., 2017), cognitive decline over a period of 24 months among patients with Alzheimer's disease (Zhu et al., 2016), and the onset of post-traumatic stress disorder among children 3 months after hospital discharge (Saxe et al., 2017), with the aim of predicting patient prognosis and treatment effects. Other studies have used patients' free-text responses to questions about their mental state to predict suicidal ideation and worsening of mental symptoms with a certain degree of accuracy (Cook et al., 2016).

In a study attempting to predict the therapeutic effects of CBT for depression (Tymofiyeva et al., 2019), 30 patients with adolescent depression received CBT for 3 months. Supervised learning (decision trees based on the C4.5 algorithm) was applied to the pretreatment structural connectome data and depression scores. The reduction of depressive symptoms after CBT was predicted with 83% accuracy. These results indicate that ML approaches utilizing neuroimaging can accurately predict the effectiveness of CBT for adolescent depression. Therefore, these methods can be used to predict the effectiveness of CBT for specific patients, which may be useful for treatment planning.

However, studies using supervised learning methods (e.g., random forest) on sociodemographic (e.g., age and sex) and clinical (e.g., symptom severity) data available in daily clinical practice indicate that CBT significantly predicts treatment effects, but the prediction accuracy is inadequate at 59%, indicating that its clinical usefulness remains limited (Hilbert et al., 2020). Therefore, for ML to be widely used in daily clinical practice, it is important to identify data that can be easily collected in clinical settings, can contribute to prediction, and to develop learning algorithms capable of providing highly accurate prediction performance for such data.

Limitations of machine learning approaches

While ML approaches have many advantages to as introduced so far, its limitations should be recognized. The biggest drawback is that ML-based research is apt to result in explanatory studies and post-hoc analysis studies due to the nature of knowledge acquired from "given" data sets. Of course, these outcomes still contribute to science by producing novel hypotheses. However, prospective studies are mandatory to validate these hypotheses with strong evidence. ML offers an useful methodology to us but is not capable of fully automating all research. In this sense, how to harmonize ML with our "human intelligence" is a critical factor for success in scientific research.

The second limitation concerns the quality control of data. In ML, we assume a certain regularity behind the data. Without information contributing to the prediction, the quality of the resulting prediction will never exceed chance level. Accordingly, to avoid potential artifacts, maximum attention should be paid to that all factors expected to contribute to the prediction can be measured as precisely as possible.

The benefit and danger of open sources for ML algorithms are two sides of the same coin. For example, deep learning, which is now indispensable to the fields of computer vision, machine translation, and speech recognition, will become more popular for accurate predictions in the field of mental health. However, the resulting functions transforming from inputs to outputs are generally too complicated for humans to understand. If we pursue the causality and mechanism of psychological phenomena, rather than accuracy of prediction, simpler algorithms equipped with feature selection (e.g., linear regression and logistic regression with LASSO regularization) are more suitable due to of high interpretability.

Finally, there is the "No Free Lunch theorem," known in the field of ML (Wolpert and Macready, 1997), which implies that no single ML algorithm is universally the best-performing algorithm for all problems. Thus, a suitable prior knowledge specialized for problems of interest is necessary to achieve the best prediction. Therefore, we expect that tight collaboration of domain-specialists (e.g., psychiatrists or clinical psychologists) and ML researchers will be increasingly more important for success in ML approaches to mental health studies.

Future perspectives on the use of machine learning approaches

Improvement of accessibility and verification of usefulness in clinical situations

Previous ML approaches have focused on diagnostic and predictive accuracy, especially for depression, suicide risk, and cognitive function, and have generally demonstrated good accuracy. Therefore, it is expected that these approaches will contribute greatly to the detection, diagnosis, and prediction of mental disorders when they are fully established in the future.

However, the accuracy of the analysis techniques used and the data sets are not consistent among studies thus, research to propose standardized methods that can be applied across institutions so that the techniques can be employed in clinical practice is needed. In addition, most of the studies that have achieved a certain level of accuracy used classification techniques based on supervised learning using neuroimaging data. However, in daily clinical practice, diagnosis and evaluation are generally based on interviews and questionnaires, and a highly accurate diagnostic aid using these data is desirable. Furthermore, as mentioned above, most of the previous studies examined prediction accuracy retrospectively, using only previously obtained data. Therefore, the prospective accuracy of newly obtained data in clinical practice is still unclear. Accordingly, it is necessary to accumulate evidence of clinical efficacy by using prospective studies. In clinical practice, there are many outpatients who are not screened for study (e.g., patients whose symptoms are considered to be somewhat homogeneous), but who present with various comorbidities or are intractable, and new issues may arise regarding sufficient accuracy. Additionally, numerous unresolved issues remain, such as the accuracy of prediction for other psychiatric disorders (anxiety, neurodevelopmental disorders, etc.) for which few research findings have been accumulated, and the diseases for which ML approaches may be optimal are unknown.

Although some other studies have applied supervised learning methods (support vector machines) to lifelog data to predict well-being (Yamamoto et al., 2019), only a few studies have applied ML methods to these positive psychological states (resilience, identity formation, personal growth, etc.) (Shatte et al., 2019). Therefore, applied research on ML that focuses on facilitating clients' adaptive aspects, as well as the maladaptive aspects, such as symptoms, is necessary.

Development of individualized intervention procedures based on machine-learning approaches

The "classical inferential approach" (a method of designing studies and evaluating results in terms of p-value-based tests and effect sizes for group variances and errors), which has been widely used in the past, is useful for comparing and understanding specific populations and has greatly promoted clinical psychology and psychiatric research. However, it tends to exclude complex interactions among variables and individual differences, and as a result, it has been difficult to apply research findings to individuals (Dwyer et al., 2018). Alternatively, the ML approach is characterized by its ability to make generalizable predictions for individuals. This suggests that ML approaches may contribute to the selection of the most effective and efficient interventions for clients.

Such individualized ML approaches may be useful in selecting psychotherapeutic and pharmacological treatments, and proposals and basic research have been put forth that will contribute to future applications. For example, studies such as the proposal of a user-individualized assessment and intervention system for predicting and coping with depression (Yang et al., 2018), the detection of high-risk drinking using cell phone sensor data (Bae et al., 2018), and the examination of the process of changing gambling behavior through individualized feedback on the amount of money lost (Auer and Griffiths, 2018) provide suggestions for providing interventions at optimal times for individuals.

In addition, an attempt has been made to identify the variables that can best reduce the risk of depression recurrence by utilizing supervised ML (random forest) methods on approximately 8 months of lifelog data collected from specific patients in remission from depression (Yamamoto and Yoshimoto, 2019). Below, the representative prediction results of this study (Fig. 13.4) are presented to propose an application method to use the classification rules obtained by learning for personalized treatment (note that Fig. 13.4 itself is one of several prediction results, and does not necessarily refer to the prediction result with the highest accuracy). The results of the analysis revealed that negative thoughts were the most important variable for identifying signs of relapse in this patient (Fig. 13.4A), followed by activity and rumination, which had a significant impact on relapse (Fig. 13.4B). In other words, the results suggest that variables such as negative automatic thoughts, activity, and rumination should be targeted preferentially. This prediction is consistent with the

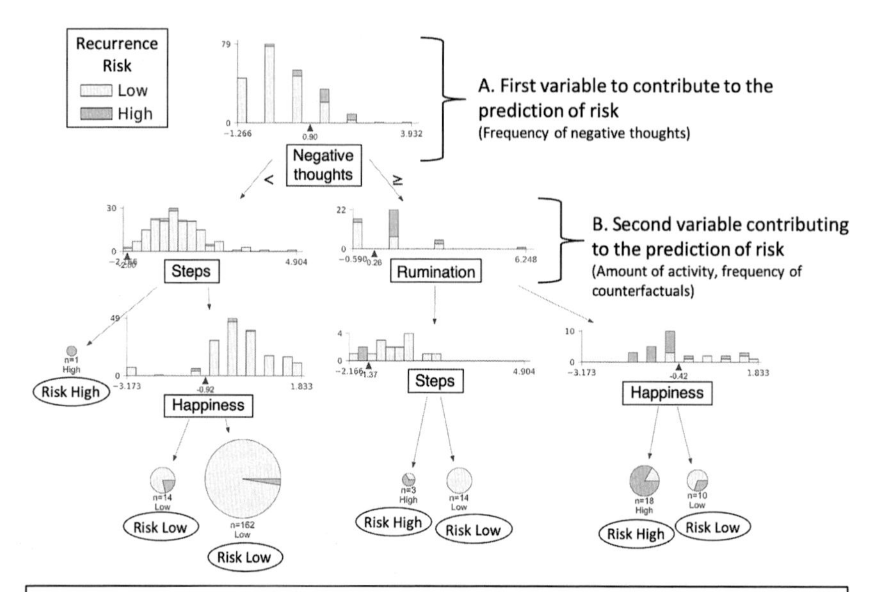

Example of interpretation: On days with high amounts of negative thinking, the tendency to ruminate should be of particular focus, and it was predicted that interventions for rumination itself, activity level, and well-being could effectively reduce recurrence risk.

Figure 13.4 Results of random forest to predict the risk of recurrence of depression and example of interpretation.

treatment model and procedures of CBT for depression. Therefore, it is presumed that an approach utilizing CBT is effective in preventing the recurrence of depression for this patient. Such a clinical approach using ML methods is still in its infancy, and more basic findings are required to improve its practicality in clinical practice. However, it has great potential for better implementation of individualized treatment for clients, similar to the trend of precision mental health care and precision medicine, which have been attracting attention recently.

Cooperation among experts and research teams

The accuracy of models built using ML is greatly affected by the quality of the data employed for training. Therefore, it is necessary to develop an environment that enables the collection and sharing of improved training data. As the No Free Lunch theorem mentioned above highlights the importance of prior knowledge, collaborative efforts between clinicians and researchers is required when collecting data. Similarly, to maximize the clinical utility of the developed model, feedback from clinicians and patients and improvement of the model by researchers in response to the feedback are necessary. In addition, it is important that the use of ML approaches be conducted by an interdisciplinary team that includes experts, decision-makers, and users, taking into account ethical considerations (Wiens et al., 2019). In light of the above, collaboration among researchers, clinicians, and patients is highly important to fully demonstrate the performance of ML approaches. Furthermore, to construct a method that can be applied to a wide range of individuals (generalizable model), it is important to examine the effects of bias due to differences in demographic attributes such as sex, age, race, and place of residence, as well as differences in measurement equipment, and to develop a method to correct for these biases (Yamashita et al., 2019). Therefore, in addition to the collection of data from different research institutes and research projects, cooperation among research groups in sharing data and a system to make this possible are also important for future research.

Conclusions

To utilize ML approaches in clinical practice, methods that can be easily implemented and to make technical improvements remain needed, and there remains a large gap between research findings and clinical practice. However, in light of the proliferation of ML approaches in the mental health field, further progress in applied research on CBT and clinical psychology is expected in the future. In the same way that "collaborative empiricism" is one of the basic principles of CBT (the therapist's attitude toward collaborative problem-solving with the client), the approach in which clients and therapists "collaborate" on problems with AI technologies such as ML may soon become commonplace. In recent years, with the advancement of information and communication technology, various intervention methods have been developed, including the practice of CBT using mobile information terminals and PCs (Yamamoto and Takebayashi, 2020), self-counseling using virtual reality (Slater et al., 2019), and game-based applications to improve cognitive function (Kollins et al., 2020), and other intervention methods utilizing various information and communication technologies have been developed. It is expected that approaches such as "clinical psychoinformatics," in which such information and communication technologies are used to understand, predict, and regulate the mind, will continue to expand. Just as many approaches have been integrated and developed under the name of CBT, new developments, including ML approaches, have the potential to further expand CBT. However, the effectiveness of CBT has been demonstrated by repeatedly testing its effects and procedures. Therefore, it is necessary to continue to monitor the effectiveness of new approaches in the future.

References

Auer, M., Griffiths, M.D., 2018. Cognitive dissonance, personalized feedback, and online gambling behavior: an exploratory study using objective tracking data and subjective self-report. Int. J. Ment. Health Addiction 16, 631—641. https://doi.org/10.1007/s11469-017-9808-1.

Bae, S., Chung, T., Ferreira, D., Dey, A.K., Suffoletto, B., 2018. Mobile phone sensors and supervised machine learning to identify alcohol use events in young adults: implications for just-in-time adaptive interventions. Addict. Behav. 83, 42—47. https://doi.org/10.1016/j.addbeh.2017.11.039.

Bak, N., Ebdrup, B.H., Oranje, B., Fagerlund, B., Jensen, M.H., Düring, S.W., Nielsen, M., GlenthØj, B.Y., Hansen, L.K., 2017. Two subgroups of antipsychotic-naive, first-episode schizophrenia patients identified with a Gaussian mixture model on cognition and electrophysiology. Transl. Psych. 7, e1087. https://doi.org/10.1038/tp.2017.59.

Bank, D., Koenigstein, N., Giryes, R., 2020. Autoencoders. ArXiv 2003, 05991. Retrieved from. http://arxiv.org/abs/2003.05991.

Bell, A.J., Sejnowski, T.J., 1995. An information-maximization approach to blind separation and blind deconvolution. Neural Comput. 7 (6), 1129—1159. https://doi.org/10.1162/neco.1995.7.6.1129.

Biau, G., 2012. Analysis of a random forests model. J. Mach. Learn. Res. 13 (1), 1063—1095.

Bishop, C., 2006. Pattern Recognition and Machine Learning. Springer-Verlag New York, Berlin, Heidelberg.

Blömer, J., Lammersen, C., Schmidt, M., Sohler, C., 2016. Theoretical analysis of the k-means algorithm — a survey. In: Kliemann, L., Sanders, P. (Eds.), Algorithm Engineering. Lecture Notes in Computer Science, vol. 9220. Springer, pp. 81—116. https://doi.org/10.1007/978-3-319-49487-6_3. LNCS.

Breiman, L., 2001. Random forests. Mach. Learn. 45, 5—32. https://doi.org/10.1023/A:1010933404324.

Cervantes, J., Garcia-Lamont, F., Rodríguez-Mazahua, L., Lopez, A., 2020. A comprehensive survey on support vector machine classification: applications, challenges and trends. Neurocomputing 408, 189—215. https://doi.org/10.1016/j.neucom.2019.10.118.

Cook, B.L., Progovac, A.M., Chen, P., Mullin, B., Hou, S., Baca-Garcia, E., 2016. Novel use of natural language processing (NLP) to predict suicidal ideation and psychiatric symptoms in a text-based mental health intervention in madrid. Comput. Math. Method. Med. 2016, 8708434. https://doi.org/10.1155/2016/8708434.

Cortes, C., Vapnik, V., 1995. Support-vector networks. Mach. Learn. 20 (3), 273—297. https://doi.org/10.1007/BF00994018.

Cox, D.R., 1958. The regression analysis of binary sequences. J. R. Stat. Soc. Ser. B 20, 215—242. https://doi.org/10.1111/j.2517-6161.1958.tb00292.x.

Dimitriadis, S.I., Liparas, D., Tsolaki, M.N., 2018. Random forest feature selection, fusion and ensemble strategy: combining multiple morphological MRI measures to discriminate among healhy elderly, MCI, cMCI and alzheimer's disease patients: from the alzheimer's disease neuroimaging initiative (ADNI) data. J. Neurosci. Method. 302, 14–23. https://doi.org/10.1016/j.jneumeth.2017.12.010.

Doan, N.T., Engvig, A., Zaske, K., Persson, K., Lund, M.J., Kaufmann, T., Cordova-Palomera, A., Alnæs, D., Moberget, T., Brækhus, A., Barca, M.L., Nordvik, J.E., Engedal, K., Agartz, I., Selbæk, G., Andreassen, O.A., Westlye, L.T., 2017. Distinguishing early and late brain aging from the Alzheimer's disease spectrum: consistent morphological patterns across independent samples. Neuroimage 158, 282–295. https://doi.org/10.1016/j.neuroimage.2017.06.070.

Drucker, H., Surges, C.J.C., Kaufman, L., Smola, A., Vapnik, V., 1997. Support vector regression machines. Adv. Neural Inf. Process. Syst. 1, 155–161.

Dwyer, D.B., Falkai, P., Koutsouleris, N., 2018. Machine learning approaches for clinical psychology and psychiatry. Annu. Rev. Clin. Psychol. 14, 91–118. https://doi.org/10.1146/annurev-clinpsy-032816-045037.

Faedda, G.L., Ohashi, K., Hernandez, M., McGreenery, C.E., Grant, M.C., Baroni, A., Polcari, A., Teicher, M.H., 2016. Actigraph measures discriminate pediatric bipolar disorder from attention-deficit/hyperactivity disorder and typically developing controls. J. Child Psychol. Psych. Allied Discip. 57, 706–716. https://doi.org/10.1111/jcpp.12520.

Fomby, T.B., Johnson, S.R., Hill, R.C., 1984. Review of ordinary least squares and generalized least squares. In: Fomby, T.B., Johnson, S.R., Hill, R.C. (Eds.), Advanced Econometric Methods. Springer New York, New York, NY, pp. 7–25. https://doi.org/10.1007/978-1-4419-8746-4_2.

Friedman, J.H., 2001. Greedy function approximation: a gradient boosting machine. Ann. Stat. 29 (5), 1189–1232. https://doi.org/10.1214/aos/1013203451.

Fu, C.H.Y., Mourao-Miranda, J., Costafreda, S.G., Khanna, A., Marquand, A.F., Williams, S.C.R., Brammer, M.J., 2008. Pattern classification of sad facial processing: toward the development of neurobiological markers in depression. Biol. Psychiatr. 63, 656–662. https://doi.org/10.1016/j.biopsych.2007.08.020.

Hastie, T., Tibshirani, R., Friedman, J., 2009. The Elements of Statistical Learning. Springer New York, New York, NY. https://doi.org/10.1007/978-0-387-84858-7.

Hilbert, K., Kunas, S.L., Lueken, U., Kathmann, N., Fydrich, T., Fehm, L., 2020. Predicting cognitive behavioral therapy outcome in the outpatient sector based on clinical routine data: a machine learning approach. Behav. Res. Ther. 124, 103530. https://doi.org/10.1016/j.brat.2019.103530.

Hinton, G.E., 2006. Reducing the dimensionality of data with neural networks. Science 313 (5786), 504–507. https://doi.org/10.1126/science.1127647.

Hyvarinen, A., 1999. Survey on independent component analysis. Neural Comput. Surv. 2, 94–128.

Jiang, L., Wang, D., Cai, Z., Yan, X., 2007. Survey of improving naive bayes for classification. In: Alhajj, R., Gao, H., Li, J., Li, X., Zaïane, O.R. (Eds.), Advanced Data Mining and Applications. Springer Berlin Heidelberg, Berlin, Heidelberg, pp. 134–145.

Johnstone, I.M., Titterington, D.M., 2009. Statistical challenges of high-dimensional data. Philos. Trans. R. Soc. A Math. Phys. Eng. Sci. 367, 4237–4253. https://doi.org/10.1098/rsta.2009.0159.

Kambeitz, J., Cabral, C., Sacchet, M.D., Gotlib, I.H., Zahn, R., Serpa, M.H., Walter, M., Falkai, P., Koutsouleris, N., 2017. Detecting neuroimaging biomarkers for depression: a meta-analysis of multivariate pattern recognition studies. Biol. Psychiatr. 82, 330–338. https://doi.org/10.1016/j.biopsych.2016.10.028.

Kambeitz, J., Kambeitz-Ilankovic, L., Leucht, S., Wood, S., Davatzikos, C., Malchow, B., Falkai, P., Koutsouleris, N., 2015. Detecting neuroimaging biomarkers for schizophrenia: a meta-analysis of multivariate pattern recognition studies. Neuropsychopharmacology 40, 1742–1751. https://doi.org/10.1038/npp.2015.22.

Kliper, R., Portuguese, S., Weinshall, D., 2016. Prosodic analysis of speech and the underlying mental state. In: Serino, S., Matic, A., Giakoumis, D., Lopez, G., Cipersso, P. (Eds.), Pervasive Computing Paradigms for Mental Health. MindCare 2015. Communications in Computer and Information Science. Springer, Cham, pp. 52–62. https://doi.org/10.1007/978-3-319-32270-4_6.

Kollins, S.H., DeLoss, D.J., Cañadas, E., Lutz, J., Findling, R.L., Keefe, R.S.E., Epstein, J.N., Cutler, A.J., Faraone, S.V., 2020. A novel digital intervention for actively reducing severity of paediatric ADHD (STARS-ADHD): a randomised controlled trial. Lancet Digit. Heal. 2, e168–e178. https://doi.org/10.1016/S2589-7500(20)30017-0.

König, A., Satt, A., Sorin, A., Hoory, R., Toledo-Ronen, O., Derreumaux, A., Manera, V., Verhey, F., Aalten, P., Robert, P.H., David, R., 2015. Automatic speech analysis for the assessment of patients with predementia and Alzheimer's disease. Alzheimer's Dement. Diagnosis, Assess. Dis. Monit. 1, 112–124. https://doi.org/10.1016/j.dadm.2014.11.012.

Koutsouleris, N., Borgwardt, S., Meisenzahl, E.M., Bottlender, R., Möller, H.J., Riecher-Rössler, A., 2012. Disease prediction in the at-risk mental state for psychosis using neuroanatomical biomarkers: results from the fepsy study. Schizophr. Bull. 38, 1234–1246. https://doi.org/10.1093/schbul/sbr145.

Lachenbruch, P.A., Goldstein, M., 1979. Discriminant analysis. Biometrics 35, 69–85.

Lavagnino, L., Amianto, F., Mwangi, B., D'Agata, F., Spalatro, A., Zunta-Soares, G.B., Abbate Daga, G., Mortara, P., Fassino, S., Soares, J.C., 2015. Identifying neuroanatomical signatures of anorexia nervosa: a multivariate machine learning approach. Psychol. Med. 45, 2805–2812. https://doi.org/10.1017/S0033291715000768.

Lee, D.D., Seung, H.S., 1999. Learning the parts of objects by non-negative matrix factorization. Nature 401 (6755), 788–791. https://doi.org/10.1038/44565.

Lueken, U., Straube, B., Yang, Y., Hahn, T., Beesdo-Baum, K., Wittchen, H.U., Konrad, C., Ströhle, A., Wittmann, A., Gerlach, A.L., Pfleiderer, B., Arolt, V., Kircher, T., 2015. Separating depressive comorbidity from panic disorder: a combined functional magnetic resonance imaging and machine learning approach. J. Affect. Disord. 184, 182–192. https://doi.org/10.1016/j.jad.2015.05.052.

MacQueen, J., 1967. Some methods for classification and analysis of multivariate observations. Proc. Fifth Berkeley Symp. Mathemat. Statist. Probab. 1, 281–297.

Melnykov, V., Maitra, R., 2010. Finite mixture models and model-based clustering. Stat. Surv. 4, 80–116. https://doi.org/10.1214/09-SS053.

Murphy, K.P., 2012. Machine Learning: A Probabilistic Perspective. The MIT Press.

Murtagh, F., Contreras, P., 2012. Algorithms for hierarchical clustering: an overview. WIREs Data Mining Knowledge Discovery 2 (1), 86–97. https://doi.org/10.1002/widm.53.

Natekin, A., Knoll, A., 2013. Gradient boosting machines, a tutorial. Front. Neurorob. 7, 21. https://doi.org/10.3389/fnbot.2013.00021.

Neal, R., 1996. Bayesian Learning for Neural Networks. Ph.D. Thesis. University of Toronto.

O'Brien, W.H., Haynes, S.N., Kaholokula, J.K., 2016. Behavioral assessment and the functional analysis. In: Nezu, C.M.E., Nezu, A.M. (Eds.), The Oxford Handbook of Cognitive and Behavioral Therapies. Oxford University Press, New York, NY, pp. 44–61.

Oseguera, O., Rinaldi, A., Tuazon, J., Cruz, A.C., 2017. Automatic quantification of the veracity of suicidal ideation in counseling transcripts. In: Stephanidis, C. (Ed.), HCI International 2017 – Posters' Extended Abstracts. HCI 2017. Communications in Computer and Information Science. Springer, Cham, pp. 473–479. https://doi.org/10.1007/978-3-319-58750-9_66.

Pestian, J.P., Grupp-Phelan, J., Bretonnel Cohen, K., Meyers, G., Richey, L.A., Matykiewicz, P., Sorter, M.T., 2016. A controlled trial using natural language processing to examine the language of suicidal adolescents in the emergency department. Suicide Life-Threatening Behav. 46. https://doi.org/10.1111/sltb.12180.

Peterson, L., 2009. K-nearest neighbor. Scholarpedia 4 (2), 1883. https://doi.org/10.4249/scholarpedia.1883.

Rosipal, R., Krämer, N., 2006. Overview and recent advances in partial least squares. In: Saunders, C., Grobelnik, M., Gunn, S., Shawe-Taylor, J. (Eds.), Subspace, Latent Structure and Feature Selection. Springer Berlin Heidelberg, Berlin, Heidelberg, pp. 34–51. https://doi.org/10.1007/11752790_2.

Rummery, G.A., Niranjan, M., 1994. On-line Q-Learning Using Connectionist Systems (Tr 166), Technical Report. Engineering Department, Cambridge University.

Saeed, N., Nam, H., Haq, M.I.U., Muhammad Saqib, D.B., 2018. A survey on multidimensional scaling. ACM Comput. Surv. 51 (3), 1–25. https://doi.org/10.1145/3178155.

Safavian, S.R., Landgrebe, D., 1991. A survey of decision tree classifier methodology. IEEE Trans. Syst., Man, Cybernet. 21 (3), 660–674. https://doi.org/10.1109/21.97458.

Samuel, A.L., 1959. Some studies in machine learning using the game of checkers. IBM J. Res. Dev. 3, 210–229.

Saxe, G.N., Ma, S., Ren, J., Aliferis, C., 2017. Machine learning methods to predict child posttraumatic stress: a proof of concept study. BMC Psychiatr. 17, 223. https://doi.org/10.1186/s12888-017-1384-1.

Schmaal, L., Marquand, A.F., Rhebergen, D., Van Tol, M.J., Ruhé, H.G., Van Der Wee, N.J.A., Veltman, D.J., Penninx, B.W.J.H., 2015. Predicting the naturalistic course of major depressive disorder using clinical and multimodal neuroimaging information: a multivariate pattern recognition study. Biol. Psychiatr. 78, 278–286. https://doi.org/10.1016/j.biopsych.2014.11.018.

Shlens, J., 2014. A tutorial on principal component analysis. ArXiv, 1404.1100. Retrieved from. http://arxiv.org/abs/1404.1100.

Schmidhuber, J., 2015. Deep learning in neural networks: an overview. Neural Network. 61, 85–117. https://doi.org/10.1016/j.neunet.2014.09.003.

Shatte, A.B.R., Hutchinson, D.M., Teague, S.J., 2019. Machine learning in mental health: a scoping review of methods and applications. Psychol. Med. 49, 1426–1448. https://doi.org/10.1017/S0033291719000151.

Skåtun, K.C., Kaufmann, T., Doan, N.T., Alnæs, D., Córdova-Palomera, A., Jönsson, E.G., Fatouros-Bergman, H., Flyckt, L., Melle, I., Andreassen, O.A., Agartz, I., Westlye, L.T., 2017. Consistent functional connectivity alterations in schizophrenia spectrum disorder: a multisite study. Schizophr. Bull. 43, 914–924. https://doi.org/10.1093/schbul/sbw145.

Slater, M., Neyret, S., Johnston, T., Iruretagoyena, G., Crespo, M.Á. de la C., Alabèrnia-Segura, M., Spanlang, B., Feixas, G., 2019. An experimental study of a virtual reality counselling paradigm using embodied self-dialogue. Sci. Rep. 9, 10903. https://doi.org/10.1038/s41598-019-46877-3.

Strous, R.D., Koppel, M., Fine, J., Nachliel, S., Shaked, G., Zivotofsky, A.Z., 2009. Automated characterization and identification of schizophrenia in writing. J. Nerv. Ment. Dis. 197, 585–588. https://doi.org/10.1097/NMD.0b013e3181b09068.

Sutton, R.S., Barto, A.G., 2018. Reinforcement Learning: An Introduction, second ed. The MIT Press.

Tibshirani, R., 2011. Regression shrinkage and selection via the lasso: a retrospective. J. R. Stat. Soc. Ser. B Stat. Methodol. 73, 273–282. https://doi.org/10.1111/j.1467-9868.2011.00771.x.

Tibshirani, R., 1996. Regression shrinkage and selection via the lasso. J. R. Stat. Soc. Ser. B 58, 267–288. https://doi.org/10.1111/j.2517-6161.1996.tb02080.x.

Tron, T., Peled, A., Grinsphoon, A., Weinshall, D., 2016. Automated facial expressions analysis in schizophrenia: a continuous dynamic approach. In: Serino, S., Matic, A., Giakoumis, D., Lopez, G., Cipersso, P. (Eds.), Pervasive Computing Paradigms for Mental Health. MindCare 2015. Communications in Computer and Information Science. Springer, Cham, pp. 72–81. https://doi.org/10.1007/978-3-319-32270-4_8.

Tymofiyeva, O., Yuan, J.P., Huang, C.Y., Connolly, C.G., Henje Blom, E., Xu, D., Yang, T.T., 2019. Application of machine learning to structural connectome to predict symptom reduction in depressed adolescents with cognitive behavioral therapy (CBT). NeuroImage Clin. 23, 101914. https://doi.org/10.1016/j.nicl.2019.101914.

Wang, Y.-X., Zhang, Y.-J., 2013. Nonnegative matrix factorization: a comprehensive review. IEEE Trans. Knowl. Data Eng. 25 (6), 1336–1353. https://doi.org/10.1109/TKDE.2012.51.

Watkins, C.J.C.H., Dayan, P., 1992. Q-learning. Mach. Learn. 8, 279–292. https://doi.org/10.1007/bf00992698.

Wiens, J., Saria, S., Sendak, M., Ghassemi, M., Liu, V.X., Doshi-Velez, F., Jung, K., Heller, K., Kale, D., Saeed, M., Ossorio, P.N., Thadaney-Israni, S., Goldenberg, A., 2019. Do no harm: a roadmap for responsible machine learning for health care. Nat. Med. 25, 1337–1340. https://doi.org/10.1038/s41591-019-0548-6.

Wolpert, D.H., Macready, W.G., 1997. No free lunch theorems for optimization. IEEE Trans. Evol. Comput. 1, 67–82. https://doi.org/10.1109/4235.585893.

Yamamoto, T., 2018. Neurocognitive therapy - neuroscience enhances cognitive behavioral therapy. Japanese J. Cogn. Ther. 11, 13–22.

Yamamoto, T., Takebayashi, Y., 2020. Web-based support. In: Takebayashi, Y., Maeda, M. (Eds.), Telepsychological Support - Tips for Continuity of Care beyond Physical Distance. Seishin Shobo,Ltd, pp. 126–145.

Yamamoto, T., Uchiumi, C., Suzuki, N., Yoshimoto, J., Murillo-Rodriguez, E., 2020. The psychological impact of "mild lockdown" in Japan during the COVID-19 pandemic: a nationwide survey under a declared state of emergency. Int. J. Environ. Res. Publ. Health 17, 9382. https://doi.org/10.3390/ijerph17249382.

Yamamoto, T., Yoshimoto, J., 2019. Development of individual optimized techniques for early detection of signs of depression recurrence: application of machine learning methods to psychological, social and biological data. In: The 16th Annual Meeting of the Japanese Society of Mood Disorders, p. 173.

Yamamoto, T., Yoshimoto, J., 2018. Artificial intelligence-based approaches for health behavior change. In: Association for Behavioral and Cognitive Therapies 52nd Annual Convention. Washington, DC.

Yamamoto, T., Yoshimoto, J., Murillo-rodriguez, E., Machado, S., 2019. Prediction of daily happiness using supervised learning of multimodal lifelog data. Rev. Psicol. e Saúde 11, 145–152. https://doi.org/10.20435/pssa.v11i2.823.

Yamashita, A., Yahata, N., Itahashi, T., Lisi, G., Yamada, T., Ichikawa, N., Takamura, M., Yoshihara, Y., Kunimatsu, A., Okada, N., Yamagata, H., Matsuo, K., Hashimoto, R., Okada, G., Sakai, Y., Morimoto, J., Narumoto, J., Shimada, Y., Kasai, K., Kato, N., Takahashi, H., Okamoto, Y., Tanaka, S.C., Kawato, M., Yamashita, O., Imamizu, H., 2019. Harmonization of resting-state functional MRI data across multiple imaging sites via the separation of site differences into sampling bias and measurement bias. PLoS Biol. 17, e3000042. https://doi.org/10.1371/journal.pbio.3000042.

Yamashita, O., Sato, M.A., Yoshioka, T., Tong, F., Kamitani, Y., 2008. Sparse estimation automatically selects voxels relevant for the decoding of fMRI activity patterns. Neuroimage 42, 1414–1429. https://doi.org/10.1016/j.neuroimage.2008.05.050.

Yang, S., Zhou, P., Duan, K., Hossain, M.S., Alhamid, M.F., 2018. emHealth: towards emotion health through depression prediction and intelligent health recommender system. Mobile Network. Appl. 23, 216–226. https://doi.org/10.1007/s11036-017-0929-3.

Zhu, F., Panwar, B., Dodge, H.H., Li, H., Hampstead, B.M., Albin, R.L., Paulson, H.L., Guan, Y., 2016. COMPASS: a computational model to predict changes in MMSE scores 24-months after initial assessment of Alzheimer's disease. Sci. Rep. 6, 34567. https://doi.org/10.1038/srep34567.

Index

Note: 'Page numbers followed by "f" indicate figures and "t" indicate tables.'

A

Actigraphy, 205
Aging, 82–85
Alpha anteriorization, 7
Anesthesia
  anesthetic-induced loss of consciousness, 6
  autonomic function, 9
  behavioral assessment, 6
  electroencephalographic signatures, 6–8
Artificial intelligence (AI), 256

B

Biologging, 179–181
Black box, 207
Blood pressure (BP), 4–5
Bluetooth, 178–179
Brain potentials, 172

C

Cardiac dynamics, 63f
Cardiorespiratory coupling-cardiorespiratory phase synchronization (CRPS), 80–82, 83f–84f, 86, 87f
Cell markers, sleep, 156–157, 156t
Channelrhodopsins (ChRs), 136–137
Chemical/electrical stimulation techniques, 138
Chronic fatigue syndrome (CFS), 120–121
Cognitive-behavioral therapy (CBT), 255–256
Coherence values, 114
Complex temporal organization, 60f–61f
Consumer accessibility, 204–205
Content validity, 241–242, 241t
Corticocortical connectivity, 220–221
Cross-frequency coupling, 27, 28f
Cumulative variation amplitude analysis (CVAA), 62

D

Deep brain stimulation (DBS)
  histone deacetylase (HDAC) enzymes, 105–106
  intracellular proteomic interaction, 106f
  intralaminar thalamus (ILT), 102f
  neuronal mechanisms, 103–105, 104f
  Parkinson's disease (PD), 101
  pedunculopontine nucleus (PPN), 101, 102f
  proteomic and electrophysiological studies, 105–106
  sleep disorder, 101–103
  thalamocortical dysrhythmia, 101–103
Deep-learning methods, 176
Default-mode network (DMN), 221–222
Design-based stereology, 160
Designer receptorsexclusively activated by designer drugs (DREADDS), 161
Detrended fluctuation analysis (DFA) method, 69
Digitalization, 172–173
Dorsal ParagigantocellularReticular Nucleus (DPGi), 157–158
Dorsal raphe (DR), 45–46
Dorsolateral prefrontal (DLPF), 218–220

E

Electrocorticogram signals complexity, 27–30, 29f
Electroencephalography, 171–173
  functional connectivity, 113–114
  power spectra, 112–113
Electrophysiology, 173
  electroencephalogram (EEG), 33

Electrophysiology (*Continued*)
electromyogram (EMG), 33
electrooculogram, 33
extracellular recording, 36–48
electrical stimulation, 44, 45f
findings, 47–48
multisite extracellular unit recording, 44–46, 46f
optogenetics, 46–48
single-site extracellular unit recording, 37–42
total sleep-deprivation, 40
unit recording and analysis, 38–42
Juxtacellular recording
labeling, 48–49
procedure, 49–51, 50f
sleep-wake regulatory systems, 34–36
Erratic fluctuations, 59
Exact low-resolution brain electromagnetic tomography (eLORETA) software, 115–117, 115f–116f
Extracellular recording, 36–48
electrical stimulation, 44, 45f
findings, 47–48
multisite extracellular unit recording, 44–46, 46f
optogenetics, 46–48
single-site extracellular unit recording, 37–42
total sleep-deprivation, 40
unit recording and analysis, 38–42
Extracellular unit recording-electrical stimulation, 44, 45f

F

Fast Fourier Transform (FFT), 112
Freezing behavior, 147

G

GABAergic neurons, 52–53, 157
Global System for Mobile (GSM) communication protocol, 179
Gold standard, 205

H

Halorhodopsin (NpHR), 136–137
Heartbeat dynamics changes, temporal correlations, 76–79
Heartbeat fluctuations, 59–63, 69–76
Hearth rate (HR), 4–5
Heart rate variability (HRV), 4–5, 65–67
High-frequency oscillations (HFOs), 3–4
Histone deacetylase (HDAC) enzymes, 105–106, 106f
Human polysomnography, 173–174
Huntington's disease (HD), 125
Hypothalamus, 33–34

I

Immediate-early genes (IEGs), 158–159
Immunohistochemical (IHC), 37–38
Immunohistochemical analysis
cell markers and sleep, 156–157, 156t
immunoreactive cells, statistical considerations, 161–163
pseudoreplication, 163–165, 164f
sleep studies, 157–161
Immunoreactive cells, statistical considerations, 161–163
Infrared thermography, 175–176
Intracellular proteomic interaction, 106f
Intracellular recording, 34–35
Intralaminar thalamus (ILT), 102f
Intrinsic neuroautonomic regulation, 75–76

J

Juxtacellular recording
labeling, 48–49
procedure, 49–51, 50f

L

Laterodorsal tegmental nucleus (LDT), 45–46
Light-emitting diode (LED), 137
Local field potentials (LFPs), 5

M

Machine learning (ML)
artificial intelligence (AI), 256
clinical applications, 265–268

cognitive-behavioral therapy (CBT), 255–256
definition, 256
future perspectives, 270–273
clinical situations accessibility and verification, 270
cooperation, 272–273
individualized intervention procedures, 271–272, 272f
limitations of, 269
mental disorders, 265–266
paradigm shift in science, 264
prognosis and treatment effects, 268
psychological interventions, 255–256
reinforcement learning, 259, 260t–263t
standard statistics, 259–264
supervised learning, 257–258, 257f
symptoms and risks detection, 266–268
unsupervised learning, 258–259, 258f
Magnitude and sign analysis (MSA), 72–74
Medial prefrontal cortex (MPFC), 217–218
Mental disorders, 265–266
Multielectrode microdrive system, 45
Myoclonic movements, 173–174

N

Network interactions among, 89–93, 90f
Neuroautonomic regulation, 91f
Neurodegenerative diseases, 125–127
Neuropsychiatric diseases, 125–127
Nonlinear oscillatory systems, 82
Nonrapid eye movement (NREM) sleep, 1–3, 171
default-mode network (DMN), 221–222
dorsolateral prefrontal (DLPF), 218–220
functional cerebral connectivity, 220–221
medial prefrontal cortex (MPFC), 217–218
neuronal correlates of, 217–222, 219t–220t
pedunculopontine tegmental nucleus (PPT), 218–220
regional cerebral blood flow (rCBF), 218–220
sleep-oscillations, 222–225
slow waves, 224–225
spindles, 222–224
slow wave sleep (SWS), 217–218
Normal sleep, 117–119

O

Obstructive sleepapnea syndrome (OSAS), 119–120, 122
Open-source software, 176
Optic cannulas, design/construction of, 140–142
Optogenetic light-emitting diode light delivery system, 146
Optogenetics
applications, 139–142
basics of, 136–137
channelrhodopsins (ChRs), 136–137
chemical/electrical stimulation techniques, 138
disadvantages, 138–139
experimental procedures, 146–147
experimental variants, 149–150
freezing behavior, 147
halorhodopsin (NpHR), 136–137
light-emitting diode (LED), 137
Natronomonas pharaonic, 136–137
opsin, 136–137
expression at injection site, 148–149
optic cannulas, design/construction of, 140–142
optogenetic light-emitting diode light delivery system, 146
sample protocol for, 146–149
surgical procedures, 142–145
operative procedures, 143–145
postoperative procedures, 145
postoperative recovery period, 145
preoperative procedures, 143
viral vector preparation, 142–143
telemetric data, 147

P

Paradigm shift in science, 264
Paradoxical sleep, 3
Parkinson's disease (PD), 101
Pedunculopontine nucleus (PPN), 101, 102f
Pedunculopontine tegmental nucleus (PPT), 45–46, 218–220
Phase synchronization, 114
Physiologic coupling, coexisting forms of, 85–88
Physiologic dynamics systems
  cardiac dynamics, 63f
  cardiorespiratory coupling-cardiorespiratory phase synchronization (CRPS), 80–82, 86, 87f
  circadian influence on, 78f–79f
  complex temporal organization, 60f–61f
  cumulative variation amplitude analysis (CVAA), 62
  detrended fluctuation analysis (DFA) method, 69
  erratic fluctuations, 59
  heartbeat dynamics changes, temporal correlations, 76–79
  heartbeat fluctuations, 69–76
  heartbeat fluctuations, 59–63
  heart rate variability (HRV), 65–67
  interactions, wake and sleep, 80–93
  intrinsic neuroautonomic regulation, 75–76
  magnitude and sign analysis (MSA), 72–74
  network interactions among, 89–93, 90f
  neuroautonomic regulation, 91f
  physiologic coupling, coexisting forms of, 85–88
  physiologic network reorganization, 91–93, 92f
  rapid eye movement (REM), 63–64
  respiratory sinus arrhythmia (RSA), 87–88, 88f
  scale-invariant fractal processes, 72–74
  scaling behavior, 59–61
  sleep-stage stratification, 73f
  sympathetic and parasympathetic activation, 63f
  sympathetic nerve activity measurements, 65
  sympathovagal balance stage related modulation, 89–91
  time delay stability (TDS) method, 89
Physiologic network reorganization, 91–93, 92f
Polysomnography (PSG), 2, 203
  animal models
    data analysis, 23–24
    procedures, 21–23
  electroencephalogram(EEG), 17
  humans, 18–21
    procedures, 18–20
    sleep characteristics, 20–21, 22f
  limitations, 206–207
  quantitative electroencephalogram analysis (qEEG), 24–30
    cross-frequency coupling, 27, 28f
    electrocorticogram signals complexity, 27–30, 29f
    spectral coherence, 25–26
    spectral power, 24–25, 25f
Ponto-geniculo-occipital (PGO) waves, 215
Power spectral analysis (PSA), 112
Primary insomnia, 120–122
Pseudoreplication, 163–165, 164f

Q

Quantitative electroencephalogram analysis (qEEG), 24–30
  cross-frequency coupling, 27, 28f
  electrocorticogram signals complexity, 27–30, 29f
  spectral coherence, 25–26
  spectral power, 24–25, 25f
Quantitative electroencephalographic (QEEG)
  abnormal sleep, 119–128
    future directions, 127–128
    neurodegenerative diseases, 125–127
    neuropsychiatric diseases, 125–127
    primary insomnia, 120–122
    sleep parasomnias, 124–125

sleep-related breathing disorders, 122–123
coherence values, 114
electroencephalography functional connectivity, 113–114
electroencephalography power spectra, 112–113
exact low-resolution brain electromagnetic tomography (eLORETA) software, 115–117, 115f–116f
normal sleep, 117–119
phase synchronization, 114
rapid-eye-movement (REM), 117
source localization methods, 114–117

R

Radio frequency technology, 176–177
Rapid eye movement (REM) sleep, 1–2, 63–64, 117, 171
neuronal correlates of, 225–228
neurophysiological mechanisms, 227
sleep-activity, 228–230
Recovery sleep (RS), 40, 41f
Regional cerebral blood flow (rCBF), 218–220
Reinforcement learning, 259, 260t–263t
Reliability, 202–207
definitions, 242–244
indices of, 245–246
standard error of measurement, 247–248
standards for, 246–247
types of, 244–245
REM sleep behavior disorder (RBD), 124
Respiratory sinus arrhythmia (RSA), 80–81, 87–88, 88f

S

Saw-tooth waveforms, 3
Scale construction
content validity, 241–242, 241t
items checking, 238–241
items devising, 237–238
required, 236–237
test theory soupçon, 242
Scale-invariant fractal processes, 72–74
Scaling behavior, 59–61
Sleep
abnormalities, 126
disorder, 101–103
deep brain stimulation, 101–103, 102f
histone deacetylase (HDAC) enzymes, 105–106, 106f
nonrapid eye movement sleep. *See* Nonrapid eye movement sleep
normal human sleep imaging, 215–217
parasomnias, 124–125
rapid eye movement sleep. *See* Rapid eye movement sleep
Sleep-inducing factors (SIFs), 157
Sleep-related breathing disorders, 122–123
Sleep-stage stratification, 73f
Sleep-wake cycle
autonomic function, 4–5
blood pressure (BP), 4–5
electroencephalographic signatures, 2–4
hearth rate (HR), 4–5
heart rate variability (HRV), 4–5
hypometabolism, 9–10
invertebrate species
identification, 5
quantification, 5
regulatory systems, 34–36
Slow wave sleep (SWS), 120–121
Source localization methods, 114–117
Spectral coherence, 25–26
Spectral power, 24–25, 25f
Spindle-related activity, 223–224
Standard error of measurement, 247–248
Standard statistics, 259–264
Supervised learning, 257–258, 257f
Sympathetic nerve activity measurements, 65
Sympathetic/parasympathetic activation, 63f
Sympathovagal balance stage related modulation, 89–91
Systematic and random sampling (SRS), 160
SystemDrive, 45–46

T

Telemetric data, 147
Telemetric electrophysiological devices, 177–178
Telemetry, 176–179
Test theory soupçon, 242
Thalamocortical dysrhythmia, 101–103
Thermoregulation, 9
Time delay stability (TDS) method, 89
Total sleep-deprivation studies, 40
Transmitter, 178–179

U

Unsupervised learning, 258–259, 258f

V

Validity, 248–251
Video, 175–176
  recorders, 175–176

W

Wearable/nonwearable sleep-tracking devices
  accuracy, 202–207
    requirements, 204
  actigraphy, 205
  applications, 208
  black box, 207
  consumer accessibility, 204–205
  definition, 192–193
  diseases, 209–210
  gold standard, 205
  hardware, 193–201, 196t–199t
    connectivity, 199–200
    physical features, 193
    sensors, 194
    software, 193–201
    types of, 200–201
  health conditions, 209–210
  interventions, 210
  lifespan, 211
  limitation of, 205–206
  polysomnography (PSG), 203
    limitations, 206–207
  recommended validation practices, 206
  reliability, 202–207
  sleep and circadian assessment, actigraphy for, 203–204
  sleep health parameters, 208–209
  sleep trackers, 192
  software
    algorithm, 201–202
    engagement, 201–202
  utility, 208
  validity, 202–207
Wireless vigilance state monitoring
  biologging, 179–181
  brain potentials, 172
  electroencephalography, 171–173
  electromyogram (EMG), 173
  electrooculogram (EOG), 173
  electrophysiology, 173
  limitations, 181–182
  myoclonic movements, 173–174
  nonrapid eye movement (NREM) sleep, 171
  opportunities, 183
  rapid eye movement (REM), 171
  telemetry, 176–179
  video, 175–176

Printed in the United States
by Baker & Taylor Publisher Services